Ozone in Food Processing

Ozone in Food Processing

Edited by

Colm O'Donnell
School of Biosystems Engineering
University College Dublin
Dublin
Ireland

B.K. Tiwari
Hollings Faculty
Manchester Metropolitan University
Manchester
UK

P.J. Cullen
School of Food Science and Environmental Health
Dublin Institute of Technology
Dublin
Ireland

Rip G. Rice
RICE International Consulting Enterprises
Sandy Spring
MD
USA

WILEY-BLACKWELL
A John Wiley & Sons, Ltd., Publication

Blackwell Publishing was acquired by John Wiley & Sons in February 2007. Blackwell's publishing program has been merged with Wiley's global Scientific, Technical and Medical business to form Wiley-Blackwell.

Registered Office
John Wiley & Sons, Ltd, The Atrium, Southern Gate, Chichester, West Sussex, PO19 8SQ, UK

Editorial Offices
9600 Garsington Road, Oxford, OX4 2DQ, UK
The Atrium, Southern Gate, Chichester, West Sussex, PO19 8SQ, UK
2121 State Avenue, Ames, Iowa 50014-8300, USA

For details of our global editorial offices, for customer services and for information about how to apply for permission to reuse the copyright material in this book please see our website at www.wiley.com/wiley-blackwell.

Library of Congress Cataloging-in-Publication Data

Ozone in food processing / edited by Colm O'Donnell ... [et al.].
p. cm.
Includes bibliographical references and index.
ISBN 978-1-4443-3442-5 (hard cover : alk. paper)
1. Ozone. 2. Food industry and trade. I. O'Donnell, C. P. (Colm P.)
QD181.O1O96 2012
664'.0286–dc23

2011035803

A catalogue record for this book is available from the British Library.

Wiley also publishes its books in a variety of electronic formats. Some content that appears in print may not be available in electronic books.

Set in 9.5/12.5pt Palatino by SPi Publisher Services, Pondicherry, India
Printed and bound in Malaysia by Vivar Printing Sdn Bhd

1 2012

Contents

A colour plate section falls between pages 180 and 181

Contributors

Hee-Jung An
Department of Food Science, Louisiana State University Agricultural Center, 111 Food Science Building, Louisiana State University, Baton Rouge, LA, USA

Ioannis S. Arvanitoyannis
Department of Agriculture Crop and Animal Production, Ichthyology and Aquatic Environment, School of Agricultural Sciences, University of Thessaly, Volos, Greece

Gilbert Y.S. Chan
Department of Applied Biology and Chemical Technology, The Hong Kong Polytechnic University, Hung Hom, Kowloon, Hong Kong

P.J. Cullen
School of Food Science & Environmental Health, Dublin Institute of Technology, Dublin, Ireland

Annel K. Greene
Department of Animal and Veterinary Sciences, Clemson University, Clemson, SC, USA

Zeynep B. Güzel-Seydim
Food Science and Human Nutrition, Clemson University, Clemson, SC, USA

Joan M. King
Department of Food Science, Louisiana State University Agricultural Center, Louisiana State University, Baton Rouge, LA, USA

Paula Misiewicz
Engineering Department, Harper Adams University College, Newport, Shropshire, UK

K. Muthukumarappan
Agricultural and Biosystems Engineering Department, South Dakota State University, Brookings, SD, USA

Shigezou Naito
Department of Nutrition and Food Science, Aichi Gakusen College, Aichi, Japan

Tomás Norton
Engineering Department, Harper Adams University College, Newport, Shropshire, UK

Colm O'Donnell
School of Biosystems Engineering, University College Dublin, Dublin, Ireland

V. Lullien-Pellerin
INRA, UMR 1208 Ingénierie des Agropolymères et Technologies Emergentes, Montpellier, France

Fred W. Pohlman
Department of Animal Science, University of Arkansas, Fayetteville, AR, USA

Alfredo Prudente
Rutgers The State University, Department of Food Science, New Brunswick, NJ, USA

Rip G. Rice
RICE International Consulting Enterprise, Sandy Spring, MD, USA

Seung-wook Seo
Nongshim, 203-1, DangJeong-Dong, Gunpo-SiGyeonggi-Do, Korea

Atıf Can Seydim
Department of Food Engineering, Faculty of Engineering, Süleyman Demirel University, Isparta, Turkey

Cameron Tapp
Clearwater Tech, San Luis Obispo, CA, USA

B.K. Tiwari
Manchester Food Research Centre, Hollings Faculty, Manchester Metropolitan University, Manchester, UK

Vasilis P. Valdramidis
Department of Food Science, Louisiana State University Agricultural Center, Louisiana State University, Baton Rouge, LA, USA

J.G. Wu
Department of Applied Biology and Chemical Technology, The Hong Kong Polytechnic University, Hung Hom, Kowloon, Hong Kong

1 Status and Trends of Ozone in Food Processing

Colm O'Donnell, B.K. Tiwari, P.J. Cullen and Rip G. Rice

1.1 Why ozone?

Interest in ozone has expanded in recent years in response to consumer demands for 'greener' food additives, regulatory approval and the increasing acceptance that ozone is an environmentally friendly technology. The multifunctionality of ozone makes it a promising food processing agent. Excess ozone autodecomposes rapidly to produce oxygen and thus leaves no residues in foods from its decomposition. In particular, the US Food and Drug Administration (FDA)'s rulings on ozone usage in food have resulted in increased interest in potential food applications worldwide. Ozone as an oxidant is used in water treatment, sanitising, washing and disinfection of equipment, odour removal, and fruit, vegetable, meat and seafood processing.

1.2 Drivers of ozone in the food industry

1.2.1 Regulation

While food safety assurance is a global concern, approaches to regulation differ throughout the world. Globally the regulatory status of ozone for food processing applications is still in an evolving state of flux, and in some countries has not been addressed to date. Legislation governing ozonation for treating, handling, processing and storage of foods has typically developed in response to the evolving use of ozone from initial applications for water treatment, through surface and equipment cleaning, food produce washes, and finally to use as a direct food additive. The use of ozone in food processing has become increasingly important as a result of the affirmation of ozone as a GRAS (Generally Recognised as Safe) chemical in 1997 (Graham et al. 1997) and its subsequent approval by the US FDA as an antimicrobial additive for direct contact with foods of all types (FDA 2001). The use of ozone in food processing has been approved to various degrees

Ozone in Food Processing, First Edition. Edited by Colm O'Donnell, B.K. Tiwari, P.J. Cullen, and Rip G. Rice.

in many countries, including the USA, Japan, Australia, France and Canada. Given the complexities of food matrices and the range of foods produced, demonstrating process validation is a challenge for industry. However, more expedited validation processes are likely with validation of comparable products.

1.2.2 Surface cleaning and disinfection

The need to develop nonresidual and validated cleaning approaches for the food industry has been clearly indentified. Ozone offers the food industry an alternative or complementary cleaning and sanitising agent. The efficacy of ozone for physical, chemical and biological cleaning within food processing units has been reported. The potential inclusion of ozone-containing water within the clean-in-place (CIP) cycle offers significant opportunities for food processors. Comparisons of treatment efficacy against traditional approaches are discussed in Chapter 10. Applications of ozone for sanitising various food processing equipment items are reviewed. The use of ozone for cleaning within comparable process industries, such as pharmaceuticals, is also outlined.

1.2.3 Food safety and shelf life extension

Food treatment approaches include ozone applications in both the gaseous and aqueous phases. Washes in ozone-containing water, storage in ozone-rich atmospheres and direct addition of ozone in fluid foods are reviewed. The antimicrobial efficacies of ozone for control of pathogenic micro-organisms of concern in the food industry are reviewed in Chapter 3. The effectiveness of ozone against microorganisms present in food systems depends on several factors including the amount of ozone applied, the residual ozone in the medium and various environmental factors such as medium pH, temperature, relative humidity, additives and the amount of organic matter surrounding the cells.

Storage grains are susceptible to a number of insects, which cause considerable damage to the stored grains and could potentially develop resistance to the currently employed insecticides. Increasing environmental problems and new legislation have tended to reduce the permitted pesticide amounts or even prohibited their use. Ozone use in fumigation is an alternative to chemicals in controlling insect development. The use of ozone for the control of fungi and mycotoxins in grains is discussed in Chapter 6.

The potential of ozone for the degradation of pesticide residues found in food is discussed in Chapter 13. The proposed mechanisms for degradation of pesticides, including organophosphates and organochlorinated compounds, are outlined. The efficacy of both gaseous and aqueous ozone for degradation and the processing parameters governing the process are reviewed.

1.2.4 Nutrient and sensory aspects

Taste and sensory properties are consistently rated as the most important factors driving consumption and repeat purchase of food products. The principal driver for industrial adoption of new processing technology is to meet consumers' demands for improved taste and nutrition. Ozone is a strong oxidising agent and its effect on such parameters must be considered prior to any potential food application. Effects will be dependent upon the mode of application, the dose, food composition and so on. Each application chapter discusses the reported effects on such parameters.

1.2.5 Consumer and processor acceptability

Consumers are not only concerned about the ingredients within the foods they consume, but also about the processes that are employed in bringing food 'from farm to fork'. A growing body of consumer research suggests that consumers are increasingly conscious of the food supply chain, which will continue to influence their perceptions of emerging food processes. Paradoxically, consumers are demanding foods which are minimally processed, meet their nutritional and taste desires yet require minimal preparation. Understanding and addressing consumer issues related to any novel food process are some of the most important challenges facing the developers of innovative food products. Research suggests that acceptance of new technologies is based to a great extent on public perceptions of the associated risks, and that perceptions of risk are influenced by trust in information and the source that provides it (Frewer et al. 2003). Several consumer research studies have consistently shown that consumers have poor knowledge and awareness levels towards most novel food processing techniques, which serves as a major impediment to their acceptance. Thus, effective communication regarding details of the technologies and their benefits becomes essential for successful marketing of these products. If a novel technology allows the introduction of new products with tangible benefits, consumers are most likely to accept it.

For the processor, it is critical that any process adopted is safe for the production staff. Chapter 15 reviews the precautions dealing with the release of gaseous ozone in amounts that might cause discomfort or injury to plant workers. The health and safety issues associated with ozone are reviewed, followed by a discussion of the commonly accepted worker safety regulations for breathing gas-phase ozone.

1.2.6 Technology advances

There have been significant developments in the methodologies of ozone production, including corona discharge/plasma and UV radiation, which make ozonation a more attractive approach for food processing. Economic and technical aspects of ozone production are outlined in Chapter 4, including process controls, production scales, application approaches and the limitations of each procedure. Challenges encountered in the industrial production of ozone are addressed, along with future trends. Novel systems for generation of ozone within sealed packages under air or modified atmospheres are described. Such approaches are suitable for many food applications, from fresh produce to meat products.

1.2.7 Environmental impact

To achieve the full potential of commercial exploitation of novel technologies, issues related to environmental impacts, such as wastewater and gas emissions, the conservation of nonrenewable resources and energy consumption, must be investigated and understood by food processors, since they can represent significant potential reductions in processing costs (Pereira and Vicente 2010). The food industry is a significant consumer of energy, with the principal type of energy used for traditional thermal processing being fossil fuel. Water is a key ingredient in the food industry, playing a fundamental role in many of the common food processing methods and unit operations, such as soaking, washing, rinsing, blanching, heating, pasteurising, chilling, cooling and steam production, acting as an ingredient, and being used for general cleaning, sanitation and disinfection purposes. However, the industry is not so well known for its use of water-saving devices and practices. While ozone has been a globally successful water treatment method, the literature has shown that it has not been largely employed as yet by the food industry, even though it was approved for application in the reconditioning of recycled poultry chilling water by the US Department of Agriculture in 1997 (Güzel-Seydim et al. 2004). Chapter 11 discusses the potentials of ozone as an alternative for potable water treatment, wastewater treatment and water reuse in the food industry. Applications are identified in the fruit and vegetable, meat and dairy sectors. The efficacy of ozone for physically, chemically and microbiologically safe reuse of water in the food industry is discussed.

1.3 The hurdle concept

Combining a number of preservation methods may enhance the overall antimicrobial effect so that lower process intensities can be employed. This approach, known as 'hurdle technology', has already been applied successfully using traditional techniques of food preservation (Leistner and Gorris 1995). Combining ozone methods with other food preservation techniques can (1) enhance the lethal effects, (2) reduce the severity of treatment required to obtain a given level of microbial inactivation and

(3) prevent the proliferation of survivors following treatment. The choice of hurdles – several combinations of either novel thermal, novel nonthermal or conventional processing technologies – is generally made to maximise the synergistic effect on the microbial inactivation kinetics. Food preservation using combined methods involves successive or simultaneous applications of various individual treatments. Combined treatments are advantageous, principally because many individual treatments alone are not adequate to ensure food safety or stability.

1.4 Challenges

Despite significant scientific advances and the demonstrated industrial potential of ozone in seafood, meat and decontamination of pesticide residues in the food chain, there is a paucity of reported studies in this area in general. Also, the potential for reduced processing costs through the use of ozone technologies has not been widely disseminated. Awareness and understanding of ozone applications for foods is key to improved uptake of ozone technology by industry. Increased clarity of the regulatory status of ozone for food applications would facilitate increased global adoption by the food industry.

1.5 Objective

The objective of this book is to demonstrate the potential technoeconomic benefits of employing ozone in the food industry to facilitate increased industry adoption. This book provides an insight into the current state of the art and reviews emerging applications of ozone processing. The principles of ozonation, process control parameters, microbial inactivation mechanisms and the effects on food nutritional and quality parameters are outlined. Separate chapters are dedicated to covering different food processing applications. Finally, health and safety aspects of ozone as used in food processing plants and future trends in industry adoption of ozonation are discussed.

References

FDA (2001) Hazard analysis and critical control point (HACCP): procedures for the safe and sanitary processing and importing of juice; final rule, *Federal Register*, **66**: 6137–6202.

Frewer, L., Scholderer, J. and Lambert, N. (2003) Consumer acceptance of functional foods: issues for the future, *British Food Journal*, **105**: 714–31.

Graham, D.M., Pariza, M.W., Glaze, W.H., Erdman, J.W., Newell, G.W. and Borzelleca, J.F. (1997) Use of ozone for food processing, *Food Technology*, **51**(6): 72–6.

Güzel-Seydim, Z.B., Greene, A.K. and Seydim, A.C. (2004) Use of ozone in the food industry, *Lebensm.-Wiss. Technol.-Food Sci. Technol.*, **37**(4): 453–460.
Leistner, L. and Gorris, G.M. (1995) Food preservation by hurdle technology, *Trends in Food Science and Technology*, **6**: 41–46.
Pereira, R.N. and Vicente, A.A. (2010) Environmental impact of novel thermal and non-thermal technologies in food processing, *Food Research International*, **43**: 1936–43.

2 Regulatory and Legislative Issues

B.K. Tiwari and Rip G. Rice

2.1 Introduction

Ozone has been used commercially for the treatment of drinking water since 1906 Nice, France (Hill and Rice 1982) and is increasingly employed in the food industry for produce preservation and sanitising of food-contact surfaces. Demand for new preservation approaches arises from growing consumer preference for minimally processed foods, frequent outbreaks of food-related illnesses, identification of new food pathogens and the passage of legislation governing food quality and safety. The World Health Organization (WHO) identified foodborne diseases as a considerable threat to human health and the global economy which requires a concerted effort on the part of three principal partners, namely governments, the food industry and consumers. For sanitising applications, ozone may be preferred over traditional sanitisers such as chlorine because of the relatively low inactivation rate of chlorine at concentrations which are limited by regulation, combined with consumer concerns over chemical residues and potential environmental impacts.

The legislation governing ozonation typically has been developed in response to the evolving use of ozone, from initial applications for water treatment, through surface and equipment cleaning, to food produce washes and more recently as a direct food additive. It is likely that the introduction of legislation governing ozonation applications in food processing will encourage the adoption of ozonation processes in industry. However, globally the regulatory status of ozone for food processing applications is still in an evolving state, and in some countries it has not yet been addressed. Food processors who wish to employ ozonation in their plants should consult with their regulatory agencies to ascertain what regulatory constraints, if any, exist that impact on their proposed process or product involving the use of ozone. This chapter outlines the current legislative and regulatory status of ozone for food processing applications where it has been developed.

Ozone in Food Processing, First Edition. Edited by Colm O'Donnell, B.K. Tiwari, P.J. Cullen, and Rip G. Rice.

Table 2.1 History of ozone application and regulation.

Year	Achievements
1839	Discovery of ozone by Schönbein.
1895	The molecular formula of ozone determined by Soret.
1886	The potential of ozone to disinfect polluted water recognised in Europe.
1891	Test results from Germany show that ozone is effective against bacteria.
1893	First full-scale application using ozone for drinking water in the Netherlands.
1906	France commissions its first municipal ozone plant for drinking water.
1909	Ozone employed for preservation of meat in Germany.
1914	Research leads to the production of inexpensive chlorine gas and interest in ozone for water treatment begins to decline.
1936	Ozone used to depurate shellfish in France.
1939	Ozone found to prevent the growth of yeast and mould during the storage of fruits.
1942	Ozone used in egg-storage rooms and in cheese-storage facilities in the USA.
1957	Ozone implemented for oxidation of iron and manganese in German drinking water.
1964	Spontaneous flocculation in ozone contact chambers leads to France constructing an ozone plant to enhance particulate removal.
1965	Ozone employed for colour control of surface water in Ireland and the UK. Ozone used to oxidize micropollutants such as phenolic compounds and several pesticides in Switzerland.
1970	Ozone exploited for algae control in France.
1982	US FDA grants GRAS (generally recognised as safe) status for ozone disinfection of bottled water.
1987	600 MGD (million gallons per day) ozonation plant comes on line in Los Angeles after seven years of pilot testing.
1995	FDA GRAS approval for ozone disinfection of bottled water renewed.
1997	Expert Panel convened by the Electric Power Research Institute (EPRI) affirms ozone as GRAS for direct contact with foods. FDA does not object to this GRAS affirmation. Regulators have the option to later add controls on ozone use.
1999	United States Department of Agriculture (USDA) rejects an ozone protocol for meat application, citing the 1982 GRAS declaration for disinfection of bottled water in which the FDA stated 'any other use must be regulated by a Food Additive Petition'.
2000	A Food Additive Petition (FAP) filed by the EPRI requests FDA approval of ozone for direct contact with foods.
2001	FDA approves ozone as a secondary direct food additive, antimicrobial agent (Federal Register, Vol. 66, no. 123, June 26). The American Meat Institute files a letter with the US FSIS (Food Safety and Inspection Service of the FDA) asking for an interpretation on the scope of FDA rule. In its response, FSIS determines that, 'The use of ozone on meat and poultry products, including treatment of ready-to-eat meat and poultry products just prior to packaging, is acceptable', and that there are 'no labelling issues in regard to treated product'.
2004	FDA issues industrial guidance and recommendations to processors of apple juice or cider on the use of ozone for pathogen reduction purposes.

2.2 History of ozone application and regulation

Ozone was first discovered in 1839 by Schönbein, who observed that the electrolysis of water produced an odorous gas. Table 2.1 shows a brief history of ozone application and regulation. Ozone was first used commercially as a disinfectant of drinking water in France early in the 1900s (Hill and Rice 1982). Currently there are an estimated several thousand drinking water treatment plants in the world using ozone (authors' estimate based on industry contacts).

2.3 Ozone regulation

2.3.1 Overview of US regulations

Although ozone and its oxidising properties were first discovered as early as 1840, application of ozone was relatively recent in the USA. Ozone is approved by the US Food and Drug Administration (FDA) for use in the USA and has been employed successfully for applications including surface decontamination to extend the shelf life of cheeses and fresh produce, decontamination of packaging materials, disinfection of process water and sanitisation of processing equipment and food storage areas, among other things (Mahapatra et al. 2005).

Ozone is Generally Recognised as Safe (GRAS) in the USA for disinfection of bottled water and as a sanitiser for process trains in bottled water plants (FDA 1995). In 1997, ozone was affirmed as having GRAS status for direct contact with foods by an independent panel of experts, sponsored by the Electric Power Research Institute (EPRI) (Graham et al. 1997). The FDA released a final ruling in June 2001, in response to an EPRI food additive petition, amending previous food additive regulations and granting regulatory approval of ozone as an antimicrobial agent for direct contact with foods (FDA 2001). An antimicrobial agent is defined as an agent that will provide 2log reductions in microbial levels. The amendment to the FDA's food additive regulations (Title 21 of the Code of Federal Regulations, part 173) allows the use of ozone when used as a gas or dissolved in water as an antimicrobial agent for food. This federal ruling cleared the way for the potential use of ozone in the US food processing industry (Hampson 2001). Ozone is also approved in the USA for use on all meat and poultry products by the US Department of Agriculture (USDA FSIS 2001) when applied in accordance with current industry standards of good manufacturing practice (21 CFR 173.368; FDA 2003).

Enforcement of the Federal Insecticide, Fungicide, and Rodenticide Act (FIFRA) is under the administrative jurisdiction of the US Environmental Protection Agency (EPA). The provisions of the FIFRA are written to regulate the use of chemicals that are applied as insecticides, fungicides and/or rodenticides. Under the FIFRA, 'chemicals' are defined as materials that are manufactured, packaged, transported, stored and applied. Because of the characteristics of ozone (an unstable gas that must be generated and used on site as it is needed), it cannot be packaged, stored or shipped from a manufacturing point miles away from its point of use. Therefore, neither ozone nor ultraviolet (UV) radiation can be regulated as a chemical under the FIFRA. However, a provision was included in the FIFRA whereby manufacturers of ozone- and UV-generating machines are required to register these generators as 'devices' that are manufactured, sold and/or used in the USA. The intent of this FIFRA provision is to provide the EPA with a register of (ozone and UV) generator manufacturers. In exchange for registering their products and for agreeing not to make claims for products (ozone and UV) that are not supported by the technical literature, (ozone and UV) generator manufacturers are issued a special EPA Establishment Registration decal that must be mounted on each generator. The decal simply states that the manufacturer of the labelled (ozone or UV) generator has registered his/her manufacturing facility with the EPA. The fact that an ozone or UV generator carries an EPA Establishment Registration Label does not in any way denote approval of ozone by the EPA for any application. On the other hand, if an EPA inspector visits a food processing plant and find that an (ozone or UV) generating device does *not* carry an EPA Establishment Registration Label, that inspector has the authority to order the unlabeled device to be taken off line. Therefore, any food processor using ozone within the USA should ensure that whoever supplies their facility with ozone generation equipment has registered that product with the EPA as a device under the FIFRA. Non-US manufacturers of ozone (or UV) generators supplying those devices inside the USA are subject to the device registration requirements of the FIFRA. From the worker safety point of view, processors must also ensure that plant workers are not exposed to concentrations of ozone higher than: (1) 0.1 ppm by volume (0.2 mg/m^3 NTP) on a time-weighted average over an 8 h/d basis and (2) 0.3 ppm by volume (0.6 mg/m^3 NTP) as a limit for a maximum exposure time of 15 minutes, not to be exceeded more than four times daily, according to US Occupational Safety and Health Administration (OSHA) regulations (please see Chapter 14) (CFR 1997).

Ozonation of fresh produce

The definition of the product used to disinfect washwater depends on the type of product to be washed, and in some cases on the location where the disinfectant is used (IFPA 2001). In the USA, the washwater disinfectants used for fresh-cut produce are regulated by the FDA as secondary direct food additives, unless they have been affirmed to be GRAS. Where a raw agricultural commodity is washed in a food processing facility, such as a fresh-cut facility, both the EPA and the FDA have regulatory jurisdiction, and the disinfecting agents (except for ozone and UV radiation) must be registered as pesticides with the EPA.

Ozonation of apple cider/juice

Several incidents of foodborne disease have been associated with juices. In 1991, an outbreak of *Escherichia coli* O157:H7 infections and haemolytic uremic syndrome was linked to traditionally pressed apple cider. In the USA 21 juice-associated outbreaks were reported to the Centers for Disease Control and Prevention (CDC) between 1995 and 2005 (Vojdani et al. 2008). *E. coli* O157:H7 is an enteric pathogen with a low infectious dose, which usually causes hemorrhagic colitis but also has the potential to cause haemolytic uremic syndrome in young children and the immunocompromised (Boyce et al. 1995).

These outbreaks led the FDA to issue hazard analysis and critical control point (HACCP) regulations for safe and sanitary processing of juice (FDA 2001). A primary performance standard is a minimum 5 log reduction of the pathogens of concern in the juice being processed (FDA 2001). The FDA's approval of ozone as a direct food additive in 2001 triggered interest in ozone applications. A number of commercial fruit juice processors in the USA and Europe began employing ozone for pasteurisation, resulting in the issuance of industry guidelines. However, these guidelines (FDA 2004) highlight gaps in the literature with respect to the critical control parameters of ozone during microbial inactivation in liquid systems.

2.3.2 Overview of European regulations

Application of ozonation in food processing commenced soon after it was first used for water treatment in the early 1900s. The interest in ozone as an antimicrobial agent for food processing is due to several advantages it has over chlorine and other chemical disinfectants presently and previously used in cleaning and disinfection operations. These advantages are generally overlooked by food processors, but the new environmental legislation emerging in Europe, especially IPPC Directive 96/61/EC, is driving changes in the food industry.

Within the EU, among several processing techniques employed in food processing only irradiation and ionisation have specific regulation and labelling rules which also vary within member states (national legislation for irradiation); for other techniques, including ozone, no specific legislation applies. However, there are two general rules which also apply to ozone:

(1) **Labelling Directive (2000/13)** Indication on the label of the specific treatment undergone by a product – if the absence of this indication would mislead the consumer.

(2) **Novel Food Regulation (258/97)** Premarket authorisation for 'novel foods', including those which have undergone a novel treatment process.

The European Council of Ministers has now adopted a proposal which permits the treatment of natural mineral water, but not spring water, with ozone, provided that the treatment information is carried on the label. The use of ozone is now permitted 'to separate unstable elements from natural mineral waters which will ensure that the composition of the water as regards its essential ingredients is not affected'. The unstable elements referred to include iron, manganese and sulfur compounds. An important side effect of this new European permission for mineral waters will be the disinfecting capacity of ozone treatment, which in many ways is superior to that of chlorine.

Before the Council vote, the European Union's Scientific Committee for Food (SCF) handed down an opinion on the use of ozone to treat natural mineral waters. The opinion recognised that ozone treatment may lead to the formation of undesirable byproducts and recommended that producers should comply with several conditions to minimise side reactions. The SCF concluded that residual ozone and the concentration of undesirable byproducts (bromate and bromoform) should be undetectable by the best available analytical methodology.

Regulations in France

In 2003 and 2004, the French Food Safety Agency rendered two Opinions as to the safety of the use of ozone as an auxiliary technology to treat wheat grains before grinding. The first (AFSSA 2003) stated that 'an ozone dose of 12 g (at standard temperature and pressure) per kg of grain, intended for the preparation of flour for pastries containing simple sugars added to a level of 7 to 50% of dry weight, does not present any health risk to the consumer'. The second Opinion (AFSSA 2004) extended the use of ozone from wheat grain treatment before grinding to 'the preparation of flour destined for bread and baked products containing up to 7% of added sugars at a concentration of 8 g ozone per kg of grain at standard temperature and pressure, exclusive of traditional French bread', which it described as posing no health risk to the consumer.

Ozonation of fresh produce

Hammond (2004) outlined the situation in Europe and discussed possible changes in regulations that may be introduced for fresh produce washing. The European Council Directive (89/107/EEC) on food additives lists the substances which legally may be added to food if they perform a useful purpose, are safe and do not mislead the consumer. The detailed controls made under the Framework Directive are implemented into the national law of each EU member state and stipulate which food additives are permitted for use, the specific purity criteria and the conditions of use, including maximum levels for specific additives; however, ozone currently is not on this list. There are opportunities to use other substances for produce decontamination, providing that they function as 'processing aids', which are defined as: 'any substance not consumed as a food itself, intentionally used in the processing of raw materials, foods or their ingredients to fulfil a certain technological purpose during treatment and processing and which may result in the unintentional but technically unavoidable presence of residues of the substance or its derivatives in the final product, provided that these residues do not present any health risk and do not have any technological effect on the finished product'. Chlorine and chlorine dioxide, which are used for fruit and vegetable washing, are regarded as 'processing aids'. Thus, they would appear to be outside the scope of the biocide controls as they are 'defined' in Directive 89/107/EEC.

Whether a washwater chemical is an additive or processing aid is of significance, since it is unlikely that a 'natural' agricultural product (such as leafy salad) which carries the name of a chemical additive on the label will appeal to consumers. Therefore, in practice, washwater decontaminants must be able to be classed as processing aids, which requires they have no lasting technological effect on the produce, a key challenge for the chemical sciences (RSC 2009).

The European Commission is planning to develop more detailed regulations governing the use of processing aids. Although it is at a very early stage of development, one possibility being considered is that the definition of a processing aid will be tightened, so that residues of processing aids in a final food will no longer be acceptable, unless the substance in question is specifically authorised for food use. Legislation on processing aids is not yet harmonised at the European Community level, and so processing aids that may be used legally in the UK and France might not be permitted in other member states. A global approach to processing aids is needed to control the agents which are essential for the minimisation of the potential transmission of pathogens from water sources to produce. The risk from pathogens is not eliminated by using large quantities of water; the risk of pathogen cross-contamination is only avoided by using processing aids.

2.3.3 Overview of Canadian regulations

Ozone is permitted for use in Canada as a food additive according to certain provisions that are listed in the Health Canada Food and Drug Regulations (Table VIII, section B.16.100). Ozone may be used as a maturating agent in cider and wine and as a chemosterilant in packaged mineral or spring waters. All of these uses should be consistent with a level of use defined by Good Manufacturing Practice (GMP).

Health Canada has not objected to the use of weakly ozonated water (up to 2 ppm) for fresh fruit/vegetable processing (e.g. flume water, transportation, water in tanks for temporary storage, etc.). Such applications of ozone are aimed at sanitising water rather than acting as a preservative on vegetables. Some petitioners carry out research to determine exact parameters of ozone concentration on fruits and vegetables. As a function of the system design these concentrations may vary; that is, they may be lower than 2 ppm.

Health Canada also has not objected to the use of ozone for the purpose of sanitising water in general in food industry premises (without direct food contact). Egg shell may be treated with ozonated water (up to 2 ppm) for decontamination. Very weak concentrations of ozone, below 1 ppm, may be used as a pesticide, against plant decay, in cold storage facilities for fresh vegetables. Such applications of ozone are considered by the Pest Management Regulatory Agency (PMRA) of Health Canada. The Food Packaging and Incidental Services Section responds to petitioners wishing to obtain 'no-objection opinions' from Health Canada, which are issued for application of ozone on food-contact surfaces and air in contact with food. That section also enables certain concentrations of ozone as a means to decontaminate the hands of food plant personnel.

The application of ozone to the water supply, including recirculated washwater, is permitted by Health Canada as an acceptable water treatment provided that the following conditions are met:

(1) The amount of ozone added to the water does not exceed the minimum level required to effectively reduce the microbial levels in the water (including water to make ice), in accordance with GMPs. A processor and the manufacturer of the ozone-generating equipment should determine and validate the amount of ozone needed to achieve disinfection and no more than that amount should be added.
(2) The concentration of residual (remaining) ozone in the water that may come into direct contact with the fresh food is negligible. In other words, GMPs are applied, and no more ozone other than that which is needed for disinfection is applied to the water, resulting in minimal or no residual ozone.
(3) If present, residual (remaining) ozone in recirculated washwater should not bring about a change in the characteristics of the fresh food and will be removed (e.g. filtered) from the washwater prior to its contact with produce or poultry carcasses/parts.
(4) The ozone in the system is not used for the purpose of preservation of the fresh food.
(5) The ozone generator does not generate ozone into the air, incidental to its normal operation, at a level in excess of 0.05 ppm (Health Canada 2007).

The current Health Canada regulations for bottled water require that producers who add ozone to mineral water or spring water must include a statement to this effect on the principal display panel of the product label. Also, rules governing food additives require that the added ozone be listed

Table 2.2 The main disinfecting compounds currently permitted by FSANZ Standard A16 that may be used as washing agents/processing aids.

Disinfecting agent	Standard A16 permission
Chlorine Chlorine dioxide Calcium hypochlorite Sodium chlorite Sodium hypochlorite Hydrogen peroxide Peracetic acid Ozone	Group II – bleaching agents, washing and peeling agents
Sodium hydroxide Phosphoric and sulfuric acids	Generally permitted processing aids

as an ingredient. So under the current regulations, ozone, when added to spring water or mineral water, must be listed on the product label twice: in the products list of ingredients and in a separate statement on the principal display panel.

2.3.4 Overview of Australian and New Zealand regulations

In Australia, ozone treatment is regarded as a processing aid in the Food Standards Code (FSANZ 2006) Standard 1.3.1, Clause 11. There are currently no restrictions on its use, as long as GMP is followed. In the Australian Food Standards Code, mineral water is defined as 'ground water obtained from subterranean water-bearing strata that, in its natural state, contains soluble matter'. It is used synonymously with the term 'spring water'. The Code permits various treatments of mineral water, including UV sterilisation, pasteurisation and 'ozone treatment'. Table 2.2 lists the main disinfecting compounds currently permitted by Standard A16 of FSANZ for use as washing agents/processing aids.

2.3.5 Overview of Japanese regulations

In the mid-1990s, ozone was approved for food processing in Japan. As of 2006, there were more than 500 ozone-based gas or water treatment installations in the food industry throughout Japan, and more than 100 000 food treatment plants processing a wide variety of foods and food products (Naito and Takahara 2006). This widespread application is a clear indication of the efficacy and usefulness of ozone in the food industry.

2.4 Global harmonisation of food safety regulations

The Global Harmonization Initiative (GHI) was initiated in 2004 as a joint activity of the US-based Institute of Food Technologists (IFT) International Division and the European Federation of Food Science and Technology (EFFoST) as a network of scientific organisations and individual scientists with the goal of working together to promote harmonisation of global food safety regulations and legislation. GHI aims at 'Achieving consensus on the science of food regulations and legislation to ensure the global availability of safe and wholesome food products for all consumers'. GHI facilitates global discussion about the scientific issues that support decisions made by national governments and international regulatory bodies by:

(1) Providing the foundation for sound, sensible, science-based regulations.
(2) Creating a forum for scientists and technologists to interact with regulatory authorities, globally.
(3) Providing industry, regulators and consumers an independent, authoritative information resource.

This global initiative should facilitate the continued adoption of ozone as a safe and environmentally friendly technique for the food processing industries.

References

AFSSA (2003) Opinion on the use of ozone as an auxiliary technology for the treatment of wheat grains before grinding for flours incorporated into pastries containing simple sugars, AFSSA, Saisine No. 2003-SA-0055, 24 July 2003.

AFSSA (2004) Opinion on the use of ozone as an auxiliary technology for the treatment of wheat grains entering into the composition of bread and baked products, excluding traditional French bread, AFSSA, Saisine No. 2004-SA-0161, 21 Sept. 2004.

Boyce, T.G., Swerdlow, D.L. and Griffin, P.M. (1995) *Escherichia coli* O157:H7 and the hemolytic–uremic syndrome, *New England Journal of Medicine*, **333**: 364–8.

CFR (1997) Air Contaminants, Title 13, Part 1910, Washington, DC: Office of Federal Register.

FDA (1995) Beverages: bottled water; final rule, Food and Drug Administration, Fed. Reg., **60**: 57 075–130.

FDA (2001) Hazard analysis and critical point (HACCP): procedures for the safe and sanitary processing and importing of juice; final rule, *Federal Register*, **66**: 6137–6202.

FDA (2003) Code of Federal Regulations, Title 21. USA: Government Printing Office.

FDA (2004) FDA Guidance to Industry, 2004: Recommendations to processors of apple juice or cider on the use of ozone for pathogen reduction purposes. Retrieved from http://www.cfsan.fda.gov/~dms/juicgu13.html.

Graham, D.M., Pariza, M.W., Glaze, W.H., Erdman, J.W., Newell, G.W. and Borzelleca, J.F. (1997) Use of ozone for food processing, *Food Technology*, **51**(6): 72–6.

Hammond, J. (2004) The legislative control of substances used for produce decontamination, *Washing and Decontamination of Fresh Produce Forum*, Chippin Campden, Gloucestershire, UK: Campden and Chorleywood Food Research Association Group, *Newsletter*, 5(December).

Hampson, B.C. (2001) Emerging technology: ozone. Presented at the IFT Annual Meeting, New Orleans, LA, 2001.

Health Canada (2007) Proposed Regulations for Intentional Ozone Generators. Retrieved from http://www.hc-sc.gc.ca/cps-spc/legislation/consultation/ozone-eng.php.

Hill, A.G. and Rice, R.G. (1982) Historical background, properties and applications, in Rice, R.G. and Netzer, A. (eds) *Handbook of Ozone Technology and Applications*, Vol. 1, Ann Arbor, MI: Ann Arbor Science Publishers, pp. 1–37.

IFPA (2001) *Food Safety Guidelines for Fresh-cut Produce Industry*, 4 edn, Alexandria, VA: IFPA.

Mahapatra, A.K., Muthukumarappan, K. and Julson, J.L. (2005) Applications of ozone, bacteriocins and irradiation in food processing: a review, *Critical Reviews in Food Science and Nutrition*, **45**: 447.

Naito, S. and Takahara, H. (2006) Ozone contribution in food industry in Japan. *Ozone: Science & Engineering*, **v28**(6): 425–9. Retrieved from http://www.informaworld.com/smpp/title~db=all~content=t713610645~tab=issueslist~branches=28.

USDA FSIS (2001) Letter from Robert C. Post (FSIS, Washington, DC) to Mark D. Dopp (Am. Meat Institute, Arlington, VA) dated Dec. 21.

Vojdani, J.D., Beuchat, L.R. and Tauxe, R.V. (2008) Juice-associated outbreaks of human illness in the United States, 1995 through 2005, *Journal of Food Protection*, **71**(2): 356–64.

3 Chemical and Physical Properties of Ozone

Annel K. Greene, Zeynep B. Güzel-Seydim
and Atıf Can Seydim

3.1 Introduction

Ozone is formed photochemically in the stratosphere, in high-voltage electrical arcs, in photochemical smog and by ultraviolet (UV) sterilisation lamps and gamma radiation plants (Mustafa 1990). The characteristic fresh, clean smell of air following a thunderstorm represents freshly generated ozone in our atmosphere. The passage of new regulations (please see Chapter 2) and broad spectrum application may make ozone a greener alternative to traditional approaches for various food applications. Generally Recognised as Safe (GRAS) status and US Food and Drug Administration (FDA) approval of ozone as an antimicrobial agent for direct food contact (FDA 2001) have allowed ozone to be used in food processing. Because of residual compounds and reaction byproducts, chemical sanitising agents have come under scrutiny. For example, chlorination byproducts such as trihalomethanes and chloramine compounds are potentially carcinogenic (Pascual et al. 2007). Ozone reaction products from oxidation of organic compounds, such as aldehydes, ketones or carboxylic acids, have not been reported to have adverse health consequences (Pascual et al. 2007). Ozone is also considered as an alternative to chlorine to prevent the formation of halogenated organic compounds. However, the efficacy of ozone on food and food products within the food industry depends on its physicochemical properties. This chapter discusses the physical, chemical and antimicrobial properties of ozone as they apply to the food industry.

3.2 The molecular structure of ozone

The strong reactivity of ozone is due to the structure of the molecule. The ozone molecule consists of three oxygen atoms. In the valence shell of each oxygen atom are two unpaired electrons, each one occupying one $2p$ orbital. This means that during its formation, three oxygen atoms are combined,

Ozone in Food Processing, First Edition. Edited by Colm O'Donnell, B.K. Tiwari, P.J. Cullen, and Rip G. Rice.

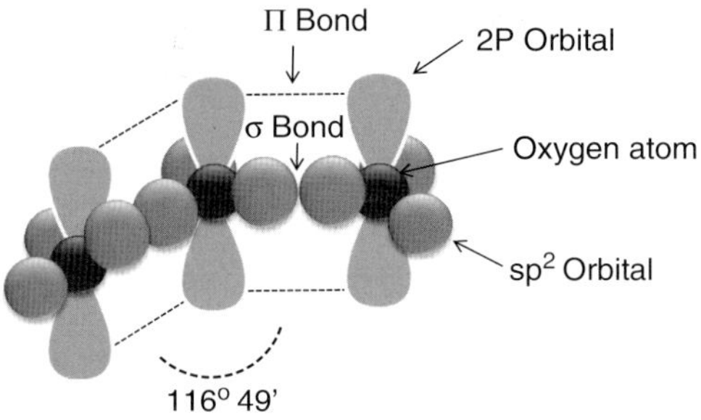

Figure 3.1 The molecular structure of ozone.

with the central oxygen rearranged in a plane sp^2 from the $2s$ and two $2p$ atomic orbitals of the valence band. With this rearrangement the three new sp^2 hybrid orbitals form a triangle with an oxygen nucleus at its centre; that is, with an angle of 116° 49′, as shown in Figure 3.1 (Beltran 2004).

The configuration of the ozone molecule arises from the way the sp^2 and $2p^2$ orbitals are combined, which results in two 9-molecular orbitals that move throughout the ozone molecule. As a consequence, the ozone molecule represents a hybrid formed by the four possible structures (Guzel-Seydim et al. 2004; von Gunten 2003). The high reactivity of ozone can then be attributed to the electronic configuration of the molecule, with the absence of electrons in one of the terminal oxygen atoms in some of the resonance structures confirming the electrophilic character of ozone, while the excess negative charge present in some other oxygen atom imparts a nucleophilic character (Beltran 2004).

3.3 The chemical and physical properties of ozone

Ozone (O_3) is the triatomic oxygen formed by addition of a free radical of oxygen to molecular oxygen. In 1781, Van Marum first described the pungent odour of ozone (Evans 1972). Later, in 1840, Schönbein named the substance 'ozone', based on the Greek word 'ozein' for 'smell' (Manley and Niegowksi 1967; Rice and Bollyky 1981; Kogelschatz 1988). Ozone is a blue gas at ordinary temperature when generated from dried air, but colourless when generated from high-purity oxygen. Regardless of how it is generated, at normal production concentrations for most applications, including food processing, the colour is not noticeable. At −112 °C, ozone condenses to a dark blue liquid. Liquid ozone can be detonated if >20% ozone–oxygen mixtures occur. The three atoms of oxygen in the ozone molecule are arranged at an obtuse angle, whereby a central oxygen atom is attached to two equidistant oxygen atoms; the included angle is approximately 116° 49′ and the bond length is 1.278 Å. The boiling point of ozone is −111.9 ± 0.3 °C, the melting point is −192.5 ± 0.4 °C, the critical temperature is −12.1 °C and the critical pressure is 54.6 atm (Manley and Niegowski

1967). Ozone is slightly denser (2.14 g/L) compared to air (1.28 g/L) at 0 °C and atmospheric pressure.

Ozone solubility in water is affected by temperature, with solubility decreasing with increasing temperature. At 0 °C, ozone solubility is 0.6401 ozone/L water, whereas at 60 °C it is insoluble in water (Hill and Rice 1982). Ozone has a high oxidising potential of 2.07 V (Manley and Niegowski 1967; Pehkonen 2001) compared to chlorine (1.36 V) and oxygen (1.23 V). The mechanism of ozone decomposition in water has not been completely resolved but some scenarios suggest decomposition into hydroxyl radicals, oxygen and hydroxide ions, and at pH ranges above 7.5, the formation of hydroxyl radicals is increased (Pehkonen 2001). Hydroxyl free radicals have a higher oxidation potential (2.80 V) than ozone (Pehkonen 2001).

At room temperature, ozone is an unstable gas. Ozone readily degrades (Manley and Niegowski 1967) but has a longer half-life in the gaseous state than in aqueous solution (Rice 1986). Purity of water usually affects ozone stability. Although ozone in pure water degrades rather quickly to oxygen, it degrades even more rapidly in impure solutions. Hill and Rice (1982) reported that approximately 50% of ozone is destroyed in 20 minutes at 20 °C in distilled or tap water, whereas only 10% of ozone breaks down in 85 minutes in 20 °C double-distilled water. Ozone solubility in water is 13 times that of oxygen at 0–30 °C and it is progressively more soluble in colder water (Rice 1986). Ozone decomposition is faster in higher water temperatures.

In order to generate ozone, a diatomic oxygen molecule must first be split. The resulting free radical oxygen is thereby free to react with diatomic oxygen to form the triatomic ozone molecule. However, in order to break the O-O bond, a great deal of energy is required.

UV radiation (188 nm wavelength) and corona discharge methods can be used to initiate free radical oxygen formation and thereby generate ozone. In order to generate commercial levels of ozone, the corona discharge method is usually used. In the corona discharge design, two electrodes are separated by a narrow discharge gap. Ozone is formed by the addition of the free radical of oxygen to a molecule of oxygen. If air is passed through the generator as a feed gas, 1–3 % ozone is made; using high-purity oxygen may yield as high as 16 % ozone (Rice et al. 1981).

The energetic process that can produce ozone molecules can also destroy ozone. Higher ozone concentrations result in higher rates of ozone destruction. Consequently, ozone concentration cannot be increased beyond the point where the rates of formation and destruction are equal (Manley and Niegowski 1967). Because ozone degrades spontaneously, ozone gas cannot be stored (Kogelschatz 1988; Wickramanayake 1991; Coke 1993).

Ozone is a toxic gas; toxicity is dependent on concentration and length of exposure (Pascual et al. 2007). At short-term exposure rates of 0.1–1.0 ppm, symptoms include headaches, nosebleeds, eye irritation, dry throat and respiratory irritation. At higher exposure levels (1–100 ppm), symptoms become more severe and include asthma-like symptoms, tiredness and loss of appetite (see Chapter 15).

3.3.1 The chemical mechanisms of ozonation

Ozone is an unstable gas that has to be generated and applied at the point of use (Rice and Netzer 1982). In order to bring ozone into contact with a target substance in the water phase, it must first be transferred into water using a gas–liquid reactor/contactor. The decay of ozone in water is characterised by a fast initial decrease, followed by a second phase in which ozone concentration decreases with first-order kinetics (von Gunten 2003). The reactions that occur are not easy to explain, not only because many chemical reactions can occur simultaneously, but also because ozone can react either indirectly (decomposition via a chain-reaction mechanism resulting in the production of hydroxyl free radicals) or directly (via selective reactions with substances in the water matrix) (Masschelein 1992). This combined effect makes ozone a highly effective oxidiser when compared to other chemicals. Therefore, the efficacy of an ozonation process is usually based on the effects of both direct and indirect reaction mechanisms, and these largely depend on the water composition, especially its pH, the type and content of natural organic matter, and its alkalinity (Hoigné 1998).

3.3.2 Ozone reaction pathways in water

Indirect reaction

The disintegration of ozone in water into OH radicals – that is, molecules with an unpaired electron – arises from the indirect reaction pathway. It is widely known that the OH radicals resulting from this reaction are very short-lived compounds that have an even stronger oxidation potential than ozone because most radicals are highly unstable and immediately undergo a reaction with another molecule in order to reclaim the missing electron. Indirect reactions in an ozone oxidation process can be very complex and lead to the formation of several primary high-reactive species. Figure 3.2 shows a schematic diagram of intertransformations among primary high-reactive species. An indirect reaction can take place according to the following process model (Gottschalk et al. 2010; Sehested et al. 1991):

(1) **Initiation** The first reaction that takes place is accelerated ozone decomposition by a type of initiator, which can be an OH molecule. The reaction between hydroxide ions and ozone leads to the formation of an O_2^{-o} anion and a HO_2^o radical:

$$O_3 + OH^- \rightarrow O_2^{-o} + HO_2^o \tag{3.1}$$

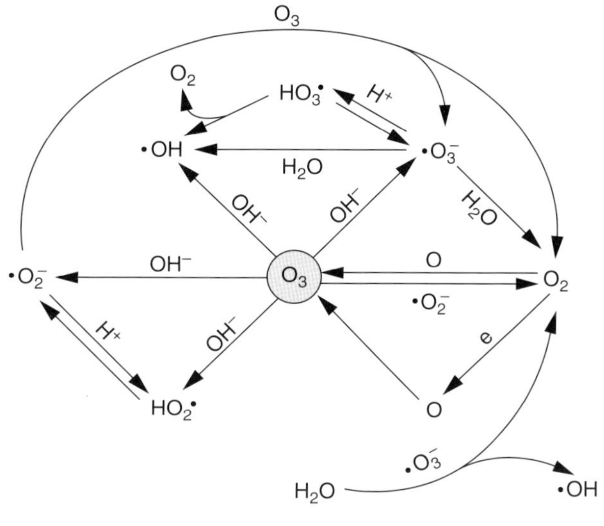

Figure 3.2 Schematic diagram of intertransformations among primary high-reactive species. (Reprinted from *Chemical Engineering Journal*, Volume 138, Issue 1–3, Jing Xue, Li Chen and Honglin Wang, Degradation mechanism of Alizarin Red in hybrid gas–liquid phase dielectric barrier discharge plasmas: experimental and theoretical examination, 120–7, 2008, with permission from Elsevier.)

This radical has an acid/base equilibrium when pKa = 4.8. However, above this value, this radical forms a O_2^{-o} radical and therefore no longer splits:

$$HO_2^o \rightarrow O_2^{-o} + H^+ \tag{3.2}$$

(2) **Radical chain reaction** During the radical chain reaction, OH radicals are formed. Once the O_2^{-o} anion reacts with ozone the resulting O_3^{-o} anion immediately decomposes via hydrogen trioxide HO_3^o to an OH^o radical as follows:

$$O_3 + O_2^{-o} \rightarrow O_3^{-o} + O_2 \tag{3.3}$$

$$HO_3^o \leftrightarrow O_3^{-o} + H^+ \tag{3.4}$$

$$HO_3^o \rightarrow OH^o + O_2 \tag{3.5}$$

The OH^o radical can then react as follows:

$$OH^o + O_3 \rightarrow HO_4^o \tag{3.6}$$

$$HO_4^o \rightarrow HO_2^o + O_2 \tag{3.7}$$

The formation of HO_2^o radicals can then initiate the reaction again. As a result, a chain reaction develops, which is maintained by promoters: substances that transform OH radicals to O_2^{-o} (superoxide) radicals, such as organic molecules (Bühler et al. 1984).

(3) **Termination** Radical catchers (or scavengers) terminate the above chain reaction and inhibit ozone decay. This is because organic and inorganic substances like CO_3^{2-} and HCO_3^- react with OH radicals to form secondary radicals that do not produce superoxide radicals. Another example of termination is the reaction of two radicals (Gottschalk et al. 2010):

$$OH^{o} + HO_2^{o} \rightarrow O_2 + H_2O \quad (3.8)$$

Direct reaction

Because of the molecular structure of ozone (Figure 3.1), it can act as an electrophilic or nucleophilic agent during reactions (von Gunten 2003), with these types of reaction occurring in solutions containing organic pollutants. Generally, electrophilic reactions will occur with organic water contaminants with a high electron density and will act faster in solutions consisting of high levels of aromatic compounds (Gottschalk et al. 2010). Nucleophilic reactions take place mainly when there is a shortage of electrons and particularly at carbon compounds that contain electron-withdrawing groups such as –COOH and $-NO_2$. However, for these groups the reaction speed is much lower. Overall, the direct oxidation of organic matter by ozone involves a quite selective reaction mechanism (von Gunten 2003). Moreover, it is important to note how the pH value of the water system can influence ozone decomposition; with $pH>7$ causing an increase in the rate of ozone decomposition. Also, in strongly acidic solutions ($pH<3$) the OH radicals do not influence the decomposition of ozone.

Advanced oxidation processes

Owing to the selectiveness of molecular ozone, not all compounds in a water solution are rapidly oxidised by it, and certainly not at the same rates. Moreover, in addition to the fact that ozone has relatively low solubility and stability in water, sometimes the water itself may contain ozone-resistant compounds (Hoigné 1998), such as many pesticides and chlorinated solvents. Therefore, as a result of both the high cost of ozone production and the only partial oxidation of most organic compounds present in water, the application of ozonation might not be feasible from an economic point of view. One way of overcoming this problem is to transform the ozone into very reactive OH radicals as quickly as possible, in a procedure known as 'advanced oxidation'. This term simply defines an oxidation process that generates OH radicals in large quantities in order to enhance the level of water treatment. These radicals can be initiated in the water containing ozone by means of an activator, such as hydrogen peroxide (H_2O_2) or UV light, or in the absence of ozone, by hydrogen peroxide and UV light. Then, due to the oxidisation processes, harmful substances are decomposed to less harmful substances without the formation of particulate matter (von Gunten 2003; Huang et al. 1993).

Over the years, advanced oxidation processes (AOPs) have been used to produce OH radicals in order to successfully degrade most organic compounds present in polluted water (Fahmi et al. 2003; Ghaly et al. 2000). However, because OH radicals are more powerful oxidants than molecular ozone, and because of their nonselective reactions, they are also likely to be utilised by competitive reactions in addition to the reaction(s) desired (Gottschalk et al. 2010). Consequently, new methods based on molecular ozone reactions have been used to study how ozone dissolution and stability in water can be improved. The main examples are ozonation in the presence of nonpolar bonded alumina phases (Kasprzyk-Hordern et al. 2004) and two-phase ozonation (Kasprzyk-Hordern et al. 2003), both of which have been successful in reducing the competitive strain on the ozonation process. The efficiency of these methods is mainly due to the immobilisation of perfluorinated hydrocarbons on the surface of alumina, which prevents the dissolution of perfluorinated molecules into the aqueous phase, using a technique involving the liquid–liquid extraction of organic substances from the aqueous phase into the organic phase and subsequent oxidation by molecular ozone dissolved in the organic phase (nonpolar perfluorinated hydrocarbon solvent saturated with ozone) (Kasprzyk-Hordern et al. 2003).

3.4 Ozone action on macromolecules

In the presence of organic compounds, ozone reacts in a variety of complex reactions due to the formation of various reactive species (Figure 3.2). The C-H and S-H bonds of alkanes, alkenes, amines and sulfhydryl compounds are attacked by ozone (Adachi 2001). Amino acids may be attacked directly at the primary amine nitrogen atom or on the R group (Mustafa 1990; Adachi 2001). Ozonation of lipids leads to formation of peroxides (Adachi 2001). Purines and pyrimidines were reported as being more resistant to ozone than many other organic compounds (Adachi 2001). Cataldo (2003) studied the reactivity of ozone with five different proteins, as shown in Figure 3.3. Using electron microscopy to monitor the reaction of invertase, pectinase, papain, trypsine and gelatin in solution with ozone, Cataldo determined that only cysteine and the aromatic amino acids (phenylalanine, tryptophan and tyrosine) were oxidised. The main polyamide bond of the protein was not affected by ozone. In the dry state, no ozone attack was noted on proteins even after prolonged ozone treatment.

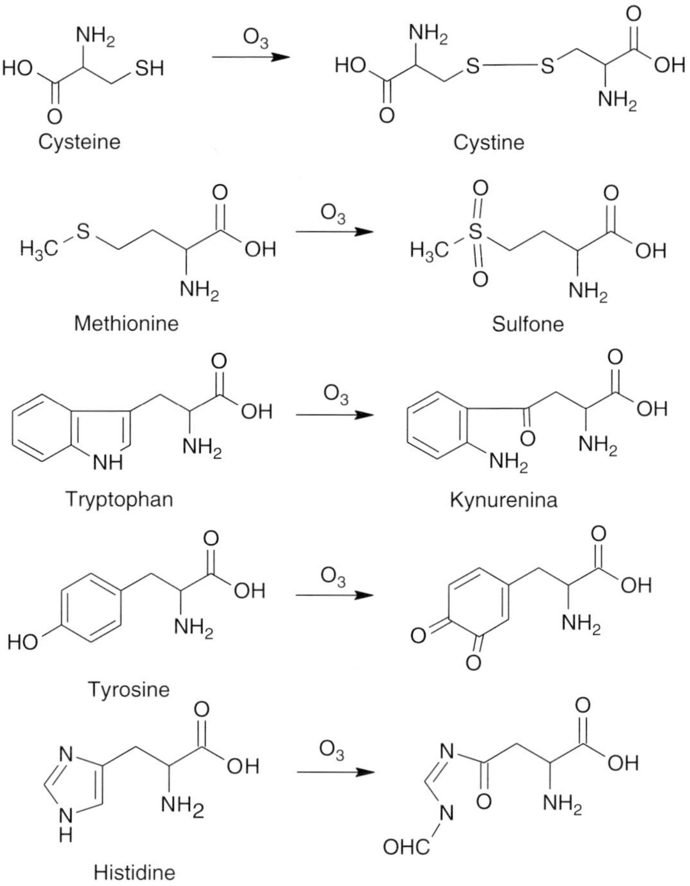

Figure 3.3 Reaction of amino acids with ozone (Cataldo 2003). (Reprinted from *Polymer Degradation and Stability*, Volume 88, Issue 1, Franco Cataldo, On the action of ozone on proteins, 105–14, 2003, with permission from Elsevier.)

3.5 Mechanisms of microbial inactivation

Ozone has a broad spectrum of antimicrobial activity in water and wastewater; accordingly, it is widely regarded as a potent bactericidal and virucidal agent. In the USA, the Surface Water Treatment Rule of 1989 requires that water suppliers must apply sufficient concentration (C) of disinfectant for sufficient time (t) in order to destroy a defined quantity of microorganisms (Botzenhart et al. 1993). Based on the results of a study on inactivation of *Bacillus subtilus* spores, Botzenhart et al. (1993) concluded that ozone is a more effective sanitiser than chlorine dioxide. Ozone is very unstable both in the gaseous phase and in solution, decomposing into hydroxyl (HO°), hydroperoxy ($\cdot HO_2$) and superoxide radicals ($\cdot O_2^-$).

The high reactivity of ozone is attributed to the oxidising power of these free radicals (Manousaridis et al. 2005).

The bactericidal effects of ozone have been studied on a variety of organisms, including Gram-positive and Gram-negative bacteria as well as spores and vegetative cells (Ishizaki et al. 1986; Restaino et al. 1995). The antimicrobial efficacy of ozone against food-related microorganisms has been studied for Gram-positive bacteria (*Listeria monocytogenes, Staphylococcus aureus, Bacillus cereus* and *Enterococcus faecalis*) and Gram-negative bacteria (*Pseudomonas aeruginosa* and *Yersinia enterocolitica*, yeasts; and *Candida albicans, Zygosaccharomyces bacilli* and spores of *Aspergillus niger*) (Restaino et al. 1995).

Inactivation by ozone is a complex process which involves ozone acting upon various cell-membrane and wall constituents (e.g. unsaturated fats) along with cell-content constituents (e.g. enzymes and nucleic acids). Ozone may affect membrane glycoproteins and/or glycolipids (Guzel-Seydim et al. 2004), membrane-bound enzymes (Murray et al. 1965) and oxidation of double bonds by singlet oxygen found in the cell (Scott 1975), and possibly may damage proteins and DNA (Komanapalli and Lau 1996). Two major mechanisms of ozone destruction of the target organisms were identified: (1) ozone oxidises sulfhydryl groups and amino acids of enzymes, peptides and proteins to smaller peptides; and (2) ozone oxidises polyunsaturated fatty acids to acid peroxides (Victorin 1992). Micro-organisms are inactivated by disruption of the cell envelope or disintegration leading to cell lysis. Transmission electron microscopic micrographs of *Bacillus* spores treated with ozone (Figure 5.3) suggest that ozone inactivates spores by degrading the outer spore component (spore coat layers make up approximately 50% of the spore volume), thus exposing the cortex and core to the action of ozone (Foegeding 1985; Khadre et al. 2001). Young and Setlow (2004) determined that ozone does not kill spores by DNA damage but rather by damaging the ability of the spores to germinate. The researchers hypothesised that damage to the inner membrane of spores causes defects in spore germination.

Both molecular ozone and the free radicals produced by its breakdown play a part in this inactivation mechanism but there is no consensus on which is more decisive. It has not been well established whether molecular ozone or the radical species are responsible for inactivation of microorganisms. Some researchers have suggested that direct reaction with molecular ozone is the predominant mechanism for inactivation of microorganisms (Finch et al. 1992; Labatiuk et al. 1994; Hunt and Mariñas 1997), while others suggest indirect reactions with radicals are responsible for inactivation (Bancroft et al. 1984). It is likely that the relative importance of direct and indirect reactions with ozone in determining microbial inactivation responses will vary between microorganisms (Blatchley and Hunt 2002).

Ozone decomposition has been explained as occurring in three stages; initiation, promotion and inhibition. During the initiation step, free radicals are generated, such as superoxide radical ions and hydroperoxide radicals, which lead to formation of the highly reactive hydroxyl radical. These hydroxyl radicals are one of the factors contributing to ozone decomposition. The promotion step involves regeneration of the hydroperoxide and superoxide radicals through reactions involving participation of promotors such as formic acid, glyoxylic acid, primary alcohols and aryl groups. In contrast, in the inhibition step the consumption of hydroxyl radicals occurs via ions like bicarbonate, carbonate, tertiary alcohols and alkyl groups, without regeneration of the superoxide radical ion (Staehelin and Hoigné 1985; Khadre et al. 2001). Bicarbonate ions are generally present in microbial cells, which could act as scavengers of radicals otherwise responsible for inactivation of microorganisms. Additionally, factors promoting ozone decomposition in the system can lead to faster dissipation of ozone, resulting in a requirement for increased ozone concentration in order to achieve the desired inactivation level (Zuma et al. 2009).

The resultant disruption or lysis of cell walls (probably by oxidative destruction) associated with ozone is a faster inactivation mechanism than that of other disinfectants, which require the disinfecting agent to permeate through the cell membrane in order to be effective (Pascual et al. 2007). Scott and Lesher (1963) reported that ozone caused leakage of cell contents into the medium and lysis of some cells. Therefore, ozone-demanding substances are generated during the ozone inactivation process. Finch and others (1988) found that *E. coli* cells demanded 0.06 mg/L ozone after lysis and attributed the second phase of inactivation to this ozone-created demand (Kim and Yousef 2000). Generally, with regard to the spectrum of microbial action, each microorganism has an inherent sensitivity to ozone. Bacteria are more sensitive than yeasts and fungi. Gram-positive bacteria are more sensitive to ozone than Gram-negative organisms, and spores are more resistant than vegetative cells. Due to the mechanism of ozone action, which destroys the microorganism through cell lysis, the development of resistance to ozone disinfection is not found (Pascual et al. 2007).

3.6 Ozone reactions against virus

Several researchers reported that ozone inactivates enveloped and nonenveloped viruses in water (Bolton et al. 1982; Roy et al. 1982; Akey and Walton 1985). Ozone was effective against bacteriophage f2 (Kim et al. 1980), enveloped viral species (including vesicular stomatitis viral species, influenza A virus (WSN strain), infectious bovine rhinotracheitis virus) and nonenveloped viruses (including polio type I and infectious canine hepatitis virus) (Bolton et al. 1982). Research indicated that enveloped viruses were more susceptible to ozone inactivation than those lacking lipid envelopes (Bolton et al. 1982). Hall and Sobsey (1993) determined that hepatitis A virus was inactivated in 5 seconds with 0.4 ppm ozone dose. Shin and

Sobsey (2003) studied methods of controlling Norwalk virus in drinking water. Ozone (0.37 ppm) at pH 7 was used to treat water for 5 minutes at 5 °C. Norwalk virus concentration was reduced by greater than 3 logs during a contact time of 10 seconds. Similar results were obtained for poliovirus 1 and bacteriophage MS2 (Shin and Sobsey 2003).

3.7 Ozone reaction on biofilms

Stewart and Raquepas (1995) developed a method for analysing the efficacy of antimicrobial agents when applied to microbial biofilms. Using an adaptation of reaction-diffusion theory, the researchers studied the rate of antimicrobial penetration into a biofilm and the limitations of transport on overall biofilm disinfection rates. The researchers noted that in the case of a strong oxidising agent such as ozone or chlorine, the antimicrobial agent eventually penetrates into the biofilm because it depletes the reactive biomass components in the biofilm. However, the time to penetrate the biofilm is considerably longer than would be expected by diffusion. The researchers theorised that this was the basis for observations that biofilms offer microorganisms resistance to chemical disinfectant agents (Stewart and Raquepas 1995). Tachikawa et al. (2009) studied the disinfection and removal of biofilms of *Pseudomonas fluorescens* and *Pseudomonas aeroginosa* by ozone-containing water using an artificial microbial biofilm system. Results indicated that ozone was effective in destroying the organisms and was capable of removing exopolysaccharides in the biofilm matrices.

Acknowledgments

The authors wish to thank Mr Steven D. Chambers, Ms M. Melissa Hayes and Ms Rayleen A. Hendrix for their kind assistance in preparing and reviewing this chapter.

References

Adachi, D. (2001) Virus inactivation by ozone, Master of Science thesis, University of Toronto.

Akey, D.H. and Walton, T.E. (1985) Liquid-phase study of ozone inactivation of Venezuelan Equine encephalomyelitis virus, *Appl Environ Micro*, **50**(4): 882–6.

Beltran, F.J. (2004) *Ozone Reaction Kinetics for Water and Wastewater System*, New York, NY: CRC Press CLL.

Blatchley, ER, and Hunt, NK. (2002) Ozone disinfection in drinking water, *American Society of Civil Engineers*, **2002**: 241–71.

Bolton, D.C., Zee, Y.C. and Osebold, J.W. (1982) The biological effects of ozone on representative member of five groups of animal viruses, *Environmental Research*, **27**: 476–84.

Botzenhart, K., Tarcson, G.M. and Ostruschka, M. (1993) Inactivation of bacteria and coliphages by ozone and chlorine dioxide in a continuous flow reactor, *Water Sci Technol*, **27**: 363–70.

Bühler, R.E., Staehelin, J. and Hoigné, J. (1984) Ozone decomposition in water studied by pulse radiolysis. 1. HO_2/O_2 and HO_3/O_3 as intermediates, *Journal of Physical Chemistry*, **88**: 2560–4.

Cataldo, F. (2003) On the action of ozone on proteins, *Polymer Degradation and Stability*, **82**: 105–14.

Coke, A.L. (1993) Mother nature's best remedy: ozone, *Water Conditioning and Purification*, **Oct.**: 48–51.

Evans, F.L. (1972) *Ozone in Water and Wastewater Treatment*, Ann Arbor, MI: Ann Arbor Science Publishers.

Fahmi, Nishijima, W. and Okada, M. (2003) Improvement of DOC removal by multi-stage AOP-biological treatment, *Chemosphere*, **50**(8): 1043–8.

FDA (2001) Secondary direct food additives permitted in food for human consumption, *Federal Register*, **66**(123): 33 829–30.

Finch, G.R., Smith, D.W. and Stiles, M.E. (1988) Dose-response of *Escherichia coli* in ozone demand-free phosphate buffer, *Water Res*, **22**: 1563–70.

Finch, G.R., Yuen, W.C. and Uibel, B.J. (1992) Inactivation of *Escherichia coli* using ozone and ozone hydrogen peroxide, *Environ. Technol.*, **13**(6): 571–8.

Foegeding, P.M. (1985) Ozone inactivation of *Bacillus* and *Clostridium* spore populations and the importance of the spore coat to resistance, *Food Microbiology*, 2: 123–34.

Ghaly, M.Y., Hartel, G., Mayer R. and Haseneder, R. (2000) Photochemical oxidation of p-chlorophenol by UV/H_2O_2 and photo-Fenton process: a comparative study, *Waste Management*, **21**(1): 41–7.

Gottschalk, C., Libra, J.A. and Saupe, A. (2010) *Ozonation of Water and Waste Water*, 2 edn, Weinheim, Germany: Wiley-VCH Verlag GmbH & Co. KGaA.

Guzel-Seydim, Z.B., Greene, A.K. and Seydim, A.C. (2004) Use of ozone in the food industry, *Lebensmittel-Wissenschaft und Technologie*, **37**(4): 453–60.

Hall, R.M. and Sobsey, M.D. (1993) Inactivation of hepatitis A virus and MS2 by ozone and ozone –hydrogen peroxide in buffered water, *Water Sci Technol*, **27**: 371–8.

Hill, A.G. and Rice, R.G. (1982) Historical background, properties and applications, in Rice, R.G. and Netzer, A. (eds) *Handbook of Ozone Technology and Applications*, Vol. 1, Ann Arbor, MI: Ann Arbor Science Publishers, pp. 1–37.

Hoigné, J. (1998) Chemistry of aqueous ozone, and transformation of pollutants by ozonation and advanced oxidation processes, in Hubrec, J. (ed) *The Handbook of Environmental Chemistry Quality and Treatment of Drinking Water*, Berlin: Springer.

Huang, C.P., Dong, C. and Tang, W.Z. (1993) Advanced chemical oxidation: its present role and future potential in hazardous waste treatment, *Waste Management*, **13**(5/7): 361–77.

Hunt, N.K. and Mariñas, B.J. (1997) Kinetics of *Escherichia coli* inactivation with ozone, *Water Res.*, **31**(6): 1355–62.

Ishizaki, K., Shinriki, N. and Matsuyama, H. (1986) Inactivation of *Bacillus* spores by gaseous ozone, *J. Appl. Bacteriol.*, **60**(1): 67–72.

Kasprzyk-Hordern, B., Andrzejewski, P. and Nawrocki, J. (2003) Catalytic ozonation and methods of enhancing molecular ozone reactions in water treatment, *Applied Catalysis B: Environmental*, **46**: 639–69.

Kasprzyk-Hordern, B., Andrzejewski, P. and Nawrocki, J. (2004) MTBE, DIPE, ETBE and TAME degradation in water using perfluorinated phases as catalysts for ozonation process, *Applied Catalysis B: Environmental*, **51**(1): 51–66.

Khadre, M.A., Yousef, A.E. and Kim, J.-G. (2001) Microbiological aspects of ozone applications in food: a review, *Journal of Food Science*, **66**(9): 1242–52.

Kim, J.G. and Yousef, A.E. (2000) Inactivation kinetics of foodborne spoilage and pathogenic bacteria by ozone, *Journal of Food Science*, **65**: 521–8.

Kim, C.K., Gentile, D.M. and Sproul, O.J. (1980) Mechanism of ozone inactivation of bacteriophage f2, *Appl Environ Micro*, **39**: 210–18.

Kim, J.-G., Yousef, A.E. and Dave, S. (1999) Application of ozone for enhancing the microbiological safety and quality of foods: a review, *Journal of Food Protection*, **62**(9): 1071–87.

Kogelschatz, U. (1988) Advanced ozone generation, in Stucki, S. (ed) *Process Technologies for Water Treatment*, New York, NY: Plenum Publishers, pp. 87–120.

Komanapalli, I.R. and Lau, B.H.S. (1996) Ozone-induced damage of *Escherichia coli* K-12, *Applied Microbiology and Biotechnology*, **46**: 610–14.

Labatiuk, C.W., Belosevic, M. and Finch, G.R. (1994) Inactivation of *Giardia muris* using ozone and ozone-hydrogen peroxide, *Ozone Sci. Eng.*, **16**(1): 67–78.

Manley, T.C. and Niegowski, S.J. (1967) Ozone, in *Encyclopedia of Chemical Technology*, 2 edn, New York, NY: John Wiley & Sons, pp. 410–32.

Manousaridis, G., Nerantzaki, A., Paleologos, E.K., Tsiotsias, A., Savvaidis, I.N. and Kontominas, M.G. (2005) Effect of ozone on microbial, chemical and sensory attributes of shucked mussels, *Food Microbiology*, **22**: 1–9.

Masschelein, W.J. (1992) *Unit Processes in Drinking Water Treatment*, New York, NY: Marcel Dekker.

Murray, R.G.E., Steed, P. and Elson, H.E. (1965) Location of mucopeptide in sections of cell wall of *Escherichia coli* and other gram-negative bacteria, *Can. J. Microbiol.*, **11**(3): 547–60.

Mustafa, M.G. (1990) Biochemical basis of ozone toxicity, *Free Radical Biology and Medicine*, **9**: 245–65.

Pascual, A., Llorca I. and Canut, A. (2007) Use of ozone in food industries for reducing the environmental impact of cleaning and disinfection activities, *Trends Food Sci & Tech*, **18**: S29–S35.

Pehkonen, A. (2001) The effect of dissolved ozone on the corrosion behavior of some stainless steels, Dissertation, Helsinki University of Technology, Department of Materials Science and Rock Engineering.

Restaino, L., Frampton, E.W., Hemphill, J.B. and Palnikar, P. (1995) Efficacy of ozonated water against various food-related microorganisms, *Applied and Environmental Microbiology*, **61**(9): 3471–5.

Rice, R.G. (1986) Application of ozone in water and wastewater treatment, in Rice, R.G., Bollyky L.J. and Lacy, W.J. (eds) *Analytical Aspects of Ozone Treatment of Water and Wastewater*, Chelsea, MI: Lewis Publishers, pp. 7–26.

Rice, R.G. and Bollyky, L.J. (1981) Fundamental aspects of ozone technology, in Rice, R.G. (ed) *Ozone Treatment of Water for Cooling Applications*, Vienna, VA: The International Ozone Association, pp. 1–19.

Rice, R.G. and Netzer, A. (1982) *Handbook of Ozone Technology and Applications*, Ann Arbor, MI: Ann Arbor Science Publishers.

Rice, R.G., Robson, C.M., Miller, G.W. and Hill, A.G. (1981) Uses of ozone in drinking water treatment, *Journal of the American Water Works Association*, **73**(1): 44–57.

Roy, D., Englebrecht, R.S. and Chian, E.S.K. (1982) Comparative inactivation of six enteroviruses by ozone, *J American Water Works Association*, **74**: 660–4.

Scott, D.B.M. (1975) The effect of ozone on nucleic acids and their derivatives, in Rice, R.G. (ed.) *Aquatic Applications of Ozone*, Syracuse, NY, USA: International Ozone Association, Pan American Group, pp. 1–15.

Scott, D.B.M. and Lesher, E.C. (1963) Effect of ozone on survival and permeability of *Escherichia coli*, *Journal of Bacteriology*, **85**: 567–76.

Sehested, K., Corfitzen, H., Holcman, J., Fischer, C.H. and Hart, E.J. (1991) The primary reaction in the decomposition of ozone in acidic aqueous solutions, *Environmental Science and Technology*, **25**: 1589–96.

Shin, G.-A. and Sobsey, M.D. (2003) Reduction of Norwalk virus, poliovirus 1, and bacteriophage MS2 by ozone disinfection of water, *Appl Environ Micro*, **69**(7): 3975–8.

Staehelin, J. and Hoigné, J. (1985) Decomposition of ozone in water in the presence of organic solutes acting as promoters and inhibitors of radical chain reactions, *Environ. Sci. Technol.*, **19**(12): 1206–13.

Stewart, P.S. and Raquepas, J.B. (1995) Implications of reaction-diffusion theory for the disinfection of microbial biofilms by reactive antimicrobial agents, *Chem. Eng. Sci.*, **50**: 3099–104.

Tachikawa, M., Yamanaka, K. and Nakamuro, K. (2009) Studies on the disinfection and removal of biofilms by ozone water using an artificial microbial biofilm system, *Ozone Sci Eng*, **31**: 3–9.

Victorin, K. (1992) Review of the genotoxicity of ozone, *Mutation Research*, **277**: 221–38.

von Gunten, U. (2003) Ozonation of drinking water. Part I. Oxidation kinetics and product formation, *Water Research*, **37**(7): 1443–67.

Wickramanayake, G.B. (1991) Disinfection and sterilization by ozone, in Seymour, S.B. (ed) *Disinfection, Sterilization and Preservation*, 4 edn, Malvern, PA: Lea and Febiyer, pp. 182–90.

Young, S.B. and Setlow, P. (2004) Mechanisms of *Bacillus subtilis* spore resistance to and killing by aqueous ozone, *J Appl Micro*, **96**: 1133–42.

Zuma, F., Lin, J. and Jonnalagadda, S.B. (2009) Ozone-initiated disinfection kinetics of *Escherichia coli* in water, *J Environ Sci, Health Part A – Toxic/Hazard, Subst Environ Eng*, **44**(1): 48–56.

4 Generation and Control of Ozone

Cameron Tapp and Rip G. Rice

4.1 Introduction

There are several ways to produce ozone and many applications in which to apply it. Ozone is used in the food and bottled water industries, in swimming pools, spas, marine aquaria, municipal water treatment, sewage treatment and for many other industrial/commercial applications. For the purposes of this book, the authors will focus on the technology of generating, applying, monitoring and controlling ozone as it applies to food processing plants. In the vast majority of food processing applications the daily amounts of ozone required normally fit into the range of 1–5 pounds/day (454 g to 2.27 kg/day). While there are some applications that demand much higher quantities of ozone, such as the high volume cut greens processors, those installations are more the exception in the food industry at present.

In this chapter the typical components required for feed gas preparation, ozone generation, contacting and monitoring/control, in the food processing industry are described. Figure 4.1 schematicises the various components of an ozonation system from feed gas treatment to final application within a food processing plant.

4.2 Ozone generation

Ozone is generated commercially within the food industry by one of two generally accepted procedures: by passing an oxygen-containing gas through either a source of ultraviolet (UV) radiation or a high-energy electrical field. The first method is known as photochemical (UV), the second is corona discharge (CD), sometimes referred to as 'plasma techniques'.

There are several less commercially mainstream methods of making ozone (electrolysis, radiochemical, reaction of elemental phosphorus with water). However, these methods are either in their early stages of development or have requirements that make them not economically practical for the food

Ozone in Food Processing, First Edition. Edited by Colm O'Donnell, B.K. Tiwari, P.J. Cullen, and Rip G. Rice.

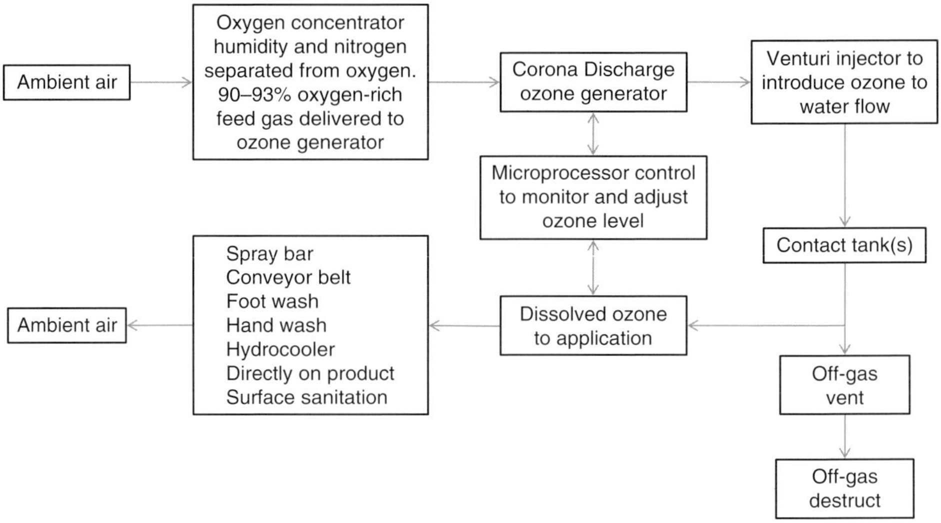

Figure 4.1 Schematic flow diagram of the various steps involving the generation, application and control of ozone in a food processing plant.

processing market at this time. For the purposes of this book, the primary focus will be on CD. Virtually all food processing applications use CD ozone generation. A few applications (air fumigation and some recent modified air packaging applications) use UV radiation. Some applications involve both CD and UV, although not applied at the same time.

Ozone is a metastable molecule produced from elemental oxygen. The overall reaction for ozone formation by any process is described by the endothermic reaction (Langlais et al. 1991, p. 101):

$$3O_2 \leftrightarrow 2O_3\,(\Delta H^\circ \text{at } 1\,\text{atm}, +284.5\,\text{kJ/mol})$$

Also, the entropy of formation is large and unfavourable:

$$\Delta S^\circ(1\,\text{atm}) = -69.9(\text{Jmol})/\text{degree}$$

Clearly, ozone cannot be generated by thermal activation of oxygen, since the standard free energy of formation of ΔG° (1 atm) = +161.3 kJ/mol. In summary, ozone can only be decomposed easily by heating, and adequate temperature control of the process (feed) gas is an important factor in ozone generation efficiency.

4.2.1 Ozone generation by corona discharge (CD)

The technologies involved in CD ozone generation are varied, but all such ozone generators operate fundamentally by passing a dried, oil-free, dust-free, oxygen-containing gas through a high-energy electrical field between two special electrodes, one of which serves as the ground electrode, and the other as the dielectric (current-bearing) medium. As oxygen molecules pass through the electrical field, they are caused to split apart, forming very active atomic oxygen radicals that can combine with intact oxygen molecules to produce molecular ozone, O_3 (Langlais et al. 1991, p. 101). A simplistic relationship is represented by the following equations:

$$O_2 + e^- \rightarrow 2O$$

$$2O + 2O_2 \rightarrow 2O_3$$

It is generally accepted that CD ozone generators are classified into three types: low frequency (50–100 Hz), medium frequency (100–1000 Hz) and high frequency (>1000 Hz). Some 85–95% of the electrical energy supplied to a CD ozone generator produces heat, inherently reducing generated O_3 output and capacity/energy efficiency. Consequently, transference and heat-removal systems are standard items for CD ozone generators. Accordingly, CD systems utilise one or more of the following cooling methods: air, water with oil or Freon, or water.

For all CD ozone generation systems, an electrical charge is diffused over a dielectric surface, creating an electrical field, or 'corona'. Many different materials in a variety of configurations are used for the dielectrics, including everything from silicone rubber to ceramics to scientific-grade glass. These dielectric materials are formed into concentric tubes or flat plates.

Any CD system requires a power supply to produce a high-voltage spark. Depending on the manufacturer, the power supply used to create the electrical field can be anything from a simple, off-the-shelf oil furnace transformer to advanced electronic technology and control subsystems capable of manipulating electrical power characteristics to enhance power generation efficiency and O_3 output concentrations.

The heart of a CD ozone generator is the reaction chamber. Depending on the technology involved – flat plate, concentric tube or otherwise – the high voltage supplied from a power supply uniformly energises the surface of the dielectric (usually made of an scientific-grade glass or ceramic). Opposing the dielectric is a grounding surface positioned to cause a uniform space between the oppositely charged surfaces. This space is referred to as the reaction chamber and is the location where a high-energy electrical field (corona) is created, through which an oxygen-containing feed gas is passed to generate ozone.

Over time, CD has taken on many different names, as listed below. They are all CD using varying techniques to create ozone with an electrical charge across an air gap.

- Silent discharge;
- dielectric barrier discharge;
- cold plasma;
- plasma block;
- cold plate discharge.

4.2.2 Ultraviolet (UV) (photochemical) ozone generation

Light is measured on a scale called an electromagnetic spectrum and its increments are referred to in terms of nanometres (nm). Plate 4.1 represents an electromagnetic scale; note the location of higher-frequency UV light relative to visible light (the range of light perceptible by the human eye).

Low-pressure mercury UV lamps are used as a means of ozone production as well as for air treatment. A major advantage lies in the emission spectrum of the mercury discharge, because mercury emits with high efficiency two resonance lines with wavelengths of 254 and 185 nm (Voronov 2008). The photons with 185 nm wavelength are responsible for ozone production, whereas photons having the 254 nm wavelength are effective in changing the DNA of microorganisms, thereby preventing their ability to reproduce.

An important point to make is that the target wavelength (254 nm) of UV lamps commonly used today for direct inactivation of microbial contaminants cannot be used to generate ozone, because ozone is actually destroyed at this wavelength. UV_{254} systems, referred to as UV sterilisers or germicidal sterilisers, inactivate microorganisms by affecting their DNA, disrupting their ability to reproduce. Water to be treated with ozone is passed by a 254 nm UV lamp separated from a process stream by a quartz glass sleeve. It is the wavelength and intensity of the light itself that impacts the organism, not ozone.

Most commercial UV lamps are made from a form of quartz that contains impurities which absorb the 185 nm emission entirely, so that they produce no ozone (see www.iuva.org). Consequently, when purchasing UV bulbs for the purpose of maximising the output of ozone, it is important that the UV equipment supplier understands the objective and provides the proper ozone-generating bulb(s).

The target wavelength of UV lamps capable of producing ozone is 185 nm. Feed gas (usually ambient air) passes through the high-energy UV_{185} irradiance field surrounding the lamp. Through photodisassociation, a small percentage of oxygen molecules (O_2) are split, resulting in unstable oxygen radical atoms (O_1). Seeking stability, O_1 radical atoms readily attach to surrounding O_2 molecules, resulting in the formation of O_3 molecules, or ozone.

For further information, see DuRon (1982) and Dohan and Masschelein (1987).

4.3 Feed gas preparation systems

4.3.1 Need for feed gas treatment

The quality of gas fed to an ozone generator can be a critical factor, particularly if ozone is produced by CD or plasma techniques. Poor-quality feed gas can dramatically affect the performance and longevity of a CD ozone generator for several reasons. Ambient air contains nitrogen, water vapour (moisture) and in many instances hydrocarbon gases (from incomplete combustion of fossil and automotive fuels). Air that has been through some treatment may also contain traces of oils picked up from improperly maintained gas cleaning equipment. Ozone can react rapidly with such impurities, causing fouling of dielectric surfaces, increased downtime and the need for increased maintenance of equipment.

Just as oxygen molecules can be split apart by strong electrical forces to produce oxygen radical atoms, nitrogen molecules can also be split apart by these same forces (although not necessarily to the same extent) to produce nitrogen radical atoms:

$$N_2 + e^- \rightarrow 2N$$

In turn, these nitrogen radical atoms can combine with oxygen molecules to form several combined nitrogen–oxygen species:

$$N + O_2 \rightarrow NO + NO_2 + N_2O_5$$

With moisture present in the feed gas, the very corrosive nitric acid (HNO_3) is readily formed. Consequently, the gas feeding an ozone generator must be very dry (*maximum* –60 °F (–54 °C) dew point), because the presence of moisture also affects ozone production, as well as leading to the formation of nitric acid. This very corrosive acid can destroy the internal parts of a CD ozone generator, which can cause premature system failure and will increase the frequency of required maintenance.

The two common types of gas preparation to feed CD ozone generators are oxygen and dry air. If an air dryer is selected to feed a CD ozone system, make sure that the air preparation equipment is matched and sized to the ozone. Each ozone generator is designed to operate at an optimal flow rate (standard cubic feet per minute (SCFM) or cubic feet per minute (CFM)) depending upon size of ozone generator, and such information should be stated in the supplier's equipment manual. Today the majority of ozone generating systems use oxygen as the feed gas because the concentrations of ozone produced are increased two to three times for the same energy expenditure. Most ozone equipment manufacturers have optimised their equipment for food processing plants to operate on oxygen feed gas.

For further information, see Langlais et al. (1991, p. 103).

4.3.2 Air preparation systems

Historically, air preparation systems supplying ozone generators for large-volume potable water treatment plants have involved low- or high-pressure air. Low-pressure air treatment also involves air compressors, heat exchangers (to remove heat of compression), refrigerant and heat-reactivated desiccant dryers. Early air-fed CD ozone generators produced ozone at concentrations between 1 and 2.5 wt%. Newer air-fed systems can produce 3–4 wt% ozone. High-pressure (about 100 psig) air-fed ozone systems involve air compressors, heat exchangers and heatless desiccant driers.

Air preparation for generating ozone in the quantities used in food processing plants comes in many sizes and several types depending on the application. In the early days of ozone, most projects used a twin-tower air dryer and filter system. Through the years there has been ongoing debate concerning the types of air preparation used to feed a CD ozone generation system. There is no one single right way, each air preparation method having its own pros and cons. However, there is no debate about how important proper air preparation is to CD ozone systems.

There are several types of air drying systems. The first, pressure swing adsorption (PSA) drying is the most common. These devices take in ambient air, automatically filter (remove dust particles), then pass the filtered air through a desiccant to remove ambient moisture and dry the processed air to below the maximum desired dew point (–54 °C, –60 °F), all at the same time in one small device.

An air compressor pressurises the airflow and sends it through a desiccant (porous bead) bed that adsorbs or traps moisture, while providing dried air to the supply output of the air preparation device. As the desiccant adsorbent bed becomes loaded with moisture, it will desorb (purge) that moisture to waste in vapour form to the environment, recovering the adsorption capacity of the desiccant bed. Some air dryers may employ multiple-chambered designs, so that at any one time some chambers will be undergoing pressurisation and adsorption, while others will be undergoing depressurisation and desorption.

Figure 4.2 is a schematic diagram showing the components of a PSA air dryer system used to provide < –54 °C (–60 °F) dew point air to an ozone generator. PSA air drying systems deliver an ambient air level of 20–22% oxygen with continuous operation.

Some PSA air drying systems use molecular sieves (synthetic zeolites) designed to adsorb moisture from ambient air. Such materials are more efficient than the older chemical desiccants for this purpose. Many ozone manufacturers supply the molecular sieve bed for the drying process but rely on the customer to supply 'plant air' to feed airflow and pressure to the beds. Generally, dry air is used when larger volumes of ozone are required with lower concentrations, for specific applications. For example, food storage rooms require high volumes of ozone-containing air at much lower ozone concentrations than are required for aqueous applications.

A second type of heat-regenerative air drying system involves a convection drying process that relies on alternating molecular sieve beds with a heater to dry the moisture-loaded sieve. These systems are sufficient for low airflow requirements in environments with relatively low dew point ambient air. They do not require an air compressor, rather simply relying upon convention flow of air. This type of air dryer would be used for light

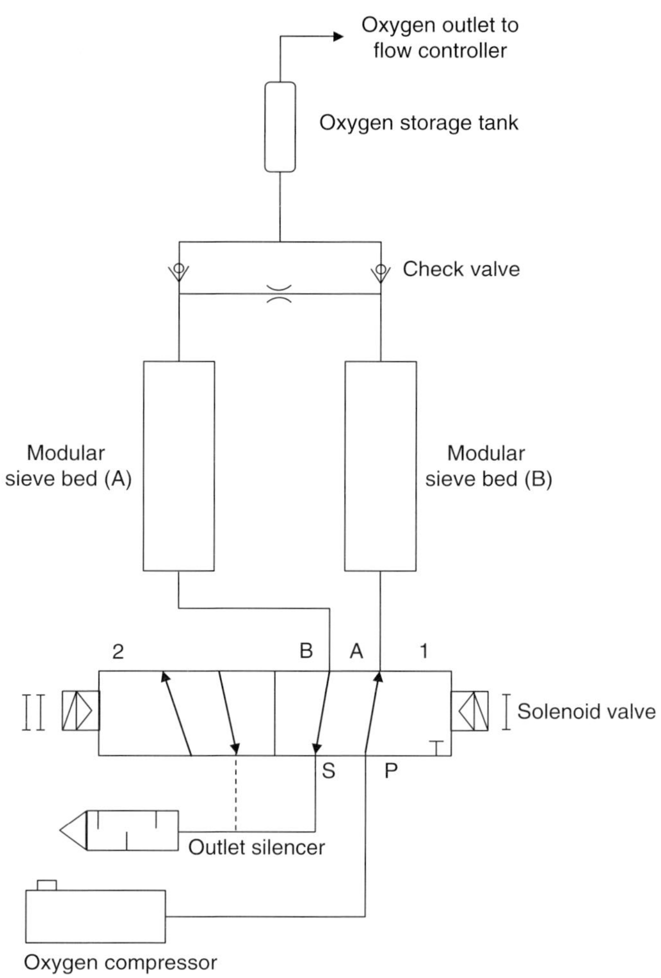

Figure 4.2 Schematic diagram of a PSA air drying system.

ozone duty applications – such as smaller fumigation applications, pilot projects or whenever 1 g/hour of ozone or less is sufficient.

As a general rule, UV generators of ozone do not require special air preparation. The ozone output is only slightly increased by increasing the oxygen content or drying the air. Although air drying is advantageous, the cost versus benefit usually does not justify the added expense of air preparation equipment when UV is used to generate ozone.

For further information, see Rakness (2005, pp. 114–32).

4.3.3 Oxygen feed gas systems

Today the majority of systems use oxygen because the concentration of ozone produced is increased two to three times over ozone concentrations in dried air for the same energy expenditure. For example, dry air-fed CD generators of the size used in food processing plants produce 1–2 wt% ozone, but when fed high-purity oxygen, ozone concentrations of 3–6 wt% are produced for the same energy expenditure. Most ozone equipment manufacturers have optimised their equipment for oxygen feed gas. In some instances, ozone generators can produce as high as 20 wt% ozone (from oxygen), but the energy and other requirements are higher to produce ozone at this concentration.

The advantage of a higher weight per cent ozone product is economics. The equipment footprint is smaller and the solubility of ozone in water is much greater, which means smaller pumps and contactors, resulting in less energy cost and space for the equipment. Even the major metropolitan drinking water plants use oxygen feed gas, even though this requires massive vacuum pressure swing adsorption (VPSA) or cryogenic systems to produce the oxygen in the large quantities necessary to produce tons/day quantities of ozone.

For food processing plants, oxygen-enriched air (>90% O_2) is provided simply and conveniently by means of oxygen concentrators. These devices take in ambient air, automatically filter (remove dust particles), then separate and remove nitrogen (thereby leaving air considerably enriched with oxygen), which is also dried to below the desired maximum dew point (–4 °C, –60 °F), all at the same time in one small device. These oxygen concentrators operate on the principle of PSA. An air compressor pressurises the airflow and sends it through a molecular sieve (microscopic porous bead) bed that adsorbs or traps nitrogen and moisture, while providing oxygen-enriched air to the supply output of the concentrator. As the molecular sieve bed becomes loaded with nitrogen and moisture, they desorb to waste in vapour form to the environment, recovering the adsorption capacity of the sieve bed. Oxygen concentrators typically employ multiple-chambered designs, so that at any one time chambers will be undergoing pressurisation and adsorption, while others will be undergoing depressurisation and desorption.

Figure 4.3 is a schematic diagram showing the components of a PSA oxygen concentrator used to provide >90 wt% oxygen to an ozone generator.

Figures 4.4 and 4.5 are front and back views of a complete ozonation system for use in food processing plants. A complete ozonation system consists of the following subunits:

- air treatment or oxygen concentrator;
- ozone generator;
- dissolved ozone controller plus vacuum, pressure and flow gauges;
- ozone/water contactor (injector), contact tank and off-gas vent;
- booster pump and lines for water in and ozone-containing water out;
- a vacuum break for backflow prevention.

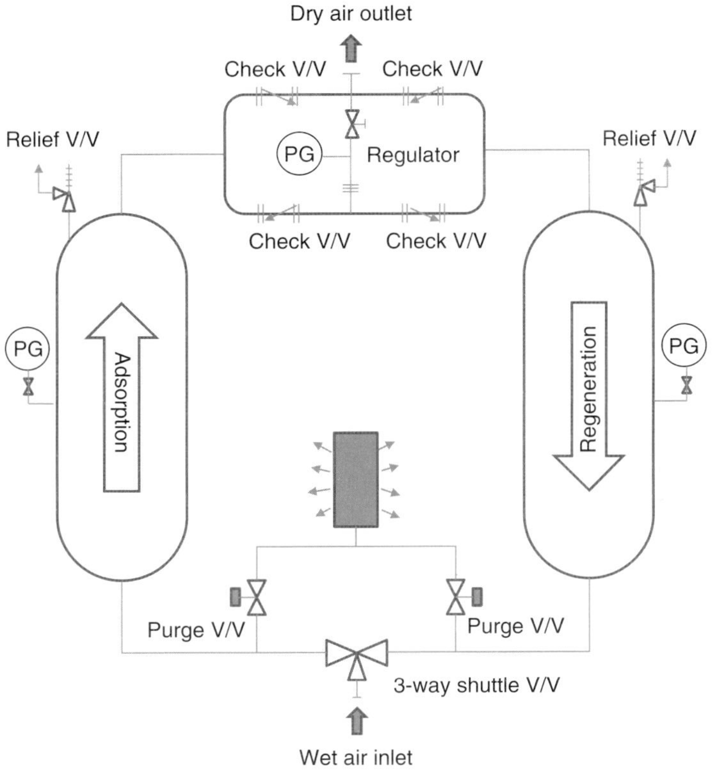

Figure 4.3 Schematic diagram of a PSA-type oxygen concentrator.

Table 4.1 shows an overall comparison of the major operating parameters of ozone generation by CD (dry air feed) and by UV radiation.

4.4 Solubility of ozone in water

Once ozone gas has been produced, the next step is to apply it, either in gas form (for storage, duct cleaning, etc.) or, more often, or by dissolving it into water. In this latter case, the controlling variable is the partial solubility of ozone in water.

The solubility of a gas in water can be described by Henry's law, which states that the partial pressure of a gas in water (solubility) is directly proportional to its partial pressure in the gas phase above that water *when the gas/liquid system is at equilibrium*. Strictly speaking, this law applies only to so-called 'ideal' gases that do not undergo chemical reactions with water.

Dalton's law states that the partial pressure of a gas is equivalent to its volumetric concentration in the gas phase multiplied by the absolute

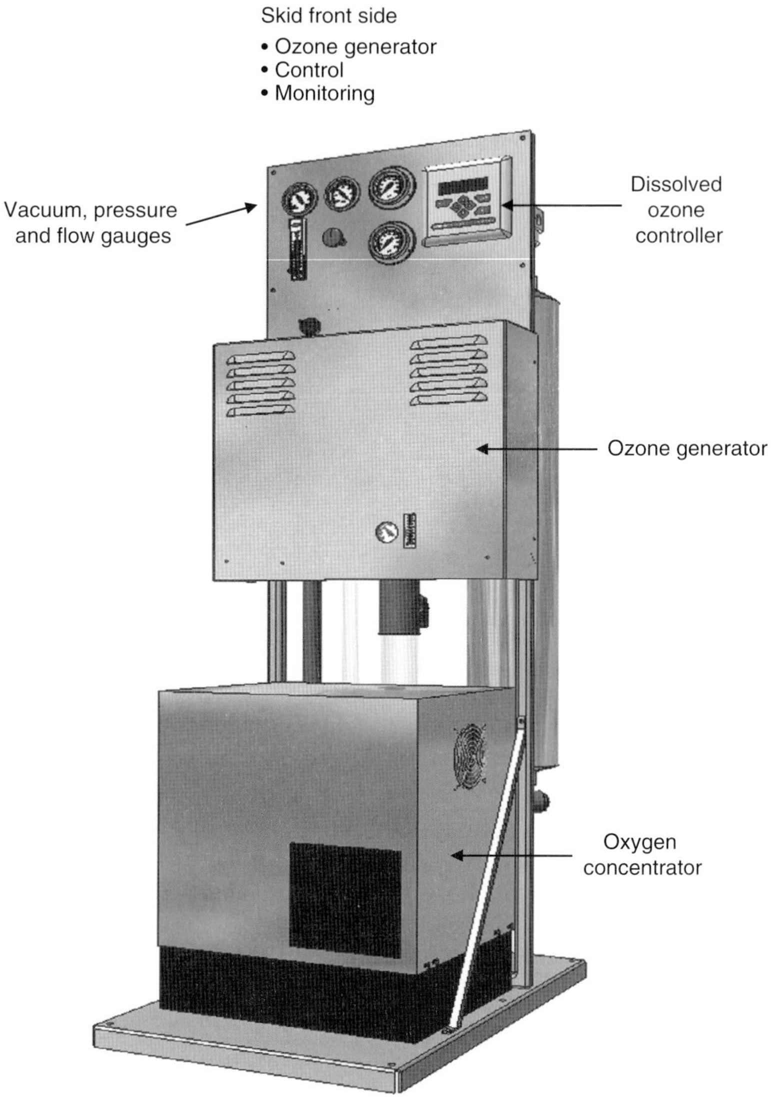

Figure 4.4 Front view of complete ozonation system for food processing plants.

pressure of the system. One form of Henry's law that expresses dissolved gas concentration in units of mg/L is as follows:

$$\mathbf{C}_s = \mathbf{B} \times \mathbf{M} \times \mathbf{P}_g$$

Where $\mathbf{C}_s$ is dissolved gas concentration, mg/L; **M** is gas phase density, mg/L; **B** is Bunsen absorption coefficient; and $\mathbf{P}_g$ is partial pressure, in atmospheres.

A number of parameters influence the solubility of ozone in water – these are: temperature, pH, ionic strength, the presence of substances readily oxidised by ozone and the autodecomposition of ozone in the gas and liquid phases.

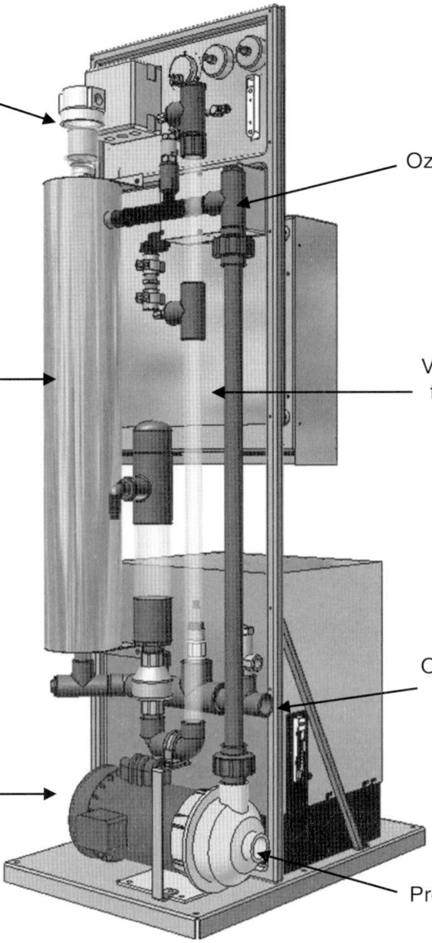

Figure 4.5 Back view of complete ozonation system for food processing plants.

Table 4.1 Comparison of ozone generation by CD (dry air feed) vs UV radiation (Water Quality Association 1997, p. 12).

Parameter	UV radiation	CD
Maximum ozone production rate	1.94 g/kWh using 185 nm lamps Lower yields are obtained from 254 nm UV bulbs	>55 g/kWh from dry air
Concentration of ozone in output gas	1.8 g/m^3 ~0.14% by weight	12–60 g/m^3 0.1–4.8% by weight
Energy required to generate 1 kg of ozone	44 kWh	6–8 kWh
Need to dry feed gas	Desirable but not critical	Critical
Consistency of ozone production	Variable	Constant
Capital costs	Relatively low	Relatively high
Operating costs (electrical energy)	High	Low

The driving force for mass transfer is the difference between the solubility of ozone and the already-dissolved ozone concentration. Gas solubility is enhanced with increase of system pressure within a water treatment train. It is important to note that this relates to water pressure versus the pressure found within the ozone gas delivery system, because the production of ozone will decrease drastically as the ozone generation cell is subjected to higher pressures. Thus it is important to ensure that the pressure within the ozone generator itself does not exceed that recommended by its manufacturer.

4.5 Contacting ozone with water: physical means of transferring ozone into water

Mass transfer of ozone into water for purposes of water purification or preparation of aqueous solutions of ozone required for surface sanitation applications is typically accomplished via two distinctly different methodologies: Venturi injection and fine bubble diffusion. Both approaches to dissolving ozone into water have their relevance, dependent upon the specific application(s) within the food processing plant.

4.5.1 *Venturi injection method*

Venturi injectors consist of two basic sections: (1) the motive fluid injection chamber nozzle, which converts the line water pressure energy to kinetic (velocity) energy; and (2) the suction chamber/diffuser section, where suction (of ozone-containing gas), entrainment and mixing (of gas and water) take place – as shown in Figure 4.6.

When the pressurised (motive) fluid (water) enters the injector inlet, it is constricted toward the injection chamber and becomes a high-velocity jet stream. The increase in velocity through the injection chamber results in a decrease in pressure, thereby enabling an additive material (ozone-containing

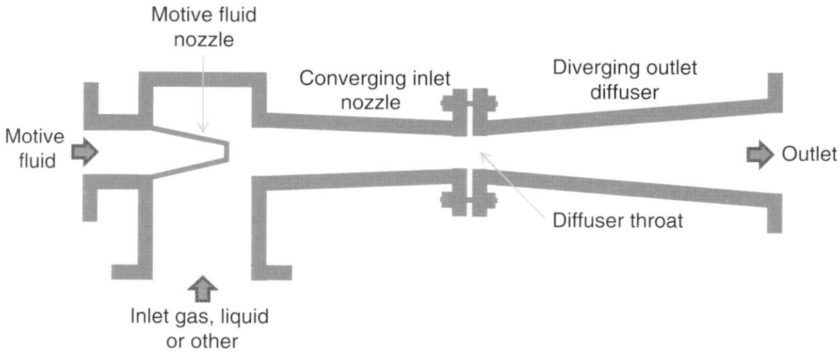

Figure 4.6 Schematic diagram of a Venturi injector contactor.

gas) to be drawn through the suction port and entrained into the motive water stream. Under these conditions, the reaction gas rapidly expands into the liquid, resulting in small bubbles of ozone gas being introduced into the liquid stream in a very turbulent environment, providing for thorough mixing of gas and water. As the jet stream is diffused toward the injector outlet, its velocity is reduced and it is reconverted into pressure energy (but at a pressure lower than the injector inlet pressure). Figure 4.7 is a photograph of a Venturi injector in operation. Motive water flow is from left to right, with air drawn in at the suction port (bottom). Note that violent mixing occurs immediately at the injector throat, with the areas of less violent mixing at the far right of the injector.

Not all Venturi-type injectors act the same. Performance characterised by the differential pressure necessary to start suction, mixing characteristics and motive pressure (energy) requirements vary widely among manufacturers. Extremely efficient injectors will operate over a wide range of pressures and require a minimal pressure differential (as low as 15%) between inlet and outlet sides to initiate a vacuum at the suction port for either liquid or gas. This more efficient performance allows for higher operational back-pressure (or water system pressure – post-ozone injection), ultimately increasing ozone mass transfer and lowering energy cost per gram of ozone transferred into solution.

A major advantage of employing Venturi contacting with ozone is that the water flow throughout the hydraulic is always under a slight negative pressure. This means that in the event of a possible leak in the hydraulic piping, room air will be drawn into the flowing water instead of liquid along with gaseous ozone leaking into the room atmosphere. At the same time, hydraulic pressure monitors/controllers will sense the change in hydraulic pressure caused by such a leakage, and shut down the system,

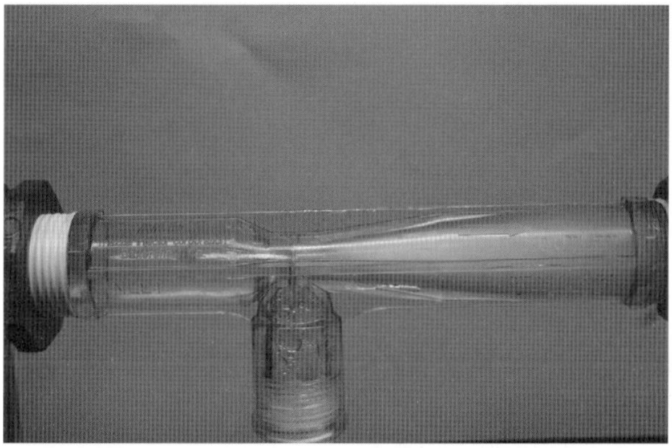

Figure 4.7 Venturi injector in operation. (Courtesy of Mazzei Injector Corp., Bakersfield, CA.)

thereby eliminating potential exposure of plant workers to ozone accidentally discharged into the workspace atmosphere.

4.5.2 *Fine bubble diffuser method*

In the fine bubble diffuser design, ozone exits the ozone generator under a pressure of a few pounds per square inch (psig) and expands through a porous stone or frit into the water being treated (Plate 4.2). This diffuser is usually made from porous stone, ceramic or stainless steel, for the purpose of distributing a fine dispersion of ozone gas bubbles into the bottom of a column of water. The pressure of the ozone gas stream must be greater than the pressure exerted by the water column above the diffuser.

Generally, the ozone gas stream is pressurised to as much as 15 psig. Higher pressures should be avoided because compressing ozone may create heat if the system is not carefully cooled, and heat destroys ozone rapidly. Diffuser assemblies are typically submerged in water columns of between 10 and 20 feet for optimum mass transfer of ozone to water. A ceramic or sintered stainless steel ozone sparger/diffuser typically may produce ozone bubbles about five microns to five millimetres in diameter, for example. Mass transfer of ozone into the water takes place as the bubbles rise toward the surface of the water. Typically, gas–water mixing is conducted in a countercurrent flow arrangement (water flows downward while ozone-containing gas flows upward), to maximise the contact time of gaseous ozone with the water, thereby increasing mass transfer of the ozone into the water.

The finer the porosity of the stone or frit, the smaller the bubble size produced, and the slower the bubbles rise through the water. The slower the bubble rise and the finer the bubbles, the greater the amount of ozone transferred to the water due to higher gas–liquid surface area interfacing as well as longer gas–liquid contact time. A well-designed countercurrent flow gas–water fine bubble contactor can have bubble rise times slower than one foot per second.

Because this method applies ozone gas under positive pressure, it is generally considered to be less safe (than Venturi injection) in the event of a potential leak in the ozone gas system between generating and contacting. With such a system under pressure, a piping leak means that the pipe contents (including ozone gas) can leak into the room. *Caution:* a barometric loop or elevated ozone gas feed line should be used between ozone generators and fine bubble diffusers to achieve a height above the maximum water column height and thereby prevent water backflow to the ozone generator.

4.6 Measuring and monitoring ozone in water

The concentration of ozone in water can be measured conveniently by means of two primary technologies, colourimetric test kits or electronic meters. In food processing applications, one primary purpose of measuring ozone in water is to ensure disinfection; that is, to determine that ozone has provided sufficiently high levels of microorganism kills/inactivations. The concentration of ozone (in mg/L) multiplied by the time of contact of ozone in a food processing water (in minutes) provides a product value known as the 'Ct value' (mg/min/L).

Ct values are related to the reduction or percentage of microbial inactivation. For example, a Ct value of 1 for a given microorganism under specific conditions of pH and temperature can be attained by providing 1.0 mg/L of residual ozone over a period of 1 minute, 0.5 mg/L dissolved ozone over 2 minutes, 0.25 mg/L of dissolved ozone over 4 minutes and so on. An ozone Ct value of 0.48 mg/min/L has been found to provide as much as a 99.9% or 3 log reduction of many microorganisms commonly encountered in food processing plants (Pascual et al. 2007).

By using one of the following methods of measuring the ozone level dissolved in water, one can accurately determine ozone's inactivation efficiency for whichever microorganisms are of concern for the specific food being processed.

4.6.1 Colourimetric method

Chemical test kits utilise ampoules of vacuum-sealed reagent to draw in a sample when the ampoule tip is snapped off underneath the ozone-containing liquid surface. After mixing, the filled ampoule is placed in the cell holder of the photometer supplied. Absorbance values obtained are converted to ppm (mg/L) of dissolved ozone with the included calibration chart. The two primary chemical substances utilised are DDPD (a methyl-substituted form of DPD, which is N,N-diethyl-p-phenylenediamine) and indigo trisulfonate.

With the DDPD procedure, potassium iodide is added to the sample before analysis. Ozone reacts with the iodide anion to liberate free iodine. The iodine then reacts with the DDPD reagent to produce a blue-violet colour, the intensity of which is then measured colourimetrically. Various free halogens and halogenating agents also produce the blue-violet colour with the DDPD reagent, and therefore interfere with the analysis for ozone. Chromate in test samples below 25 ppm will not interfere with results. Results are expressed as ppm (mg/L) of ozone.

The indigo trisulfonate reagent reacts instantly and quantitatively with ozone, bleaching the blue indigo colour in direct proportion to the amount of ozone present. Malonic acid is included in the formulation to prevent interference from chlorine. Results are expressed as ppm (mg/L) O_3.

4.6.2 Electronic method – for dissolved ozone

Electronic monitors or controllers for dissolved ozone use a membrane-covered amperometric sensor, which consists of a gas-permeable membrane stretched tightly over a gold cathode. A silver anode and electrolyte solution complete the internal circuit. During operation, ozone diffuses from the sample through the membrane. Once inside the sensor, the ozone reacts with the electrolyte solution to form an intermediate compound. A polarising voltage applied to the cathode completely reduces this intermediate compound, producing a current between the cathode and the anode, which the analyser measures. This current is directly proportional to the rate at which ozone diffuses through the membrane into the sensor, which ultimately is proportional to the concentration of ozone in the sample.

An advantage of the electronic method is that it measures the sample in real time and allows for control of the ozone generator as well as measurement of levels of dissolved ozone. Disadvantages are the higher capital expense and maintenance costs.

4.6.3 Electronic method – for ORP

ORP stands for 'oxidation-reduction potential', which is measured by a probe. In practical terms, an ORP probe is a voltmeter, measuring the voltage across a circuit formed by a reference electrode constructed of silver wire (in effect, the negative pole of the circuit) and a measuring electrode constructed of a platinum band (the positive pole), with the analysis water in between. Although ORP does not measure dissolved ozone (unless ozone is the only oxidising substance that is present in solution), it can be very useful for controlling the output of ozone generators. When a preset ORP level is exceeded, a signal is sent to the ozone generator to decrease ozone production, and *vice versa*.

The reference electrode is surrounded by salt (electrolyte) solution, which produces another tiny voltage. The voltage produced by the reference electrode is constant and stable, so it provides a reference against which the voltage generated by the platinum measuring electrode and the oxidisers in the water may be compared. The difference in voltage between the two electrodes is what is actually measured by the meter. (Note: high or low pH can alter ORP readings involving dissolved ozone due to the rapid decomposition of ozone at elevated pH. Optimal accuracy requires pH levels within 6.5–8.0.)

4.7 Measuring and monitoring ozone in air

In food processing plants it is very important to measure and monitor ozone in the gaseous phase to ensure that the appropriate concentrations of ozone are available for its intended purposes. These include food storage (before and after processing), during processing, treating (disinfecting) incoming plant air and ensuring that ozone-treated plant air that might be breathed by plant workers is in compliance with Occupational Safety and Health Administration (OSHA) regulations.

Ozone in the gaseous phase can be a very beneficial tool in the food processing industry. Storing processed foods in atmospheres containing small amounts of gaseous ozone is effective in controlling mould, mildew and many airborne contaminants on both the food products and the surfaces of storage racks and other equipment. In addition, ozone is known to destroy ethylene gas, which is released when many fruits and vegetables begin to ripen. Ultimately, ethylene is converted to carbon dioxide, carbon monoxide and water upon prolonged exposure to ozone in a humid cold storage environment (Bailey 1978).

Therefore, ozone can slow the ripening process while reducing levels of spoilage agents such as mould and mildew. For example, studies have shown that ozone exposure of oranges reduces ageing and weight loss over oranges stored in a non-ozonised environment (Di Renzo et al. 2005).

Both UV and CD ozone generators can be used in controlled atmosphere (CA) storage room applications (fruits, vegetables, grains, dairy products, meats and so on, for killing microorganisms, mildew control, destroying ethylene, stopping spore proliferation and so on, and overall – reducing the need for chemical fumigants). CA rooms are the most common application for gaseous ozone in the food industry, but there are other uses as well, notably fumigation for pests, meat ageing, barrel storage and overall plant mildew control.

For control of mould and destruction of ethylene, a gas phase ozone level of 0.3 ppm (vol) is considered sufficient. However, dosage rates vary greatly depending on desired goals and commodities being treated. The best way to truly determine the proper applied dose is to conduct a pilot test for each specific ozone need.

4.7.1 Ozone measurement equipment for food processing plant air

There are two primary methods of measuring ozone in food processing plant air: UV absorption (based on the Beer–Lambert law) and metal oxide semiconductor technology (MOS). The UV systems are widely used for determining ozone levels as low as typical atmospheric amounts (less than 0.10 ppm) up to concentrations produced by large municipal water treatment plant ozone generators operating on oxygen (12–15 wt%). The UV method is very accurate, ±1%, and therefore quite technically complex and expensive. The ozone readout is in weight per cent (wt%), parts per million (ppm) or grams per normalised cubic metre (g/Nm^3). Some analytical equipment manufacturers are starting to provide gas phase UV ozone analysers containing fewer technical features, which will lower the initial equipment costs.

Gas phase ozone analysers based on MOS technology are lower in cost than UV analysers, and are typically used in situations where precise accuracy is less critical. MOS analysers are widely used as safety devices to warn if levels of ozone exceed a predetermined concentration, such as 0.10 ppm. MOS units are available in handheld, wall-mounted and even

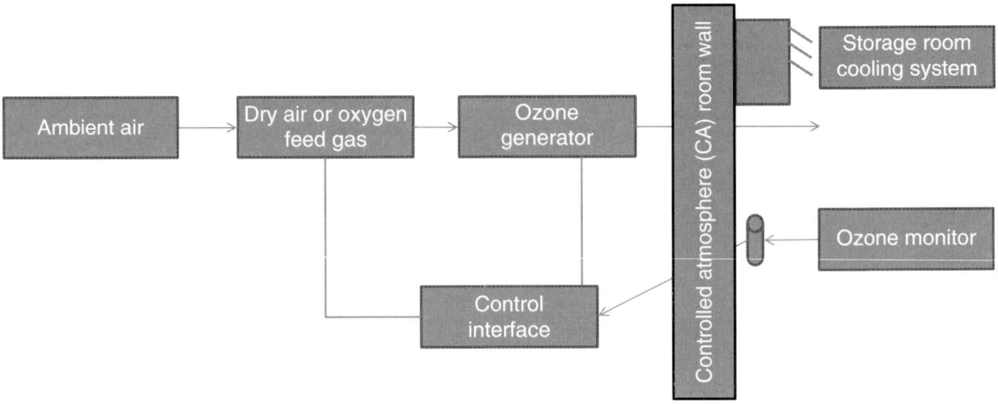

Figure 4.8 Schematic diagram of ozone applied in a food storage room.

personnel clip-on formats. The sensing unit needs to be replaced every 6–12 months, depending on the specifics of location and application conditions.

4.8 Ozonation equipment for food storage rooms

The equipment components required for generating and applying ozone in the gas phase are fairly simple compared to aqueous ozone applications, primarily because mixing a gas (ozone) with another gas (air) is much less complicated than dissolving ozone gas in water. Typical components are a feed gas, which will both dry the air and pressurise the ozone entering the storage room, an ozone generator, preferably with 4/20 mA control for real-time variable output, and finally a gas phase ozone monitor, for achieving and maintaining a desired ozone level in the room. One design note: it is advisable to interface the system with the cooling fans so there is no build-up (or localisation) of ozone without air movement anywhere in the room.

Figure 4.8 is a schematic diagram showing how ozone is generated and applied to a CA food storage room.

4.9 Ozone generator output control

For most food processing or surface sanitation applications, a predetermined amount of a disinfecting oxidiser is required to achieve microbial inactivation objectives. As an example, in the case of ozone, a specific dissolved ozone level is required to obtain a specific Ct value. Ozone generators are easily provided with the means by which to control the ozone output either manually or automatically to achieve these dissolved ozone concentration levels.

With some ozone generators, a simple manual volume control switch can be used to increase and/or decrease the ozone output level from 0 to 100%. However, for an accurate real-time approach a more sophisticated ozone generator can be provided with either an ON/OFF contact or a 4–20 mA control. With either feature the ozone generator can be controlled via a dissolved ozone monitor/controller or ORP monitor/controller. The controller can be programmed with predetermined set points; these are levels at which the dissolved ozone level or ORP reading is not to be exceeded nor fallen below. The controller can provide an ON function when the level declines to the low set point and an OFF function when the level rises to the high set point value.

A 4–20 mA signal from the controller can provide a linear output signal, relating to 0–100% ozone output. As the ozone level declines, the 4–20 mA signal given to the ozone generator increases toward 20 mA (or 100%); similarly, as the signal level increases toward the high set point, the 4–20 mA signal given to the ozone generator decreases towards 4 mA (or 0%). It is this advancement in technology which enables a food processing plant to have the ability to achieve higher quality standards when it comes to food sanitation practice. In addition, most controllers can send data to a chart recorder or history stored directly in the controller to meet sanitarians' requirements for record keeping.

4.10 Some recent novel applications for ozone generation in food processing plants

Much of the ozone food processing literature reports on studies or case applications of ozone that involve ozone as the primary or even sole treatment to attain specific product quality objectives. Many reports demonstrate clearly that it is critical to use enough ozone to do its intended job, but at the same time, the indiscriminate overdosing of ozone can damage or at least change the quality of an ozone-treated food product. A Swiss firm pioneering the use of ozone for a variety of applications has developed an approach of applying sequential and multiple applications of ozone, along with other processing steps, such as UVC radiation, electrolysed water, ultrasound washing and modified air packaging, to produce higher-quality food products with longer than customary product shelf lives.

Some of these treatment steps, termed 'Ventafresh technologies', involve the generation of ozone inside packaged products, after multistep pretreatment processing. For example, in a Swiss sushi plant (Steffen et al. 2010), ozone, ultrasound, electrolysed water and UV (185 and 254 nm) radiation are employed to sanitise all production equipment and factory space, including incoming and cooled plant air, as well as to sanitise the sushi products themselves. Fish, vegetables and rice all are washed with electrolysed water as ultrasound is applied. Sushi itself is disinfected prior to packaging by fumigation with ozone and UV radiation in a special UV disinfection tunnel. Packaging materials (film and trays) are disinfected with gaseous ozone and UV radiation.

After sealing the sushi packages with modified atmosphere packaging (MAP) (including oxygen above its atmospheric concentration of ~21%), UV radiation is again applied in another, longer UV disinfection tunnel. This converts about 12–14% of the oxygen remaining inside the packed tray into ozone, creating an ozone-containing atmosphere. By this Ventafresh technology step, in combination with the other processing steps, the shelf life of sushi products increases from three to seven days. Plant ambient temperature is maintained at 3 °C at all times during processing to provide additional improvement in microorganism control. Cost savings at this Sushi Mania plant are significant, but secondary to the prime objective of producing very high-quality sushi – only one failure and the plant is shut down. Ventafresh is, at the very least, a technological insurance policy that allows the plant manager to sleep well at night.

A recently developed in-package ionisation (plasma) technology (PK-1) has been developed at Purdue University (USA) (Klockow and Keener 2009). This technology has been shown to significantly reduce bacteria on raw and unprocessed food products. Significant amounts of ozone are generated inside sealed packages using low-current, high-voltage electrodes placed on opposite sides of the package. The efficacy of this system has been shown using several different gas mixtures.

Preliminary experiments have been conducted, verifying that:

(1) In-package ionisation generates high levels of bactericidal molecules (ozone, oxides, peroxides and ions) using low amperage (15 mA), while virtually no heat is generated inside the package (<2 °C), and very low energy (<40 W) is required.
(2) The technology could be applied to food products placed in common food packaging materials (glass, polypropylene, PETE, multilaminate and many others).
(3) Increased treatment times result in greater amounts of species generation when even small amounts of oxygen (5% or less) are present in the package. Tests have shown that bacteria on agar plates are not detected (< 5 log reductions) after treatment inside a sealed package.
(4) Treatment times range from 15 seconds to 5 minutes, depending on product, package and gas composition.
(5) In-package ionisation using unique blends of common gases mitigates oxidative effects (unlike ozone treatment), while at the same time providing bacterial reductions and quality retention.
(6) After 24 hours or less, depending on treatment conditions, the reactive molecules revert to the original gases (air, MAP, etc.).

These developments portend a promising future for the application of ozone in many food processing applications.

4.11 Helpful calculations

$$\text{grams per hour} \times 4.41 = \text{ppm or mg/L of ozone}$$

4.11.1 *Gallons per minute*

$$\text{GPM} \times \text{ozone dosage (ppm or mg/L)} \times 0.012$$
$$= \text{lbs/day of ozone } (\times\ 19 = \text{g/h})$$

4.11.2 *Metric equivalent*

$$\text{m}^3/\text{h} \times \text{ozone dosage(ppm or mg/L)} = \text{g/h}$$
$$\text{SCFH} \times \text{per cent by weight} \times 0.34 = \text{g/h of ozone}$$

References

Bailey, P.S. (1978) *Ozonation in Organic Compounds, Vol. I. Olefinic Compounds*, New York, NY: Academic Press, Inc., pp. 227–8.

Di Renzo, G.C., Altieri, G., D'Erchia, L., Lanza, G. and Strano, M.C. (2005) Effects of gaseous ozone exposure on cold stored orange fruit, *Acta Hort. (ISHS)*, **682**: 1605–10.

Dohan, J.M. and Masschelein, W.J. (1987) The photochemical generation of ozone: present state of art of technology, *Ozone Science & Engineering*, **9**: 325.

DuRon, B. (1982) Ozone generation with ultraviolet radiation, in Rice R.G. and Netzer A. (eds) *Handbook of Ozone Technology and Applications*, Ann Arbor, MI: Ann Arbor Science Publishers, Inc.

Klockow, P.A. and Keener, K.M. (2009) Safety and quality assessment of packaged spinach treated with a novel ozone-generation system. *LWT – Food Science and Technology*, **42**: 1047–53. doi:10.1016/j.lwt.2009.02.011.

Langlais, B., Reckhow, D.A. and Brink, D.R. (1991) *Ozone in Water Treatment: Application and Engineering*, Denver, CO: Am. Water Works Assoc. Research Foundation.

Pascual, A., Llorca, I. and Canut, A. (2007) Use of ozone in food industries for reducing the environmental impact of cleaning and disinfection activities, *Trends in Food Science & Technology*, **18**: S29–S35. doi:10.1016/j.tifs.2006.10.006.

Rakness, K.L. (2005) *Ozone in Drinking Water Treatment: Process Design, Operation, and Optimization*, Denver, CO: Am. Water Works Assoc.

Steffen, H., Duerst, M. and Rice, R.G. (2010) User experiences with ozone, electrolytic water (active water) and UV-C light (Ventafresh technology) in production processes and for hygiene maintenance in a Swiss sushi factory, *Ozone: Science & Engineering*, **32**(1): 71–8. doi:10.1080/01919510903489561.

Voronov, A. (2008) New generation of low pressure mercury lamps for producing ozone, *Ozone: Science & Engineering*, **30**(6): 395–7. doi:10.1080/01919510802341012.

Water Quality Association: Harrison, J.F. and Blazek, P. (1997) *Ozone for Point-of-Use, Point-of-Entry, and Small Water System Water Treatment Applications, A Reference Manual*, Lisle, IL: WQA.

5 Ozone in Fruit and Vegetable Processing

B.K. Tiwari and K. Muthukumarappan

5.1 Introduction

Ozone finds wide application in the food industry, including surface decontamination of fruits and vegetables, drinking water disinfection and wastewater treatment (Guzel-Seydim et al. 2004; Karaca and Velioglu 2007). Ozone is applied in either gaseous or aqueous form. Minimising the occurrence of pathogenic and spoilage microorganisms in fruits, vegetables and their products is a primary food-safety concern. Consumer preference for minimally processed foods and foods free of chemical preservatives, as well as recent outbreaks of foodborne pathogens, identification of new food pathogens and the passage of new legislation such as the Food Quality Protection Act in the USA, have all stimulated demand for novel food processing and preservation systems. Several incidents of foodborne disease have been associated with fruit and vegetable products. In 1991, an outbreak of *Escherichia coli* O157:H7 infections and haemolytic uremic syndrome was linked to traditionally pressed apple cider. In the USA 21 juice-associated outbreaks were reported to the CDC (Centers for Disease Control and Prevention) between 1995 and 2005 (Vojdani et al. 2008). Recent outbreaks have shown that fruit juices can be vehicles for foodborne pathogens (CDC 1996 and 1999). *E. coli* O157:H7 is an enteric pathogen with a low infectious dose, which usually causes haemorrhagic colitis, but also has the potential to cause haemolytic uremic syndrome in young children and the immunocompromised (Boyce et al. 1995).

Ozone is a powerful broad-spectrum antimicrobial agent that is active against bacteria, fungi, viruses, protozoa, and bacterial and fungal spores (Khadre et al. 2001) pertinent to fruits and vegetables and their products. Efficacy against both Gram-positive and Gram-negative bacteria and fungi is reported, as well as potential virucidal effects (Restaino et al. 1995). Apart from the wide spectrum of microbial inactivation, ozone also has the potential to kill storage pests and degrade mycotoxins. One of the potential advantages of ozone is that excess ozone autodecomposes rapidly to

Ozone in Food Processing, First Edition. Edited by Colm O'Donnell, B.K. Tiwari, P.J. Cullen, and Rip G. Rice.

produce oxygen, and thus generally leaves no residue in food. However, ozone reactions with organic compounds can lead to new, partially oxidised compounds, some of which may remain in the food.

Its efficacy against a wide range of microorganisms, including bacteria, fungi, viruses, protozoa and bacterial fungal spores, has been reported (Restaino et al. 1995; Khadre et al. 2001; Cullen et al. 2009). Such advantages make ozone attractive to the food industry and consequently it has been affirmed as Generally Recognised as Safe (GRAS) for use in food processing (Graham 1997) and was approved as an antimicrobial food additive in 2001 (FDA 2001). This chapter outlines the efficacy of ozone for the storage and preservation of fruits, vegetables and their products, the effect of ozonation on product quality, and the current status of ozone application in fruit and vegetable processing.

5.2 Applications in fruit and vegetable processing

Ozone as a disinfecting agent has widespread application to assure safety and quality. It has several advantages over conventional disinfectant agents such as chlorine, chlorine dioxide, calcium hypochlorite, sodium chlorite, peroxyacetic acid and sodium hypochlorite. Some of these agents are inefficient against specific organisms. Table 5.1 shows the advantages and disadvantages of several disinfectants used in the fruit and vegetable processing industries. Ozone is preferred over most popular disinfectants, such as chlorine, because of the relatively low inactivation rate of chlorine at concentrations limited by regulation. The main purposes of ozone application at the postharvest stage of fruit and vegetable processing are: inactivation of pathogenic and spoilage microorganisms, and destruction of pesticide and chemical residues.

5.2.1 Surface decontamination

Traditionally, ozone treatment within the fruit and vegetable processing industry has been carried out for surface decontamination of whole fruits and vegetables by either gaseous treatment or washing with ozone-containing water. Table 5.2 shows some examples of the effects of aqueous and gaseous ozone treatment on fruit and vegetable preservation and quality.

Aqueous ozone

Ozone has been used routinely for washing and storage of fruits and vegetables (Liangji 1999; Karaca and Velioglu 2007). Water containing ozone has been applied to fresh-cut vegetables for sanitation purposes, reducing microbial populations and extending shelf life (Beltrán et al. 2005a,b). Treatment of apples with ozone resulted in lower weight loss and spoilage (Achen and Yousef 2001). Increased shelf life of apples and oranges following ozone treatment has been attributed to the oxidation of ethylene.

Table 5.1 Advantages and limitations of disinfection methods proposed for fresh-cut organic vegetables (Olmez and Kretzschmar 2009). (Reprinted from *LWT – Food Science and Technology*, Volume 42, Issue Number 3, Hülya Ölmez, Ursula Kretzschmar, Potential alternative disinfection methods for organic fresh-cut industry for minimising water consumption and environmental impact, 686–93, 2009, with permission from Elsevier.)

Disinfectant agent	Advantages	Disadvantages/limitation
Chlorine (hypochlorite)	Low cost Easily available	Hazardous DBP at high levels Reacts with organic matter, in some cases leads to the production of toxic compounds Efficacy is affected by the presence of organic matter Corrosive Activity pH dependent Not allowed for organic products
Ozone	High antimicrobial activity Short contact time GRAS substance No residue problem No hazardous DBP formation No need to store hazardous substances Lower running cost Requires onsite generation	Toxic when inhaled Requires monitoring in indoor applications Corrosive above 4 ppm Higher initial investment cost
Chlorine dioxide	Higher antimicrobial efficacy at neutral pH than chlorine Effectiveness less pH dependent than that of chlorine Fewer potentially hazardous DBP formation than chlorine Less corrosive than chlorine and ozone Requires onsite generation	Not efficient at permitted levels for fresh produce Explosive Only allowed in whole produce Final water rinsing is required after treatment More iodinated DBP formation than chlorine if iodide ion is present in water Formation of specific byproducts, chlorite and chlorate Requires monitoring in indoor applications Not allowed for organic products
Organic acids	Easy to use No toxicity Allowed for organic products	Long contact time, not relevant to the industry Interferes with the sensory quality Relatively lower antimicrobial efficacy
Peroxyacetic acid	Efficacy is not affected by the organic load of water Efficacy unaffected by temperature changes No harmful DBP formation Not corrosive at permitted levels (<80 ppm)	Low antimicrobial efficacy at permitted levels for vegetables Not allowed for organic products

(*Continued*)

Table 5.1 (*continued*)

Disinfectant agent	Advantages	Disadvantages/limitation
Hydrogen peroxide	No residue problem Easy to use Low cost	Low antimicrobial efficacy Long contact time Phytotoxic, negative impact on overall quality Requires the removal of residual H_2O_2 after processing Not allowed for organic products

Table 5.2 Effect of ozone on fruit and vegetable preservation and quality.

Food product	Target microbial population	Quality attributes	Reference
Aqueous ozone treatment			
Lettuce	*Shigella sonnei* (1.8 LR)		Gil et al. (2006)
Iceberg lettuce	APC (1.1 LR)	Shelf life (↑), visual sensory quality (↑)	Garcia et al. (2003)
Apple	*E. coli* O157:H7 (3.7 LR)		Achen and Yousef (2001)
	E. coli O157:H7 (2.6 LR)		Achen and Yousef (2001)
Coriander (*Coriandrum sativum* L.)	TPC (↓)	Aroma (~), flavour (~), overall quality (↑)	Feng et al. (2004)
Lettuce	*E. coli* O157:H7 (1.42 LR)	Colour (↓)	Singh et al. (2002)
Baby carrot	*E. coli* O157:H7 (1.8 LR)		Singh et al. (2002)
Watermelon	APC (1–1.5 LR)	Colour (↓), overall quality (↓)	Fonseca and Rushing (2006)
Celery	Total bacteria (1.15 LR) PPO (↓)	Total sugar (~), colour (~)	Zhang et al. (2005)
Lettuce	PPO (↓)	Antioxidants (~), vitamin C (↓), visual appearance (~)	Beltran et al. (2005a)
Fresh-cut potato strips	LAB(↓), coliforms (↓) and anaerobic bacteria (↓)	Shelf life (↑), non-enzymatic browning (↑)	Beltran et al. (2005b)
Blueberries	*E. coli* O157:H7 (3.0 LR)	Colour (~)	Bialka and Demirci (2007)
Gaseous ozone treatment			
Lettuce	*E. coli* O157:H7 (1.84 LR)	Colour (↓)	Singh et al. (2002)
Baby carrot	*E. coli* O157:H7 (2.64 LR)		Singh et al. (2002)
Green pepper	*E. coli* O157:H7 (5 LR)		Han et al. (2002)

(*Continued*)

Table 5.2 (*continued*)

Food product	Target microbial population	Quality attributes	Reference
Pistachio (*Pistachio vera* L.)	*E. coli* (↓) and *Bacillus cereus* (↓)	pH (~), FFA (~) and peroxide values (~), colour (~), fatty acid composition (~)	Akbas and Ozdemir (2006)
Fig	Total aerobic (mesophyllic) Microorganisms 38% inactivation Coliform (TI) Yeast/mould (72% inactivation)		Oztekin et al. (2006)
Blueberries	*Salmonella enterica* (2.2 LR)	Colour (~)	Bialka and Demirci (2007)

FFA, free fatty acid; APC, aerobic plate count; AA, ascorbic acid; PPO, polyphenol oxidase; LR, log reduction. (×) significant difference; (√) insignificant; (↑) increases; (↓) decreases; (~) no change.

Fungal deterioration of blackberries and grapes was decreased by ozonation of the fruits (Beuchat 1992). Ozone-containing water was found to reduce bacterial content in shredded lettuce, blackberries, grapes, black pepper, broccoli, carrots and tomatoes (Kim et al. 1999b; Barth et al. 1995; Zhao and Cranston 1995; Sarig et al. 1996). Microbial studies typically show a 2 log reduction of total counts and significant reductions in spoilage and potentially pathogenic species most commonly associated with fruit and vegetable products.

Selma et al. (2007) reported that ozone treatments at levels of 1.6 and 2.2 ppm for 1 minute decreased *Shigella sonnei* population in water by 3.7 and 5.6 log colony forming unit (CFU)/mL, respectively. In addition, *S. sonnei* counts were reduced by 1.8 log units in lettuce treated with 5 ppm of ozone for 5 minutes (Selma et al. 2007). Oztekin et al. (2006) reported the effects of ozone treatment on the microflora of dried figs, where the application of gaseous ozone at 5 or 10 ppm for 3 to 5 hours resulted in significant reductions in total bacteria, coliform and yeast/mould counts. Najafi and Khodaparast (2009) concluded that a minimum of 1 hour of ozone treatment at 5 ppm could be successfully used to reduce both coliform and *Staphlyococcus aureus* populations on date fruits, but that longer exposure times were required for elimination of the total mesophilic bacteria as well as yeast/mould present.

Washing of fruits and vegetables was also reported to degrade pesticide residues. Wu et al. (2007) reported that rinsing at a dissolved ozone concentration of 1.4 mg/L for 15 minutes effectively removes 27–34% of residual pesticide from vegetables. However, higher degradation of pesticides residues can be obtained with an increase in ozone concentration (Ou-Yang et al. 2004; Ong et al. 1996). Inan et al. (2007) investigated the

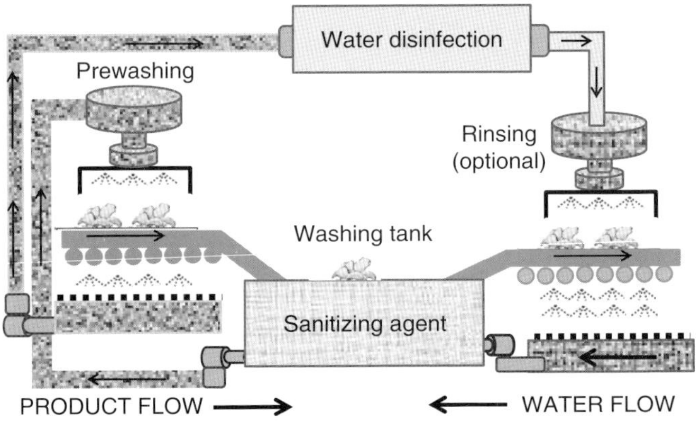

Figure 5.1 Efficient water disinfection process with a recirculation system proposed as an alternative for traditional fresh-cut washing (Gil et al. 2009). (Reprinted from *International Journal of Food Microbiology*, Volume 134, Issue 1–2, Maria I. Gil, Maria V. Selma, Francisco López-Gálvez and Ana Allende, Fresh-cut product sanitation and wash water disinfection: problems and solutions, 37–45, 2009, with permission from Elsevier.)

efficacy of ozone for the degradation of aflatoxin B1 in flaked and chopped red peppers. They observed 80 and 93% degradation of aflatoxin B1 at an ozone concentration of 33 and 66 mg/L for 60 minutes, respectively.

Wang et al. (2004) employed tap water, acidic electrolysed water (AEW), aqueous ozone, chlorinated water and aqueous ozone followed by AEW (sequential wash) for treatment of fresh-cut coriander samples. They observed that the sequential wash – that is, aqueous ozone followed by AEW – is more effective in initial microbial count reduction and maintains low microbial growth during storage at 0 °C for 14 days. However, the combination of ozone and AEW led to more tissue injury, which influences the overall quality of coriander, whereas ozone treatment alone achieved the highest overall quality of coriander during storage and maintained the typical coriander aroma.

Figure 5.1 shows an effective ozone washing system, which includes a shower as a prewashing step to remove dirt and cell exudates from the cut surfaces. This step is followed by the immersion of the product in a washing tank, which contains ozone as the sanitising agent (Gil et al. 2009). A rinse step is optional depending on the sanitising agent. It is recommended that water flows in the opposite direction to the movement of produce through the different unit operations. Thus, water in the sanitising tank could be recirculated for use in the prewashing step (Figure 5.1). The same applies to the rinse water, which could be incorporated into the sanitising tank after the shower in a continuous process. Water disinfection remains an essential activity in the fresh-cut industry and is possible with an efficient disinfection strategy such as chlorine, ozone and AOPs in a recirculated system (Figure 5.1) (Gil et al. 2009).

Gaseous ozone

The use of gaseous ozone to reduce *Bacillus* spp. and *Micrococcus* counts in peas and beans by up to 3 log units, depending on ozone concentration, temperature and relative humidity (RH) conditions, was reported by Naitoh et al. (1988). Fan et al. (2007) reported that gaseous ozone effectively inactivated *Listeria innocua* on solid media at concentrations of 50 and 100 nl/L during short exposure times at both 5 and 20 °C. The red colour of intact, whole berry fruit was optimum in 0.3 ppm ozone-treated samples during storage (Barth et al. 1995). It was also reported that the undesirable colour change from green to yellow in broccoli was significantly less pronounced for ozone-treated samples. However, ozone was reported to change the surface colour of products such as peaches (Badiani et al. 1995) and carrots (Liew and Prange 1994). The majority of gaseous ozone treatments are reported to be during storage, as discussed in section 5.2.2.

5.2.2 *Storage in ozone-rich atmospheres*

Apart from surface decontamination of fruits and vegetables, storage of fruits and vegetables in ozone-rich atmospheres has been reported to reduce or eliminate odour, and to control spoilage caused by microbial and fungal pathogens. Continuous exposure of fresh commodities to ozone during storage is reported to reduce postharvest decay and to reduce microbial spoilage of fruits and vegetables (Liew and Prange 1994; Barth et al. 1995; Sarig et al. 1996; Perez et al. 1999; Palou et al. 2002; Aguayo et al. 2006; Tzortzakis et al. 2007a,b). Applications of ozone-rich atmospheres have been studied for apples, cherries, carrots, kiwi, onions, peach, plum, potatoes, table grapes, tomatoes, blackberries and strawberries (Table 5.3). Ozone can be used as a relatively brief prestorage treatment in air or water, or it can be added continuously or intermittently to the storage room atmosphere throughout the storage period to prevent or delay fruit decay (Skog and Chu 2001; Palou et al. 2003; Cayuela et al. 2009). Hildebrand et al. (2008) observed a reduction in postharvest decay when carrots inoculated with *S. sclerotiorum* and *B. cinerea* were held in 115–530 nl/L ozone at 10 °C for 20 days.

Reported studies show that the effect of ozone during storage is variable and strongly depends on the type of microorganism, commodities and storage conditions. For example, Forney et al. (2003) observed a decay resistance towards *B. cinerea* in carrots treated with 1000 nl/L ozone for 2 or 4 days, however they did not observe a decay resistance towards *S. sclerotiorum*. Similarly, Skog and Chu (2001) reported that an ozone concentration of 0.04 μl/L has the potential to extend the storage life of broccoli and cucumbers stored at 3 °C. However, they did not observe similar effects for mushrooms stored at 4 °C or cucumbers stored at 10 °C. Ogawa et al. (1990) reported the inactivation of *B. cinerea* spores in tomato fruits after

Table 5.3 Application of ozone during storage of fruits and vegetables.

Food product	Storage conditions	Target Microbial population	Salient findings	Reference
Carrot	Ozone concentration of 0 (control), 7.5, 15, 30 or 60 μl/L. Treatment chambers were flushed with a total flow rate of 0.5 L/min (air and ozone) for 8 hours daily for 28 days. The experiment was repeated twice at storage temperatures of 2, 8 and 16 °C.	*Botrytis cinerea* Pers. and *Sclerotinia sclerotiorum* de Bary	A 50% reduction of daily growth rates of both fungi at the highest ozone concentration indicated that ozone was fungistatic. Carrot respiration rate, electrolyte leakage and total colour differences increased with ozone concentration. Ozone-treated carrots were lighter (higher L^* values) and less intense (lower chroma values) in colour than control carrots.	Liew and Prange (1994)
Carrot	Ozone concentration of 50 ± 10 nl/L ozone during storage for up to 6 months at 0.5 °C and >95% relative humidity.	*Sclerotinia sclerotiorum* and *Botrytis cinerea*	Reduced lesion size and aerial mycelium of both pathogens. Ozone-induced injury, appearing as blotches of brownish discoloured periderm. Ozone treatment had no effect on fresh weight loss, sprouting of carrot crowns or concentrations of glucose, fructose, sucrose or galactose.	Hildebrand et al. (2008)
Blackberry	Stored for 12 days at 2 °C in 0.0, 0.1 and 0.3 ppm ozone.	Fungal decay (*Botrytis cinerea*)	Ozone storage suppressed fungal development for 12 days. Ozone storage did not cause observable injury or defects. On day 12, anthocyanin content of juice was similar to initial levels for all treatments. Surface colour was better retained in 0.1 and 0.3 ppm-stored berries by 5 days and in 0.3 ppm berries by 12 days, by hue angle values. POD was greater in controls and 0.1 ppm samples, and was lowest in 0.3 ppm fruits by 12 days.	Barth et al. (1995)
Strawberry (*Fragaria × ananassa* Duch. cv. Camarosa)	Strawberry fruits were stored at 2 °C in an atmosphere containing ozone (0.35 ppm). After 3 days at 2 °C, fruits were moved to 20 °C to mimic retail conditions.	Fungal decay (*B. cinerea*)	Ozone treatment was ineffective in preventing fungal decay in strawberries after 4 days at 20 °C. Significant differences in sugars and ascorbic acid (reduced by 3 times) content were found in ozone-treated strawberries.	Perez et al. (1999)

			A detrimental effect of ozone treatment on strawberry aroma was observed, with a 40% reduced emission of volatile esters in ozonated fruits.	
Grape and Peach	Continuous ozone exposure at 0.3 ppm (v/v) for 4 weeks at 5 °C and 90% RH.	*Monilinia fructicola*, *Botrytis cinerea*, *Mucor piriformis* or *Penicillium expansum*	Ozone exposure did not significantly reduce the incidence and severity of decay caused by these fungi, with the exception of brown rot. Continuous ozone exposure at 0.3 ppm increased water loss after 5 weeks of storage. No phytotoxic injuries of fruit tissues were observed in ozonated or ambient atmosphere treatments.	Palou et al. (2002)
Whole and fresh-cut Tomato	Humidified flow of ozone-enriched air applied cyclically (4 ± 0.5 µl/L of O_3 for 30 minutes every 3 hours) stored up to 15 days at 5 °C	Mesophilic and psychrotrophic aerobic bacteria, yeasts and moulds count	Significant reduction in bacteria (1.1–1.2 $\log_{10}$ units) and fungi (0.5 $\log_{10}$ units). In whole tomatoes, O_3 maintained the tissue firmer than in control fruit, while no influence was found on slices. No significant changes in appearance and overall quality in slices but a slight reduction in aroma. No significant damage or off-flavour in slices or whole tomatoes.	Aguayo et al. (2006)
Tomato, Strawberry, Table Grape and Plum	Chilled storage (13 °C) and exposed to 'clean air' or low-level ozone enrichment (0.1 µmol/mol).	*Botrytis cinerea* (grey mould)	Ozone-enrichment resulted in a substantial decline in spore production as well as visible lesion development.	Tzortzakis et al. (2007)
Tomato	Ozone concentrations between 0.005 (control) and 5.0 µmol/mol up to 13 days at 13 °C.	*Alternaria alternata* or *Colletotrichum coccode*	Significant reduction in fungal lesion development. Concentration-specific impacts on fungal lesion development.	Tzortzakis et al. (2008)

RH, relative humidity; POD, peroxidases; L^*, lightness.

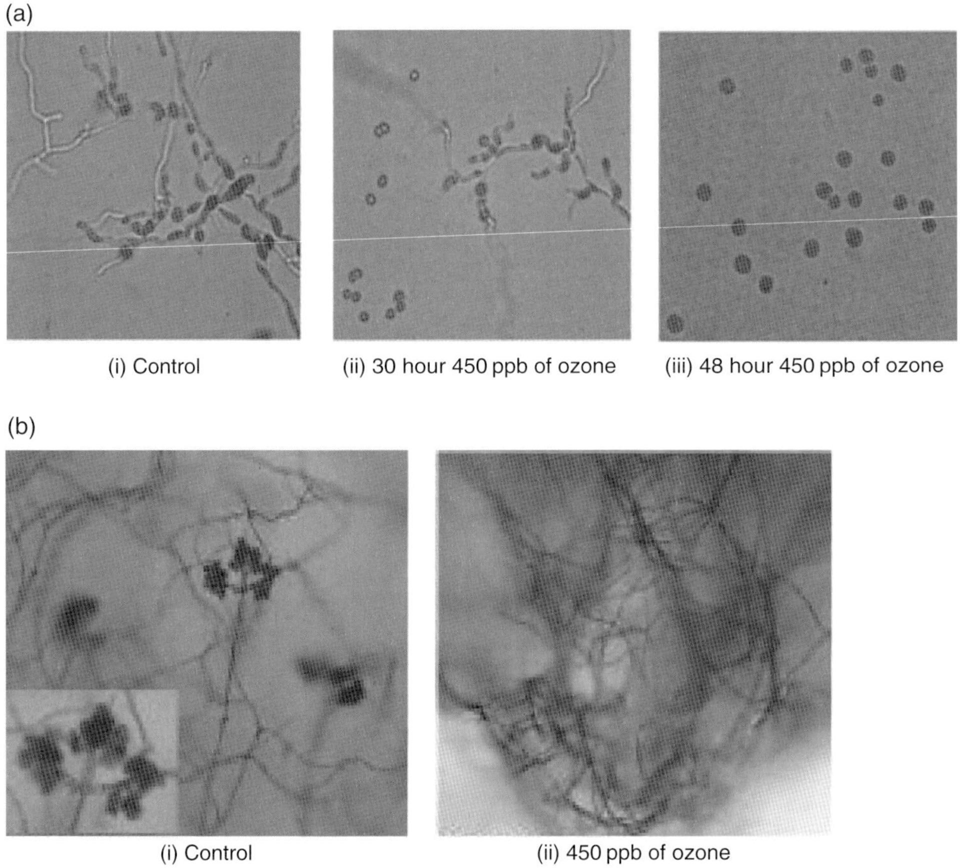

Figure 5.2 (a) Mycelial growth of *B. cinerea* after treatment with or without 450 ppb of ozone for 48 hours. (b) Micrographs of *B. cinerea* were taken after incubation for 24 hours at 20 °C using a light microscopy at 100× magnification. (i) Control with no ozone treatment, asexual conidiophores were able to develop. (ii) Ozone-treated *B. cinerea* without conidiophores, only mycelium growth was observed (Sharpe et al. 2009).

ozone treatment, whereas Liew and Prange (1994) concluded that the effect of ozone on *B. cinerea* was fungistatic but not fungicidal in treated carrots. Sharpe et al. (2009) investigated the effect of gaseous ozone on spore viability of *B. cinerea* and mycelial growth of *B. cinerea* and *S. sclerotiorum* for apples, grapes, highbush blueberries and carrots. They observed a significant reduction in spore viability of *B. cinerea* of over 99.5% and a reduction in the aerial mycelium from 4.7 mm in the control to less than 1 mm after exposure to 450 or 600 ppb ozone for 48 hours at 20 °C (Figure 5.2).

Furthermore, high ozone concentrations during storage may cause surface discolouration (blotches). Studies have shown reductions in decay of blackberries during continuous ozone treatment (Barth et al. 1995) and in mould inhibition and decay of onions. A delay of the growth of green and blue mould was observed in ozonated citrus fruit (Palou et al. 2001).

Apart from reduction in microbial decay, ozone is reported to be an effective agent in removing ethylene from the atmosphere during apple and pear storage without a significant change in quality attributes (Skog and Chu 2001). Exposure of horticulture crops to ozone can reduce postharvest decay and may be effective in reducing application of field-applied fungicides used to control these pathogens.

5.2.3 *Ozone in fruit and vegetable juice processing*

Currently the practical application of ozonation to fruit juices is still in its infancy. Ozonation of liquid phases is most frequently accomplished by injecting ozone gas (mixtures of air/ozone or oxygen/ozone) through a sparger into a liquid. Usually the studies on ozone absorption in the aqueous systems are carried out in stirred-tank reactors or bubble columns (Cullen et al. 2009). The approval of ozone as a direct food additive (FDA 2001) led to the application of ozone for processing of various fruit juices, including: apple cider (Steenstrup and Floros 2004; Choi and Nielsen 2005) orange juice (Angelino et al. 2003; Tiwari et al. 2008; Patil et al. 2009a,b), blackberry juice (Tiwari et al. 2009a) and strawberry juice (Tiwari et al. 2009b). Ozonation of fruit juices is reported to meet the FDA's requirement of a mandatory 5 log reduction of the most resistant pathogens (*E. coli, Salmonella, Listeria monocytogens*).

Apart from microbial inactivation, ozone treatment of apple juice has also been reported for destruction of mycotoxins (Cataldo 2008). Mycotoxins have been found to occur in a number of foods, including apple juice. Patulin is a predominant mycotoxin in apple juice with an FDA action level of 50 µg/L. Patulin has carcinogenic properties and it survives conventional pasteurisation processes. Cataldo (2008) reported that a moderate ozone treatment of apple juice may become a standard industrial practice to reduce or eliminate the patulin toxin from juices. Similarly, Ashirifie-Gogofio et al. (2009) reported that an apple juice model system (0.5% malic acid, buffered at pH 3.5–4.0) with an initial level of 1000 ppb patulin was degraded to <50 ppb in 20, 15 and 10 minutes for 350 ppm, 1500 ppm and 2500 ppm O_3, respectively. The efficiency of ozone in mycotoxin degradation is due to the presence of a polyketide lactone, which makes it highly vulnerable to oxidation. Hence ozone treatment effectively degrades mycotoxins.

5.3 Efficacy of ozone

Efficacy of ozone is affected by both extrinsic and intrinsic factors (Table 5.4) and it is difficult to predict ozone behaviour in the presence of specific compounds, other ingredients and environmental factors such as medium pH, temperature and humidity. Residual ozone is the concentration of

Table 5.4 Factors influencing efficacy of ozone.

Parameters	Factors
Extrinsic factors	
Water quality	pH, temperature, turbidity, organic matter, oxidizable inorganic materials (e.g. ferrous iron, manganous, sulfide, etc.)
Ozone	Concentration, contact time
Decontamination treatment	Application method (dipping, spraying and agitated, rubbed or static condition during exposure), produce/water ratio, single or multiple batches, rinse after sanitation, multiple washings
Intrinsic factors	
Microbial load	Characteristics of microbial strain, physiological states of the bacterial cells, natural or inoculated microorganisms, population size
Food product	Type of fruit and vegetable, characteristics of the product surfaces (cracks, crevices, hydrophobic tendency and texture), relation weight and surface area

ozone that can be detected in the medium after application to the target surface. Both the instability of ozone under certain conditions and the presence of ozone-consuming materials affect the level of residual ozone available in the medium. Therefore, it is important to distinguish between the concentration of applied ozone and residual ozone necessary for effective disinfection. It is advisable to monitor ozone availability during treatment (Pascual et al. 2007).

Food components are reported to interfere with the bactericidal properties of ozone (Guzel-Seydim et al. 2004). Efficacy of ozone is demonstrated more readily when targeted microorganisms are suspended and treated in pure water or simple buffers (with low ozone demand) than in complex food systems, in which it is difficult to predict how ozone reacts in the presence of organic matter (Cho et al. 2003). Organic substances with high ozone demand in a medium may compete with microorganisms for ozone (Khadre et al. 2001). Hence, the presence of organic matter or dissolved solids in water intended for washing of fruits and vegetables may increase ozone demand and may form undesirable byproducts due to reaction with ozone. The formation of these byproducts may shorten the shelf life, change the organoleptic quality or jeopardise the safety of the final product (Khadre et al. 2001).

The effectiveness of ozone against microorganisms depends not only on the amount used, but also on the residual ozone in the medium and various environmental factors such as medium pH, temperature, humidity, additives (surfactants, sugars, etc.) and the amount of organic matter surrounding the cells (Restaino et al. 1995). For example, whole-fruit bubbling of ozone in stored apples inoculated with *E. coli* O157:H7 was found to be more effective than dipping apples in ozone-containing water.

Bubbling and dipping resulted in 3.7 log and 2.6 log reductions in counts of *E. coli*, respectively (Achen and Yousef 2001). A 1.3–3.8 log reduction range for *E. coli* at ozone concentrations of 0.3–1.0 ppm (O_3 demand-free water) and pH of 5.9 was reported. Populations of *Leuconostoc mesenteroides* at similar treatment conditions were reduced by 1.3 to ~7 log CFU/mL and ozone concentrations of 0.2–1.8 ppm yielded 0.7 to ~7 log reductions in *L. monocytogenes* (Kim et al. 1999a). The effect of pH on ozone inactivation is mainly attributed to the fact that the ozone decomposition rate changes substantially with changes in pH. At high pH, the chain reactions of ozone decomposition result in the formation of numerous radical species with high oxidative capabilities.

$$O_3 + OH^- \rightarrow HO_2^- + O_2$$
$$O_3 + HO_2^- \rightarrow {}^{\bullet}OH + O^{\bullet}_2 + O_2$$

Patil et al. (2010) studied the effect of pH (at levels of 3.0, 3.5, 4.0, 4.5 and 5.0) on the microbial safety of apple juice. Apple juice inoculated with *E. coli* strains (10^6 CFU/mL) was treated with ozone at a flow rate of 0.12 L/min and ozone concentration of 0.048 mg/min/mL for different time periods (0–18 minutes). The results revealed that pH had a significant effect on the ozone inactivation kinetics. The ozone treatment duration for achieving a 5 log reduction was faster (4 minutes) at the lowest pH (3.0) than at the highest pH (5.0) (18 minutes) studied. The higher inactivation rates observed at lower pH may be due to the synergistic effect of pH on inactivation kinetics.

Factors influencing the solubility, stability and reactivity of ozone may also affect the efficacy of ozone (Table 5.4). There is no consensus on the effect of temperature on the biocidal efficacy of ozone. For example, a reduction in the temperature of an aqueous medium increases ozone solubility and stability, augmenting its availability in the medium, and consequently efficacy rises. The simultaneous contribution of these two factors (solubility/stability and reactivity) to ozone efficacy can vary with experimental conditions, making it difficult to predict the influence of temperature on a particular application (Pascual et al. 2007).

It has been widely reported since the 1930s that high RH is required for microorganisms to be inactivated by ozone gas. The optimum RH level is 90–95%. However, in general ozone loses its bactericidal efficiency below 50% RH. The strong effectiveness of ozone gas at high RH levels is beneficial for sanitation of fruits and vegetables where environmental RH levels generally are over 80% (Han et al. 2002). At high RH, ozone gas is more effective as an antimicrobial agent than gaseous disinfectants, such as ethylene oxide and propylene oxide (Wiley 1994), but may be less effective than ClO_2 gas (Han et al. 2002). Decreasing pH is reported to increase ozone efficiency (Kuscu and Pazir 2004) because the concentrations of molecular ozone molecules, responsible for providing the 'Ct value' (mg/min/L), are more stable at lower pH than at the higher pH ranges. Zhao and Cranston (1995) found that ozone gas treatment of black pepper at higher moisture content led to increased reductions in the microbial load.

The effectiveness of sanitisers can be affected by product surface features (Han et al. 2002) and ozone application methods, such as bubbling and agitation (Kim et al. 1999a). Variations in these factors will influence the antimicrobial activity of ozone. The degree of attachment of the microorganism to the food significantly influences the bactericidal effects of ozone. Microorganisms embedded in product surfaces are more resistant to ozone than those readily exposed. Application of aqueous ozone to products having smooth intact surfaces with low ozone demand (for example, fruits and vegetables) produced promising results (Kim et al. 1999b; Achen and Yousef 2001). Similarly, Kim et al. (1999b) reported that inactivation of microflora on food by ozone depends greatly on the nature and composition of the food surface, the type and load of microbial contaminant and the degree of attachment or association of microorganisms with the food. However, application must ensure direct contact of ozone with the target microbial cells. A variety of methods have been employed to accomplish this, including stirring, pumping, fluming, bubbling, sonication, abrasion and pressure washing (Kim et al. 2001).

5.4 Synergistic effects with ozone

The efficacy of ozonation may be increased by use in combination with other technologies. The disaggregating effect of ultrasound upon solid matter and on gas bubbles may improve efficacy by increasing surface area. Furthermore, ultrasound accelerates the sedimentation of oxidisable organic matter, thus reducing ozone demand. Williams et al. (2005) reported that combinations of hydrogen peroxide and ozone treatment followed by refrigerated storage caused greater than 5 log CFU/mL reduction of *E. coli* O157:H7 and *Salmonella* in apple cider and orange juice. Some microorganisms are sensitive to lower concentrations of oxidising agents when exposed to ultrasound, and the combined action of ultraviolet (UV) radiation with high-frequency ultrasound increases the rate of bacterial inactivation (Sierra and Boucher 1971). Employing such hybrid techniques can also reduce the dosage of the chemical disinfectant required. Thus by using the combination of ultrasonic sonotrode and ozone or hydrodynamic cavitation and ozone, the concentration of ozone required for disinfection may be significantly reduced to half or one-third, depending upon the type of microorganism.

The combination of hydrodynamic cavitation and ozone has proved to be an efficient method of water disinfection (Jyoti and Pandit 2004). Yuk et al. (2007) found that a combined ozone and organic acid treatment was more effective than individual application for control of *E. coli* O157:H7 on enoki mushrooms. Garcia et al. (2003) reported an improvement in microbial reduction and an extension of shelf life as determined by appearance of a ready-to-eat (RTE) salad lettuce by using a sequential combination of ozone and chlorine. They also observed a significant reduction in process water turbidity compared to chlorine treatment alone, which would facilitate increased water reuse (Strickland et al. 2010).

Ozone in combination with citric or oxalic acid as a fumigant is reported to reduce postharvest decay and pericarp browning of longan fruit. Whangchai et al. (2006) observed that longan fruit treated with ozone in combination with oxalic or citric acid reduces enzymatic browning and could be a partial alternative to sulfur dioxide fumigation for control of postharvest decay and browning. Pericarp browning has been attributed to oxidation of phenolics by polyphenol oxidase (PPO), producing brown-coloured byproducts (Ferrar and Walker 1996).

5.5 Effect of ozone on product quality and nutrition

Microbial studies to date typically show that the mandatory 5 log reductions of spoilage and potentially pathogenic species most commonly associated with fruit and vegetable juices may be achieved. A number of studies report the effects of ozone on quality parameters of treated fruits and vegetables (Zhang et al. 2005; Fonseca and Rushing 2006). The effects of ozone treatment on quality and physiology of various foods are reported in Table 5.2. Applying ozone at doses that are large enough for effective decontamination may change the sensory qualities of food. Ozone is not universally beneficial and in some cases may promote oxidative spoilage in foods. Surface oxidation, discolouration or development of undesirable odours may occur in substrates from excessive use of ozone (Khadre et al. 2001). Dock (1999) reported no detrimental change in the quality attributes of apple cider when it was treated with ozone.

5.5.1 *Chemical attributes*

No change in onion chemical composition and sensory quality was reported by Song et al. (2000). Ozone-containing water treatment resulted in no significant difference in total sugar content of celery and strawberries (Zhang et al. 2005) during storage. Ozonation is expected to lead to the loss of antioxidant constituents, because of its strong oxidising activity. However, ozone washing treatment was reported to have no effect on the final phenolic content of fresh-cut iceberg lettuce (Beltrán et al. 2005a). Contradictory reports were found in the literature regarding ascorbic acid. Decomposition of ascorbic acid in broccoli florets was reported after ozone treatment by Lewis et al. (1996), but Zhang et al. (2005) reported no significant difference in ascorbic acid content between treated and nontreated celery samples. Moreover, increases in ascorbic acid levels in spinach (Luwe et al. 1993), pumpkin leaves (Ranieri et al. 1996) and strawberries (Perez et al. 1999) were reported in response to ozone exposure. Slight decreases in vitamin C content were reported in lettuce (Beltrán et al. 2005a). Ozone treatments were reported to have minor effects on the anthocyanin contents of strawberries (Perez et al. 1999) and blackberries (Barth et al. 1995).

Beltrán et al. (2005a) compared the effect of ozone-containing water (10 mg/L/min, 20 mg/L/min, 10 mg/L/min ozone converted into a stronger oxidising moiety or moieties by UVC radiation, and chlorine 80 mg/L) and reported that ozone concentration of 10 ppm in water activated by UVC radiation aids in extending the shelf life of fresh-cut lettuce. They concluded that ozone-containing water activated with UVC radiation could be an alternative to chlorine for washing of shredded lettuce, by reducing microbial populations on the product but also by maintaining the visual quality and controlling browning without any detrimental effect on the antioxidant constituents when combining with active modified atmosphere packaging (MAP).

Recently researchers in Spain evaluated the effects of continuous and intermittent applications of ozone gas treatments, applied during cold storage to maintain postharvest quality during subsequent shelf life, on the bioactive phenolic composition of Autumn Seedless table grapes after long-term storage and simulated retail display conditions (Artes-Hernandez et al. 2007). They found that the sensory quality was preserved with both ozone treatments. Although ozone treatment did not completely inhibit fungal development, its application increased the total flavan-3-ol content at all sampling times. Continuous 0.1 µl/L O_3 application also preserved the total amount of hydroxycinnamates, while both treatments assayed maintained the flavonol content sampled at harvest. Total phenolics increased after the retail period in ozone-treated berries. Therefore, the improved techniques tested to enhance the quality of Autumn Seedless table grapes during long-term storage seem to maintain or even enhance the antioxidant compound content.

5.5.2 Visual quality

Gabler et al. (2010) investigated the efficacy of ozone in controlling postharvest decay of table grapes and for the potential replacement of sulfur dioxide, which is used as a commercial fumigant. They observed that ozone fumigation with up to 10 000 µl/L for up to 2 hours helps to control postharvest grey mould of table grapes caused by *Botrytis cinerea* (Figure 5.3). However, grapes stored in ozone-rich atmospheres may develop thin longitudinal darkened lesions (Figure 5.4). This injury is reported to be irregular and was not always associated with an ozone dose or cultivar (Gabler et al. 2010). Similarly, Tzortzakis et al. (2007b) observed a substantial decline in spore production and development of

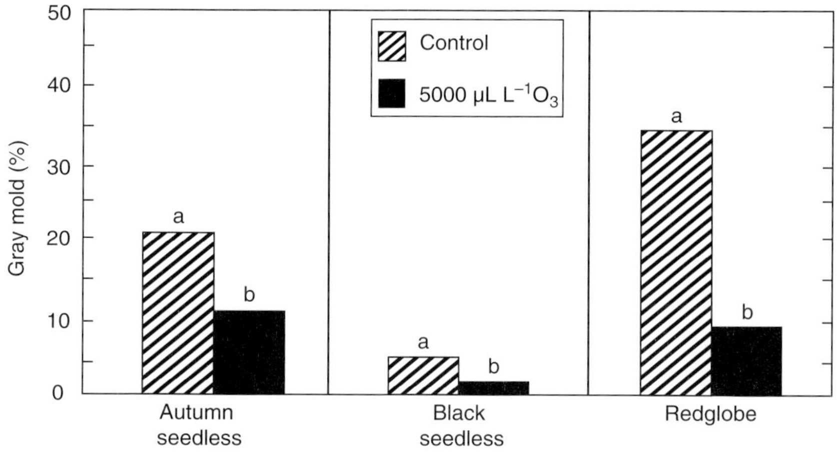

Figure 5.3 Influence of postharvest ozone fumigation on the natural incidence of postharvest grey mould among several table grape cultivars. Freshly harvested, organically grown table grapes were fumigated with 5000 µl/L ozone for 60 minutes in a commercial ozone chamber and stored for 6 weeks at 0.5 °C. Each value is the mean of four replicates per treatment and each replicate box contained nine cluster bags with 1000 g of grapes each. For each cultivar, columns with different letters differ significantly at $P \leq 0.05$ (Gabler et al. 2010). (Reprinted from *Postharvest Biology and Technology*, Volume 55, Issue 2, Franka Mlikota Gabler, Joseph L. Smilanick, Monir F. Mansour and Hakan Karaca, 85–90, 2010, with permission from Elsevier.)

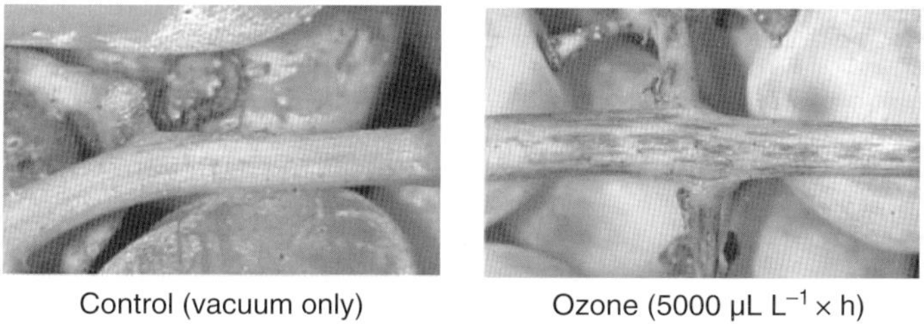

Figure 5.4 Occasional injuries to Thompson Seedless grape cluster rachis after grapes were fumigated once with 5000 µl/L ozone for 1 hour. Grapes were stored for 7 days at 15 °C (Gabler et al. 2010). (Reprinted from *Postharvest Biology and Technology*, Volume 55, Issue 2, Franka Mlikota Gabler, Joseph L. Smilanick, Monir F. Mansour and Hakan Karaca, 85–90, 2010, with permission from Elsevier.)

visible lesions mainly due to *B. cinerea* (grey mould) in tomatoes, strawberries, table grapes and plums during chilled storage (13 °C) with low-level ozone-enrichment (0.1 µmol/mol). Martínez-Sánc et al. (2006) investigated the effect of several sanitisers (Figure 5.5) on the visual quality and colour of rocket leaves during storage in air and low O_2 (1–3 kPa) + high CO_2 (11–13 kPa) for 15 days at 4 °C. They observed that ozone effects were comparable with other sanitisers, except for Purac- (lactic acid) treated samples.

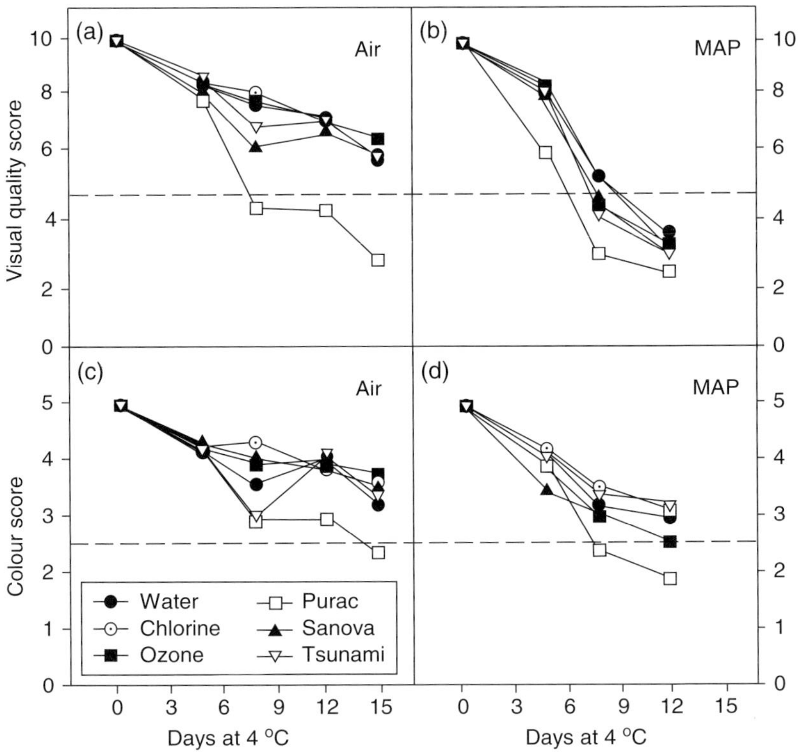

Figure 5.5 Effect of chlorine (sodium hypochlorite), 100 mg/L), ozone (10 mg/L), Purac (lactic acid, 20 mL/L), Sanova (acidified sodium chlorite, 250 mg/L) and Tsunami (peroxyacetic acid, 300 mL/L) sanitisers on the visual quality and colour of rocket leaves stored under air and low O_2 and high CO_2 (MAP). Values are the mean of three replicates (Martínez-Sánchez et al. 2006). (Reprinted from *Postharvest Biology and Technology*, Volume 42, Issue 1, Ascensión Martínez-Sánchez, Ana Allende, Richard N. Bennett, Federico Ferreres and María Isabel Gil, Microbial, nutritional and sensory quality of rocket leaves as affected by different sanitisers, 86–97, with permission from Elsevier.)

5.5.3 *Texture*

Texture or firmness is an important rheological property pertinent to fresh fruits and vegetables. Fruits and vegetables with a firm, crunchy texture are highly desirable because consumers associate these textural attributes with freshness and wholesomeness. The appearance of a soft or limp product may give rise to consumer rejection prior to consumption (Rico et al. 2007). Textural changes in fruits and vegetables could be due to various enzymatic and non-enzymatic processes. Ozone treatment of fresh fruit and vegetables either by washing or in storage consisting of ozone gas is reported to have significant effects on texture. Firmness of fresh coriander leaves was reported to decrease through washing with ozone-containing water compared to control. A decrease in firmness was also reported by washing with chlorinated water (Wang et al. 2004). Another study conducted by Selma et al. (2008) reported nonsignificant changes in firmness of fresh-cut cantaloupe irrespective of gaseous ozone concentration (5000 or 2000 ppm) for 30 minutes during storage compared to control.

Change in texture during ozonation and subsequent storage may possibly be due to postharvest changes in cellulose and hemicellulose contents due to ozone application during MAP. This could be due to polymerisation and epimerisation of cellulose and hemicelluloses contents of cell walls inducing thickening of the cell walls, causing textural changes in fresh-cut green asparagus during storage after ozone treatment (An et al. 2007). An et al. (2007) reported an increase in cellulose, hemicelluloses and lignin content during MAP storage after pretreatment with aqueous ozone.

Ozonation of fruits has been reported to enhance firmness of citrus fruits and cucumbers compared to controls (Skog and Chu 2001). Ozone is reported to delay softening in strawberries during cold-room storage and storage at room temperature (Nadas et al. 2003). Wang et al. (2004) compared five washing treatment systems including tap water, AEW, ozone-containing water, chlorinated water and aqueous ozone followed by AEW (sequential wash) on the firmness of coriander packaged in polyethylene bags and stored at 0 °C for 14 days. They observed a slight decrease in firmness of treated samples compared to the control on day 0, with no significant differences in firmness among chlorine, ozone, AEW and the sequential treatments at day 0. They observed a gradual decrease in firmness during storage, which may be attributed to the tissue injury caused by the treatments.

Ozone is a strong oxidising agent which can cause oxidation of feruloylated crosslinkages or phenolic crosslinkages among cell-wall pectin, structural proteins or other polymers, and thereby change the firmness of the product (Heun Hong and Gross 1998). Similarly, Forney et al. (2003) observed a reduction in firmness of ozone-treated carrot firmness during 8 weeks of storage. Aguayo et al. (2006) observed no significant changes in the firmness of tomato slices during treatment, whereas they observed a reduction of about 9.2, 16 and 35% in firmness after 5, 12 and 15 days of storage at 5 °C.

5.5.4 Sensory quality

The most notable effect of ozone on the sensory quality of fruits reported in the literature is the loss of aroma. Ozone-enriched cold storage of strawberries resulted in reversible losses of fruit aroma (Nadas et al. 2003; Perez et al. 1999). This behaviour is probably due to the oxidation of volatile compounds. However, Tzortzakis et al. (2007a) did not observe any significant changes in tomato fruit weight, antioxidant status, CO_2/H_2O exchange, ethylene production or in organic acid, vitamin C (pulp and seed) or total phenolic content when exposed to ozone concentrations ranging between 0.005 and 1.0 μmol/mol at 13 °C and 95% RH. Similar results were reported by Kute et al. (1995) for strawberry exposed to ozone concentrations between 0.3 and 0.7 μmol/mol for up to 1 week. Applying ozone at doses that are large enough for effective decontamination may change the sensory qualities of these products.

5.6 Conclusion

The effectiveness of ozone against microorganisms present in food systems depends on several factors, including the amount of ozone applied, the residual ozone in the medium, various environmental factors such as medium pH, temperature, humidity and additives (surfactants, sugars, etc.), and the amount of organic matter surrounding the cells (Pascual et al. 2007). To facilitate enhanced control of both quality and safety parameters of ozone-treated foods, mathematical models incorporating various independent factors governing ozone processing are required to describe biochemical reactions and microbial inactivation. Based on these modelling approaches, process optimisation can be carried out and specific safety constraints can be taken into account. A detailed study of the influence of food ingredients on both the inactivation and quality degradation kinetics is required to account for the complexity of food systems. Additionally, revisiting the mechanisms of the reactions of ozone with organic materials will contribute to establishing the impact of specific radical species on target microorganisms. Overall it can be concluded that ozonation is a potential treatment for producing safe and high-quality minimally processed fruit and vegetables but that specific treatment conditions must be developed and defined for each produce prior to ozone treatment.

References

Achen, M. and Yousef, A.E. (2001) Efficacy of ozone against *Escherichia coli* O157:H7 on apples, *J Food Sci*, **66**(9): 1380–4.

Aguayo, E., Escalona, V.H. and Artes, F. (2006) Effect of cyclic exposure to ozone gas on physicochemical, sensorial and microbial quality of whole and sliced tomatoes, *Postharvest Biology and Technology*, **39**: 169–77.

Akbas, M.Y. and Ozdemir, M. (2006). Effectiveness of ozone for inactivation of *Escherichia coli* and *Bacillus cereus* in pistachios, *International Journal of Food Science & Technology*, **41**: 513–19.

An, J., Zhang, M. and Lu, Q. (2007) Changes in some quality indexes in fresh-cut green asparagus pretreated with aqueous ozone and subsequent modified atmosphere packaging, *Journal of Food Engineering*, **78**(1): 340–4.

Angelino, P.D., Golden, A. and Mount, J.R. (2003) Effect of ozone treatment on quality of orange juice, in *IFT Annual Meeting Book of Abstract, 2003*, Abstract No. 76C–2, Chicago, IL: Institute of Food Technologists.

Artes-Hernandez, F., Aguayo, E., Artes, F. and Tomas-Barberan, F.A. (2007) Enriched ozone atmosphere enhances bioactive phenolics in seedless table grapes after prolonged shelf life, *J. Sci. Food Agric.*, **87**: 824–31.

Ashirifie-Gogofio, J., Floros, J.D. and LaBorde, L.F. (2009) Ozone degradation of patulin in model apple juice system, *Annual Meeting and Food Expo*, Anaheim, CA: IFT 09.

Badiani, M., Fuhrer, J., Paolacci, A.R. and Sermanni, G.G. (1995) Deriving critical levels for ozone effects on peach trees [Prunus persica (L) Batsch] grown in open-top chambers in central Italy, *8th International Symposium on Environmental Pollution and its Impact on Life in the Mediterranean Region*, Rhodes, Greece: Inst Lebensmitteltechnologie Analytische Chemie.

Barth, M.M., Zhou, C., Mercier, J. and Payne, F.A. (1995) Ozone storage effects on anthocyanin content and fungal growth in blackberries, *J Food Sci*, **60**(6): 1286–8.

Beltrán, D., Selma, M.V., Marin, A. and Gil, M.I. (2005a) Ozonated water extends the shelf life of fresh-cut lettuce, *Journal of Agricultural and Food Chemistry*, **53**: 5654–63.

Beltrán, D., Selma, M.V., Tudela, J.A. and Gil, M.I. (2005b) Effect of different sanitizers on microbial and sensory quality of fresh-cut potato strips stored under modified atmosphere or vacuum packaging, *Postharvest Biol Technol*, **37**: 37–46.

Beuchat, L.R. (1992) Surface disinfection of raw produce, *Dairy, Food and Environmental Sanitation*, **12**(1): 6–9.

Bialka, K.L. and Demirci, A. (2007) Decontamination of *Escherichia coli O157:H7* and Salmonella enterica on blueberries using ozone and pulsed UV-ligh, *Journal of Food Science*, **72**(9): 391–6.

Boyce, T.G., Pemberton, A.G., Wells, J.G. and Griffin, P.M. (1995) Screening for *Escherichia coli O157-H7* a nationwide survey of clinical laboratories, *J Clin Microbiol*, **33**(12): 3275–7.

Cataldo, F. (2008) Ozone decomposition of Patulin – a micotoxin and food contaminant, *Ozone Sci Eng*, **30**(3): 197–201.

Cayuela, J.A., Vazquez, A., Perez, A.G. and Garcia, J.M. (2009) Control of table grapes postharvest decay by ozone treatment and resveratrol induction, *Food Science and Technology International*, **15**(5): 495–502.

CDC (1996) Foodborne disease outbreak line listing (1996). Retrieved 14 Aug. 2009, from http://www.cdc.gov/foodborneoutbreaks/us_outb/fbo1999/fbofinal1999.pdf.

CDC (1999) Foodborne disease outbreak line listing (1999). Retrieved 14 Sep. 2009, from http://www.cdc.gov/foodborneoutbreaks/us_outb/fbo1996/fbofinal1996.pdf.

Cho, M., Chung, H. and Yoon, J. (2003) Disinfection of water containing natural organic matter by using ozone-initiated radical reactions, *Applied Environmental Microbiology*, **69**: 2284–91.

Choi, L.H. and Nielsen, S.S. (2005) The effects of thermal and nonthermal processing methods on apple cider quality and consumer acceptability, *Journal of Food Quality*, **28**: 13–29.

Cullen, P.J., Tiwari, B.K., O'Donnell, C.P. and Muthukumarappan, K. (2009) Modelling approaches to ozone processing of liquid foods, *Trends Food Sci Technol*, **20**(3–4): 125–36.

Dock, L.L. (1999) Development of thermal and non-thermal preservation methods for producing microbially safe apple cider, PhD thesis, West Lafayette, IN: Purdue University.

Fan, L., Song, J., McRae, K.B., Walker, B.A. and Sharpe, D. (2007) Gaseous ozone treatment inactivates *Listeria innocua in vitro*, *Journal of Applied Microbiology*, **103**: 2657–63.

FDA (2001) Hazard analysis and critical control point (HACCP): procedures for the safe and sanitary processing and importing of juice: final rule, *Federal Register*, **66**: 6137–202.

Ferrar, P.H. and Walker, J.R.L. (1996) Inhibition of diphenol oxidases: a comparative study, *Journal of Food Biochemistry*, **20**(1): 15–30.

Fonseca, J.M. and Rushing, J.W. (2006) Effect of ultraviolet-C light on quality and microbial population of fresh-cut watermelon, *Postharvest Biology and Technology*, **40**: 256–61.

Forney, C.F., Song, J., Fan, L.H., Hildebrand, P.D. and Jordan, M.A. (2003) Ozone and 1-methylcyclopropene alter the postharvest quality of broccoli, *Journal of the American Society for Horticultural Science*, **128**(3): 403–8.

Gabler, F.M., Mercier, J., Jimenez, J.I. and Smilanick, J.L. (2010) Integration of continuous biofumigation with *Muscodor albus* with pre-cooling fumigation with ozone or sulfur dioxide to control postharvest gray mold of table grapes, *Postharvest Biology and Technology*, **55**(2): 78–84.

Garcia, A., Mount, J.R. and Davidson, P.M. (2003) Ozone and chlorine treatment of minimally processed lettuce, *J Food Science*, **68**(9): 2747–51.

Gil, M.I., Selmaa, M.V., Beltrán, D., Allendea, A. and Verab, E.C. (2006) Elimination by ozone of *Shigella sonnei* in shredded lettuce and water, *Food Microbiology*, **35**: 345–51.

Gil, M.I., Selma, M.V., López-Gálvez, F. and Allende, A. (2009) Fresh-cut product sanitation and wash water disinfection: problems and solutions, *International Journal of Food Microbiology*, **134**(1–2): 37–45.

Graham, D.M. (1997) Use of ozone for food processing, *Food Technol*, **51**(6): 72–5.

Guzel-Seydim, Z.B., Greene, A.K. and Seydim, A.C. (2004) Use of ozone in the food industry, *Lebensm-Wiss Technol-Food Sci Technol*, **37**(4): 453–60.

Han, Y., Floros, J.D., Linton, R.H., Nielsen, S.S. and Nelson, P.E. (2002) Response surface modeling for the inactivation of *Escherichia coli* O157:H7 on green peppers (*Capsicum annuum*) by ozone gas treatment, *J Food Sci*, **67**(3): 1188–93.

Heun Hong, J. and Gross, K.C. (1998) Surface sterilization of whole tomato fruit with sodium hypochlorite influences subsequent postharvest behavior of fresh-cut slices, *Postharvest Biology and Technology*, **13**(1): 51–8.

Hildebrand, P.D., Forney, C.F., Song, J., Fan, L. and McRae, K.B. (2008) Effect of a continuous low ozone exposure (50 nL L^{-1}) on decay and quality of stored carrots, *Postharvest Biology and Technology*, **49**(3): 397–402.

Inan, F., Pala, M. and Doymaz, I. (2007) Use of ozone in detoxification of aflatoxin B1 in red pepper, *Journal of Stored Products Research*, **43**(4): 425–9.

Jyoti, K.K. and Pandit, A.B. (2004) Ozone and cavitation for water disinfection, *Biochem Eng J*, **18**(1): 9–19.

Karaca, H. and Velioglu, Y.S. (2007) Ozone applications in fruit and vegetable processing, *Food Reviews International*, **23**(1): 91–106.

Khadre, M.A., Yousef, A.E. and Kim, J. (2001) Microbiological aspects of ozone applications in food: a review, *Journal of Food Science*, **6**: 1242–52.

Kim, J.G., Yousef, A.E. and Chism, G.W. (1999a) Use of ozone to inactivate microorganisms on lettuce, *J Food Saf*, **19**(1): 17–34.

Kim, J.G., Yousef, A.E. and Dave, S. (1999b) Application of ozone for enhancing the microbiological safety and quality of foods: a review, *Journal of Food Protection*, **62**(9): 1071–87.

Kim, H.Y., Lee, Y.J., Hong, K.-H., Kwon, Y.-K., Sim, K.-C., Lee, J.-Y., Cho, H.-Y., Kim, I.-S., Han, S.-B., Lee, C.-W., Shin, I.-S. and Cho, J.S. (2001) Isolation of antimicrobial substances from natural products and their preservative effects, *Food Science Biotechnology*, **10**(1): 59–71.

Kuscu, A. and Pazir, F. (2004) Ozone applications in food industry, *Gida*, **29**(2): 123–9.

Kute, K.M., Zhou, C. and Barth, M.M. (1995) The effect of ozone exposure on total ascorbic acid activity and soluble solids contents in strawberry tissue, *Proceedings of the Annual Meeting of the Institute of Food Technologists (IFT)*, New Orleans, LA, 28 Jan.–2 Feb., p. 82.

Lewis, L., Zhuang, H., Payne, F.A. and Barth, M.M. (1996) Beta-carotene content and color assessment in ozone treated broccoli florets during modified atmosphere packaging, in *Book of Abstracts*, Abstract No 99, Chicago, IL: Institute of Food Technologists Annual Meeting.

Liangji, X. (1999) Use of ozone to improve the safety of fresh fruits and vegetables, *Food Technol*, **53**: 58–61.

Liew, C.L. and Prange, R.K. (1994) Effect of ozone and storage-temperature on postharvest diseases and physiology of carrots (*Daucus-Carota* L), *J Am Soc Hortic Sci*, **119**(3): 563–7.

Luwe, M.W.F., Takahama, U. and Heber, U. (1993) Role of ascorbate in detoxifying ozone in the apoplast of spinach (*spinacia-oleracea* l) leaves, *Plant Physiology*, **101**(3): 969–76.

Martínez-Sánchez, A., Allende, A., Bennett, R.N., Ferreres, F. and Gil, M.I. (2006) Microbial, nutritional and sensory quality of rocket leaves as affected by different sanitizers, *Postharvest Biology and Technology*, **42**(1): 86–97.

Nadas, A., Olmo, M. and Garcia, J.M. (2003) Growth of *Botrytis cinerea* and strawberry quality in ozone-enriched atmospheres, *Journal of Food Science*, **68**(5): 1798–802.

Naitoh, S., Okada, Y. and Sakai, T. (1988) Studies on utilization of ozone in food preservation. V. Changes in microflora of ozone-treated cereals, grains, peas, beans and spices during storage, *Journal of Japanese Society of Food Science and Technology*, **35**: 69–77.

Najafi, M.B.H. and Khodaparast, M.H.H. (2009) Efficacy of ozone to reduce microbial populations in date fruits, *Food Control*, **20**(1): 27–30.

Ogawa, J.M., Feliciano, A.J. and Manji, B.T. (1990) Evaluation of ozone as disinfectant in postharvest dump tank treatments for tomato, *Phytopathology*, **80**: 1020.

Ong, K.C., Cash, J.N., Zabik, M.J., Siddiq, M. and Jones, A.L. (1996) Chlorine and ozone washes for pesticide removal from apples and processed apple sauce, *Food Chemistry*, **55**(2): 153–60.

Ou-Yang, X.K., Liu, S.M. and Ying, M. (2004) Study on the mechanism of ozone reaction with parathion-methyl, *Safety and Environmental Engineering*, **11**(2): 38–41.

Oztekin, S., Zorlugenc, B. and Zorlugenc, F.K. (2006) Effects of ozone treatment on microflora of dried figs, *J Food Eng*, **75**(3): 396–9.

Palou, L., Smilanick, J.L., Crisosto, C.H. and Mansour, M. (2001) Effect of gaseous ozone exposure on the development of green and blue molds on cold stored citrus fruit, *Plant Disease*, **85**(6): 632–8.

Palou, L., Crisosto, C.H., Smilanick, J.L., Adaskaveg, J.E. and Zoffoli, J.P. (2002) Effects of continuous 0.3ppm ozone exposure on decay development and physiological responses of peaches and table grapes in cold storage, *Postharvest Biology and Technology*, **24**(1): 39–48.

Palou, L., Smilanick, J.L., Crisosto, C.H., Mansour, M. and Plaza, P. (2003) Ozone gas penetration and control of the sporulation of *Penicillium digitatum* and *Penicillium italicum* within commercial packages of oranges during cold storage, *Crop Protection*, **22**(9): 1131–4.

Pascual, A., Llorca, I. and Canut, A. (2007) Use of ozone in food industries for reducing the environmental impact of cleaning and disinfection activities, *Trends in Food Science & Technology*, **18**: S29–S35.

Patil, S., Bourke, P., Frias, J.M., Tiwari, B.K. and Cullen, P.J. (2009a) Inactivation of *Escherichia coli* in orange juice using ozone, *Innovative Food Science & Emerging Technologies*, **10**(4): 551–7.

Patil, S., Cullen, P.J. Kelly, B. Frías, J.M. and Bourke, B.P. (2009b) Extrinsic control parameters for ozone inactivation of *Escherichia coli* using a bubble column, *J Appl Microbiol*, **107**(3): 830–7.

Patil, S., Valdramidis, V.P., Cullen, P.J., Frias, J. and Bourke P. (2010) Inactivation of *Escherichia coli* by ozone treatment of apple juice at different pH levels, *Food Microbiology*, **27**: 835–90.

Perez, A.G., Sanz, C., Rios, J.J., Olias, R. and Olias, J.M. (1999) Effects of ozone treatment on postharvest strawberry quality, *Journal of Agricultural and Food Chemistry*, **47**: 1652–6.

Ranieri, A., Durso, G., Nali, C., Lorenzini, G. and Soldatini, G.F. (1996) Ozone stimulates apoplastic antioxidant systems in pumpkin leaves, *Physiologia Plantarum*, **97**(2): 381–7.

Restaino, L., Frampton, E., Hemphill, J. and Palnikar, P. (1995) Efficacy of ozonated water against various food related microorganisms, *Applied Environmental Microbiology*, **61**: 3471–5.

Rico, D., Martin-Diana, A.B., Barat, J.M. and Barry-Ryan, C. (2007) Extending and measuring the quality of fresh-cut fruit and vegetables: a review, *Trends in Food Science & Technology*, **18**(7): 373–86.

Sarig, P., Zahavi, T., Zutkhi, Y., Yannai, S., Lisker, N. and BenArie, R. (1996) Ozone for control of post-harvest decay of table grapes caused by *Rhizopus stolonifer*, *Physiol Mol Plant Pathol*, **48**(6): 403–15.

Selma, M.V., Beltrán, D., Allende, A., Chacón-Vera, E. and Gil, M.I. (2007) Elimination by ozone of Shigella sonnei in shredded lettuce and water, *Food Microbiology*, **24**(5): 492–9.

Selma, M.V., Allende, A., Lopez-Galvez, F., Conesa, M.A. and Gil, M.I. (2008) Disinfection potential of ozone, ultraviolet-C and their combination in wash water for the fresh-cut vegetable industry, *Food Microbiology*, **25**(6): 809–14.

Sharpe, D., Fan, L.H., McRae, K., Walker, B., MacKay, R. and Doucette, C. (2009) Effects of ozone treatment on *Botrytis cinerea* and *Sclerotinia sclerotiorum* in relation to horticultural product quality, *Journal of Food Science*, **74**(6): M250–M257.

Sierra, G. and Boucher, R.M.G. (I971) Ultrasonic synergistic effects in liquid-phase chemical sterilization, *Applied Microbiology*, **22**(2): 160–4.

Singh, N., Singh, R.K., Bhunia, A.K., Stroshine, R.L. (2002) Efficacy of chlorine dioxide, ozone, and thyme essential oil or a sequential washing in killing *Escherichia coli* O157:H7 on lettuce and baby carrots, *LWT – Food Sci Technol*, 35(8): 720–9.

Skog, L.J. and Chu, C.L. (2001) Effect of ozone on qualities of fruits and vegetables in cold storage, *Canadian Journal of Plant Science*, **81**(4): 773–8.

Steenstrup, L.D and Floros, J.D. (2004) Inactivation of *E. coli* 0157:H7 in apple cider by ozone at various temperatures and concentrations, *Journal of Food Processing Preservation*, **28**: 103–16.

Strickland, W., Sopher, C.D., Rice, R.G. and Battles, G.T. (2010) Six years of ozone processing of fresh cut salad mixes, *Ozone: Science & Engineering*, **32**: 1, 66–70.

Tiwari, B.K., Muthukumarappan, K., O'Donnell, C.P. and Cullen, P.J. (2008) Kinetics of freshly squeezed orange juice quality changes during ozone processing, *J Agric Food Chem*, **56**(15): 6416–22.

Tiwari, B.K., O'Donnell, C.P., Muthukumarappan, K. and Cullen, P.J. (2009a) Anthocyanin and colour degradation in ozone-treated blackberry juice, *Innovative Food Science & Emerging Technologies*, **10**(1): 70–5.

Tiwari, B.K., O'Donnell, C.P., Patras, A., Brunton, N. and Cullen, P.J. (2009b) Effect of ozone processing on anthocyanins and ascorbic acid degradation of strawberry juice, *Food Chem*, **113**(4): 1119–26.

Tzortzakis, N., Borland, A., Singleton, I. and Barnes, J. (2007a) Impact of atmospheric ozone-enrichment on quality-related attributes of tomato fruit, *Postharvest Biology and Technology*, **45**(3): 317–25.

Tzortzakis, N., Singleton, I. and Barnes, J. (2007b) Deployment of low-level ozone-enrichment for the preservation of chilled fresh produce, *Postharvest Biology and Technology*, **43**(2): 261–70.

Tzortzakis, N., Singleton, I. and Barnes, J. (2008) Impact of low-level atmospheric ozone-enrichment on black spot and anthracnose rot of tomato fruit, *Postharvest Biology and Technology*, **47**(1): 1–9.

Vojdani, J.D., Beuchat, L.R. and Tauxe, R.V. (2008) Juice-associated outbreaks of human illness in the United States, 1995 through 2005, *J Food Prot*, **71**(2): 356–64.

Wang, H., Feng, H. and Luo, Y. (2004) Microbial reduction and storage quality of fresh-cut cilantro washed with acidic electrolyzed water and aqueous ozone, *Food Research International*, **37**(10): 949–56.

Whangchai, K., Saengnil, K. and Uthaibutra, J. (2006) Effect of ozone in combination with some organic acids on the control of postharvest decay and pericarp browning of longan fruit, *Crop Protection*, **25**(8): 821–5.

Wiley, R.C. (1994) *Minimally Processed Refrigerated Fruits & Vegetables*, Springer.

Williams, R.C, Sumner, S.S. and Golden, D.A. (2005) Inactivation of *Escherichia coli* O157:H7 and *Salmonella* in apple cider and orange juice treated with combinations of ozone, dimethyl dicarbonate, and hydrogen peroxide, *Journal of Food Science*, **70**(4): 197–201.

Wu, J.G., Luan, T.G., Lan, C.Y., Lo, W.H. and Chan, G.Y.S. (2007) Removal of residual pesticides on vegetables using ozonated water, *Food Control*, **18**(5): 466–72.

Yuk, H.G., Yoo, M.Y. Yoon, J.W. Marshall, D.L. and Oh, D.H. (2007) Effect of combined ozone and organic acid treatment for control of *Escherichia coli* O157:H7 and *Listeria monocytogenes* on enoki mushroom, *Food Control*, **18**(5): 548–53.

Zhang, L.K., Lu, Z.X., Yu, Z.F. and Gao, X. (2005) Preservation of fresh-cut celery by treatment of ozonated water, *Food Control*, **16**(3): 279–83.

Zhao, J. and Cranston, P.M. (1995) Microbial decontamination of black pepper by ozone and the effect of the treatment on volatile oil constituents of the spice, *J Sci Food Agric*, **68**(1): 11–18.

6 Ozone in Grain Processing

V. Lullien-Pellerin

6.1 Introduction

Due to its properties, ozone has been used to control pest development in stored grains, to disinfect flours and to degrade potentially toxic molecules. It is easily generated onsite where it is used, by either corona discharge (CD), ultraviolet (UV) or electrolysis of water (Kim et al. 1999), eliminating the need to store dangerous chemicals. It can be applied in the gaseous or aqueous state and has been demonstrated to be at least as efficient a disinfectant as chlorine, and more convenient, because it quickly decomposes into O_2 and hence does not leave residues (Graham 1997). While it has been approved as an antimicrobial agent by the US Food and Drug Administration (FDA 2001) for direct contact with all foods, and recently approved by the French Food Safety Agency (AFSSA 2003, 2004; Pernot et al. 2007a) for grain treatment before fractionation, its oxidant properties (oxidation potential 2.07 V; Guzel-Seydim et al. 2004) and pungent odour could affect the inner composition and properties of the treated product; these effects must be evaluated to determine the suitability of ozone for different processes. Indeed, in nature, ozone has been found to lead to important metabolic changes in plant cells suffering from ozone stress (Sandermann 1996; Pell et al. 1997). Ozone has been shown to react either directly or indirectly (via the production of superoxide radical ion and its hydrogenated form, the hydroperoxide radical) with many endogenous compounds in the product (Khadre et al. 2001). Ozone can for example react with sulfhydryl-containing proteins (Cataldo 2003), which are known to play an important role in cereal product quality (Lindsay and Skerritt 1999).

This review aims to briefly overview the reported applications and investigations of ozone in cereal grain storage and processing, its benefits and potential drawbacks for the plant material and products, as well as the effects of process control parameters (e.g. gas flow rate, ozone concentration). The effects on the product characteristics will be discussed. The

Ozone in Food Processing, First Edition. Edited by Colm O'Donnell, B.K. Tiwari, P.J. Cullen, and Rip G. Rice.

potential industry relevance of reported ozone applications and the typical challenges encountered in industrial scale-up of ozone treatment processes will also be outlined.

6.2 Ozone application in grain storage and effects on grain components

The control of pest (insects and microorganisms including mould, fungi and bacteria) development in stored grains after harvest is essential as it currently leads to a grain yield loss around 3–10% in developed countries, and can reach 50% in certain countries (Jayas 1999; Fleurat-Lessard 2004; Magan and Aldred 2007).

6.2.1 Insect control

Storage grains are very susceptible to a number of insects, such as *Tribolium*, *Sithophilus* and moths, which cause considerable damage and which could potentially develop resistance to the currently used insecticides. Nevertheless, increasing environmental problems and new legislation in Europe and the USA, as well as in other countries, tend to reduce the permitted pesticide amounts (Fleurat-Lessard 2004) or even ban their use. Therefore, ozone that can be used in fumigation is an interesting alternative to applied chemicals for the control of insect development. However, few studies have been undertaken to precisely study its efficiency.

Kells et al. (2001) evaluated the efficiency of ozone fumigation in a corn grain mass (around 9 tons) against adult insects, such as the red flour beetle (*Tribolium castaneum*), maize weevil (*Sithophilus zeamais*) and larvae from the Indian meal moth (*Plodia interpunctella*). Insects were put in cages containing maize kernels and placed into a column filled with grains and positioned just below the surface. Columns were either treated or not with ozone (50 ppm for 3 days or 25 ppm for 5 days) and the number of dead insects was determined. Results demonstrated a significant insect mortality increase (92–100% compared to 3–10% in the control) when the insect species in grain samples were treated with 50 ppm for 3 days. *S. zeamais* was the most sensible to the ozone treatment. The lower dose was also significantly efficient but led to a lower insect mortality (77–99.9% depending on the insect species).

Similar results were obtained by Maier et al. (2006) using identical conditions but with insects positioned deeper in corn grain samples (0.6 m below the grain surface) and in the plenum of silos. The higher susceptibility of *S. zeamais* to ozone exposure compared with *T. castaneum* was also confirmed later by Rozado et al. (2008). These authors demonstrated that 95% of *Sitophilus* species could be controlled after 24 hours of ozone exposure, whereas an exposure of 64 hours was necessary to control *Tribolium* species.

Ozonation conditions were also pointed out to be critical for ozone efficiency. Kells et al. (2001) pointed out that an apparent air velocity of 0.03 m/s was necessary for ozone penetration through corn grain mass if a concentration of 50 ppm was used. Sousa et al. (2006) demonstrated the influence of the corn grain mass temperature over a 24-hour exposure and showed an increase in *S. zeamais* mortality when the grain mass temperature was higher than 35 °C. Indeed, an insect mortality of 12.7% was achieved at 20 °C, whereas 100% insect mortality was achieved at 40 °C. However, the effect of temperature was not observed for longer exposure periods.

A similar effect of temperature increase was also observed by Pereira et al. (2008b), who demonstrated that the required exposure time to kill 50 or 95% of adult insects of *T. castaneum* with 50 ppm of ozone was reduced if grain temperature was increased from 20 to 30 °C. Furthermore, if an increase of 5 °C led to the same efficiency as at 30 °C, increasing the temperature to 40 °C was proved to be the most efficient.

Potential differences in susceptibility to ozone treatment for several insect samples from distinct species (*T. castaneum, Rhyzopherta dominica, Oryzaephilus surinamensis*) collected in distinct places in Brazil, and potentially displaying distinct resistance to the widely used fumigant phosphine, were further explored by Sousa et al. (2008). Efficacy of ozone against *T. castaneum, O. surinamensis, R. dominica* was demonstrated with 150 ppm ozone at a flow rate of 2 L/min. Fortunately, the authors did not find any correlation between the resistance to phosphine and insect mortality following insect treatment, suggesting that the insect defence mechanisms developed for phosphine resistance are ineffective for ozone. Indeed, variability in the lethal time to kill 50% of the insect population only varied by 1.6-fold in each species evaluated, depending on the origin of the collected insect sample. Therefore, because no cross-resistance between the two gases was observed, as also reported before by Qin et al. (2003), ozone could be considered as a potential viable alternative to overcome insect resistance to chemicals. However, as it is not commonly used to treat commodities, real interest in its use remains to be assessed.

These authors also studied the relationships between the insect respiratory rate and ozone efficiency, as it was shown for phosphine that this physiological parameter is essential for the fumigant uptake. These studies also observed an increase in ozone efficiency related to higher grain mass temperature that could be linked to a higher metabolic rate of the insects and thus an increase of ozone uptake. Nevertheless, no correlation was found between ozone efficiency and insect respiration rate. However, Lu et al. (2009) demonstrated a two-phase effect of ozone on adult insect respiration (*T. castaneum, S. orysae* and *Rhyzopertha dominica*): a first phase

Table 6.1 Percentage of insect mortality reported by Işikberg et al. (2009) depending on the insect species and life stages and the ozone treatment conditions. Insects were treated either in a fumigation chamber (3 L volume) with ozone only (empty space: around 14 mg/L for 2 hours) or placed at the top or the bottom of wheat grain mass (2 kg) and treated for 5 hours with 10 pulses of the same amount of ozone every 30 minutes; n is the number of insects in each experiment.

Ozonation conditions	Adults (n = 25)	Pupae (n = 25)	Larvae (n = 25)	Eggs (n = 100)	Insect species
Empty space	100	100	100	62.5	*Ephesia kuehniella*
Top of wheat grain mass	100	100	100	100	
Bottom of wheat grain mass	100	90	92	3	
Empty space	4	14	74	4	*Tribolium confusum*
Top of wheat grain mass	5	44	86	8	
Bottom of wheat grain mass	1	22	72	6	

characterised by a lower respiration rate, reflecting the need for insects to reduce ozone toxicity, followed by a second phase where the insect respiration rate increased in parallel to ozone degradation to oxygen.

As outlined in Table 6.1, Işikberg and Öztekin (2009) pointed out large differences in ozone susceptibility of the different life stages of two distinct insect species (*Tribolium confusum* and *Ephestia kuehniella*), depending on treatment conditions (insects were treated with ozone only or placed at the top or at the bottom of wheat grains before ozone treatment). With *E. kuehniella*, eggs were the most difficult to kill, especially when placed at the bottom of the grain mass. *T. confusum* was shown as the most difficult insect species to kill, except for the larvae, which were eliminated at levels of 70–90%. For both species, the mortality rate was found to depend on the treatment conditions and the insect position in the grain mass. Therefore, further investigation is required to define adapted treatment conditions – ozone dose, temperature, exposure time – in order to penetrate deeper into the grain mass and to kill all forms of potentially contaminating insect species.

6.2.2 Microorganism control

Fungi development in grains was shown to depend on a number of factors, including cultural practices and field weather conditions, date of inoculation, resistance of the plant and storage conditions (Edwards 2004; Jouany 2007). The majority of infecting fungi types, such as *Aspergillus*, *Fusarium* and *Penicillium*, were able under certain conditions to produce mycotoxins, which have been shown to be harmful for humans and livestock, and therefore specific regulations are progressively defined in order to determine

Table 6.2 Percentage of spore survival reported by Wu et al. (2006), depending on the moisture content and temperature of wheat grains. Ozone treatment (T) was for 5 minutes with 0.33 mg/g of wheat grains or with an additional holding time of 30 minutes after ozone exposure (T + 30).

	% spore survival	
Ozonation conditions (T, grain moisture content)	**T**	**T + 30**
20 °C, 16.1%	30.1	26.5
20 °C, 19.3%	9.2	5.2
20 °C, 21.9%	3.1	0
10 °C, 16.1%	38.9	n.d.
40 °C, 16.1%	9.7	n.d.

n.d., not determined.

acceptable concentrations in raw matter and corresponding products (Murphy et al. 2006; Reddy et al. 2009). After harvest, the fungi growth and mycotoxin production can be reduced by drying grains or maintaining a moisture content below 14% and a low temperature of storage (Homdork et al. 2000; Lugauskas et al. 2007; Magan and Aldred 2007) but this will not ensure fungi removal, which can only be achieved by grain abrasion (Laca et al. 2006; Rios et al. 2009). The fungicidal efficacy of gaseous ozone for the reduction of *Aspergillus parasiticus* in corn was demonstrated by Kells et al. (2001). Indeed, a 63% reduction of the fungi load was observed in a batch of around 9 tons of grains using 50 ppm ozone for 3 days. However, inoculation was done artificially by shaking grains with a conidia suspension.

Inactivation of 96% of nonspecified spores or a mix of spores and a small amount of fungi mycelia was also demonstrated in barley after 5 minutes using 160 or 100 ppm of ozone, respectively, as mycelia were less resistant to ozone (Allen et al. 2003). Allen et al. (2003) also reported that an increase in water activity and temperature of barley grains enhanced the fungicidal activity of ozone. In naturally infected barley, Kottapalli et al. (2005) confirmed the fungicide effect of ozone on *Fusarium* fungi using higher doses of ozone (11 000 or 26 000 ppm), but for a shorter time (15-minute exposure), and found a statistically significant decrease (24–36% with $p < 0.05$) of fungi survival at either gas concentration.

An applied ozone dose of 330 ppm within 5 minutes was found to be sufficient to inactivate 96.9% of the fungal spores found in wheat grains (Wu et al. 2006). However, in this case the stored wheat grain moisture content was enhanced to around 22%. Indeed, an increase in the moisture content and thus in the water activity of wheat grains was shown to enhance the fungicidal effect of ozone on wheat grains (Table 6.2); this was also further confirmed by Raila et al. (2006). The presence of water potentially could accelerate ozone decomposition and thus the production of oxidant radicals able to react rapidly with organic compounds (Khadre et al. 2001). Wu et al. (2006) also observed an increase in ozone efficiency with a rise in temperature from 10 to 40 °C (Table 6.2) when controlling insect contamination. Additionally, ozone action was found to continue after exposure because a 30 minute resting time decreased the spore survival rate by up to 4%.

As wet and warm conditions are the most detrimental for stored grains, favouring microorganism development, the enhanced activity of ozone under the same conditions is advantageous. Therefore, interest in ozone use as a potential fungicide against the main grain contaminants has been demonstrated, as well as the importance of ozonation conditions in the gas efficiency for corn, barley and wheat. However, no clear studies on the fungicide effect in relation to the ozone coefficient of diffusion into the processed material have been developed and thus the observed effects depending on conditions, the nature of the grains or microorganisms are difficult to analyse.

Ozone (from 0.5 to 50 ppm) was also found to be effective in decreasing the level (at least by a 10-fold factor) of *Bacillus* and *Micrococcus* microorganisms which developed on cereal grains or flours during storage (Naito et al. 1988). Bacteria survival was also affected by tempering grains with ozone-containing water (1.5 or 11.5 mg/L), as demonstrated by Ibanoglu (2001). A significant decrease of total bacteria, as well as yeast and moulds, was reported in both soft and hard wheat-type grains. A more pronounced effect was observed with soft grains at a higher ozone dose, probably due to a better penetration and efficacy of ozone in that case. Microorganism levels were also found to be reduced in flours from grains that were washed for 30 minutes with ozone-containing water before drying and milling (Ibanoglu 2002).

Treatment with ozone was also found to reduce maize grain mass loss even under detrimental high-moisture storage conditions (White et al. 2010). Furthermore, ozone was reported to improve grain drying and thus could also affect microorganism growth reduction through this indirect effect (Lauva et al. 2006).

6.2.3 Reduction of toxic chemical levels

Due to its inactivating action on fungi, ozone can also be considered as helping to reduce mycotoxin accumulation during grain storage. Furthermore, its oxidant properties could also be used for mycotoxin degradation and detoxification, as demonstrated by McKenzie et al. (1997), except for fumonisin B1, which was transformed into another compound which was not proved to be safe. Degradation of aflatoxin (McKenzie et al. 1997),

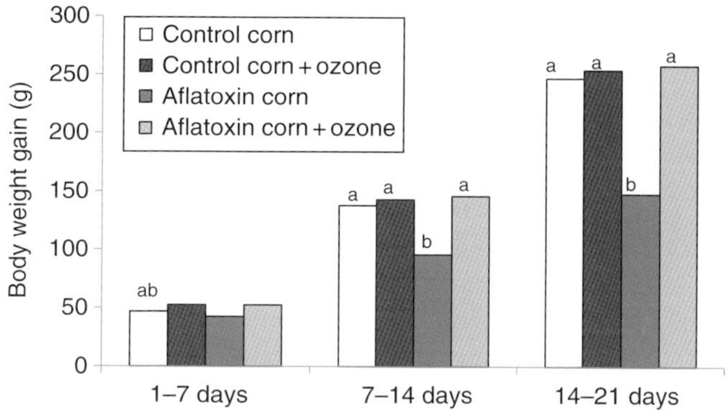

Figure 6.1 Body weight gain of turkey poultry fed with distinct grounded corn mix with soybean meal at 1–7 days old, 7–14 days old or 14–21 days old, as reported by Mc Kenzie et al. (1998). Corn samples were either noncontaminated corn (control), contaminated corn with a level of 1220 ± 73.3 ppb of aflatoxin B1, same samples after treatment with ozone (≈7 ppm/min, 92 hours). Common superscripts indicate nonsignificant mean values for each age period in days.

as well as trichothecenes (Young et al. 2006), is initiated by the attack of a double bond with addition of two oxygen atoms, which further leads to the molecule breaking apart. Young et al. (2006) furthermore pointed out that trichothecene degradation depends on ozone concentration as well as on pH.

The suitability of ozone for mycotoxin level control was claimed by McKenzie et al. (1998), who reported a 95% decrease in the aflatoxin B1 level produced by *Aspergillus* strains in naturally contaminated corn grains using around 7 mg/kg of ozone per minute for a 92-hour exposure. These results were supported by feeding poultry – which are the most sensitive animal to aflatoxicosis – with either nontreated or ozone-treated ground contaminated grains mixed with soybean. Indeed, a significant difference of weight gain between animals fed with the ozone-treated grains was observed compared to the contaminated grains (Figure 6.1). Aflatoxin-contaminated corn (1220±73.3 ppb aflatoxin B1) was found to lead to significant decrease of turkey poultry body weight gain compared with the noncontaminated control. On the other hand, no differences were observed if the grain sample was treated with ozone. Furthermore, no decreases in the relative weight of selected poultry organs such as kidney, spleen, pancreas and proventriculus were observed in poultry fed ozone-treated corn, unlike poultry fed contaminated corn, where decreases were observed.

A similar reduction of the aflatoxin B1 level in corn grains was later reported by Prudente and King (2002) in less contaminated samples (586.8±50.0 ppb) using ozone treatment. Furthermore, other studies also reported reduction of pesticides in grain following treatment with ozone (Yvin et al. 2001; Pernot et al. 2007a).

6.2.4 Effects of ozone on grain components, metabolism and physiological status

Effects on grain components

Since ozone is a powerful oxidant, a number of authors have tried to determine if its use could lead to potential negative effects on the grain compounds either by direct comparison of the molecules or via their effect on animal metabolism.

Effects of ozone exposure on fatty acids and lipids that are susceptible to oxidation were studied but contradictory results were reported. Naito (1989) pointed out that potential changes occurred preferentially above an ozone concentration of 50 ppm and were dependent on raw-matter storage parameters. On the other hand, Mendez et al. (2003) reported no significant changes after an ozone treatment (50 ppm) of wheat or corn grains for 30 days. However, these last authors pointed out that the experimental conditions did not allow ozone to penetrate deep into the grains even if the dose was sufficient to kill contaminating insects.

Absence of lipid changes was also reported by Faroni et al. (2007) in corn samples treated for 168 hours with the same ozone concentration. These authors concluded that the quality of corn oil extracted from treated versus untreated grains was similar, as demonstrated by identical fat acidity and peroxide values. Prudente and King (2002) confirmed the absence of changes in either saturated or unsaturated fatty acid levels of untreated or ozone-treated corn grains, but pointed out a significant increase in palmitic acid and a decrease in linoleic acid if treated grains were infected with fungi and also contained aflatoxins. They suggested that the damage occurring in corn grains following *Aspergillus* contamination could lead to an increase in the susceptibility of unsaturated lipids to ozone; however, the ozone concentrations used in this case were 10–12% higher than those used by Mendez et al. (2003). Dubois et al. (2006) also reported a slight decrease in linoleic acid (C18:2) content following ozone treatment of wheat grains using a higher concentration (12 g/kg). In this case, the authors also did not notice any increase of lipid oxidation compared with the water-tempered grains.

Potential oxidation of carotenoids in corn was studied by Wang et al. (2008). These authors found a significant decrease of the lutein content in ozone-treated grains (Figure 6.2) that were infected by fungi and contained aflatoxins. However, they also reported an increase in the level of extracted lutein from uninfected corn grains, which could be due to an improved extraction of lutein with ozone. Indeed, lutein could be trapped or associated with proteins from corn grains and be released by ozone treatment, which was also found to affect protein level (Figure 6.2).

Other low-molecular-weight compounds, such as B vitamins, phytic acid and ferulic acid, were compared using untreated or ozone-treated wheat

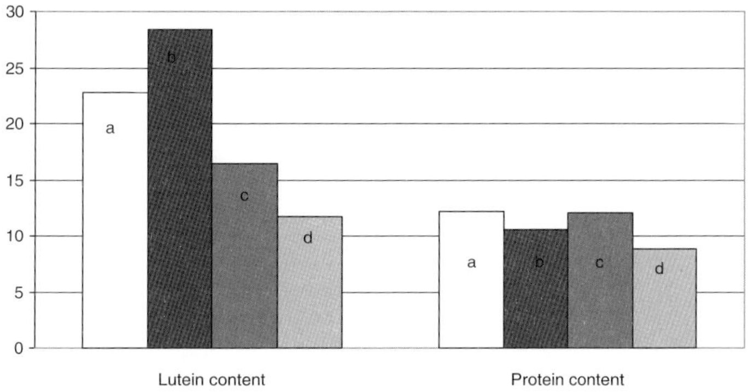

Figure 6.2 Lutein (μg/g) and protein (%) contents reported by Wang et al. (2008) in (a) clean corn grains; (b) clean corn grains treated with ozone; (c) contaminated corn grains containing aflatoxins; and (d) same as (b) after ozone treatment.

grains by Dubois et al. (2006). No distinct differences were observed, except in vitamin B5 and B6, which appeared to be lowered.

Dubois et al. (2006) also described changes in enzyme activity following ozone treatment of wheat grains using ozone concentrations between 5 and 12 g/kg of grains. These authors mainly observed an increase in protease activity and a significant decrease in lipase activity, which seems to be due mainly to grain tempering, and was amplified following ozone treatment. Protein oxidation was also observed by carbonyl group quantification and appeared to increase after 1 year of storage in the samples treated with the higher dose of ozone.

However, while these changes could alter the protein and lipid qualities of wheat grains, they did not appear to affect rat development during a 4-week feeding study (Gaou et al. 2005). Only the following few changes were observed in rats fed with either untreated or ozone-treated grains: an increase of rectal temperature in females, a slight decrease of calcium concentration in males and a slight decrease in certain blood cell numbers without clinical significances. However, the retained ozone dose used to treat wheat grains (5 g/kg) in this study was lower than those reported by Dubois et al. (2006), which showed changes in biochemical composition and enzyme activities.

Dubois et al. (2008) further studied the relationships between the effect of the ozone-treatment and the physiological state of wheat grains, pointing out that ozonation exemplified the genotoxic effects of wheat at a certain level of germination.

Effects on grain germination

No effect of ozone treatment on corn grain germination was reported by Mendez et al. (2003) using 50 ppm of ozone. These results were also later confirmed by dos Santos et al. (2007) using 100 ppm of ozone at 4.6 L/min. Also, Rozado et al. (2008) did not observe any changes in physiological quality of corn grains following ozone treatment (50 mg/kg ozone injected at a flow of 0.8 L/min). However, Allen et al. (2003) demonstrated that the effect of ozone on the germination capability of barley grains was dose-dependent. If ozone concentration was maintained under 0.98 mg/g of grains per minute, no effect on germination was observed even after 45 minutes of treatment. However, if the applied ozone dose reached 0.98 mg/g of grains per minute and was maintained in the sealed reactor, germination capability decreased after 15 minutes of treatment and was reduced by 28.5% after 45 minutes. Kottapalli et al. (2005) also pointed out that the effect of ozone on barley grain germination capability could be more pronounced if grain integrity was affected by *Fusarium* infection.

Wu et al. (2006) also studied ozonation of wheat grains, and reported that the effect of ozone on germination depends on the applied ozone dose. A similar threshold and reduction of the germination ability was found, as had been reported for barley by Allen et al. (2003).

Violleau et al. (2008) further demonstrated that short-term ozone exposure resulted in a higher germination rate of corn seeds compared to samples ozonated for longer periods, which resulted in lowering of germination levels. In this study, germination tests were undertaken immediately or after a 48 hour period following the application of ozone at 20 g/m^3 for 6, 8 or 20.5 minutes. This increase of germination rate was also reported and patented by Yvin and Coste (1997) for a number of seeds and bulbs.

6.3 Effects of ozone on grain processing, flour and product quality

Mendez et al. (2003) studied the effects of grain exposure to 50 ppm of ozone for 30 days on the popping volume of popcorn, the milling characteristics of maize and wheat, the baking of wheat flour and the stickiness of cooked rice. Contradictory results were reported for popcorn popping volume between the untreated and ozone-treated grains, depending on their position relative to the point of gas entry. However, Mendez et al. (2003) pointed out a potential effect on the pericarp integrity and grain drying, which could lead to a significant reduction of the popping volume. No differences in fraction yield between wet and dry milling of maize grains was observed.

Furthermore, Mendez et al. (2003) did not find any differences in flour moisture, ash and protein content or colour after dry fractionation of ozonated soft or hard wheat grains compared to control grains. Additionally, the bread-making properties of the flour obtained from hard red wheat grains treated with ozone were found to be similar to those of the control flour. Adhesiveness of the rice was also not significantly changed by the ozone treatment. However, the peripheral layers which are removed before cooking were found to be darker than those of the control grains and had an acidic odour. Mendez et al. (2003) thus concluded that no significant decrease of grain quality was observed for end users following ozone treatment. However, in the experimental protocol used by the researchers, layers of different types of grain were treated sequentially by the ozone flux, which could lead to variation of the received dose of ozone, and could affect conclusions as to the true impact of ozone on the grains.

Ibanoglu (2001) also reported an absence of changes in flour yield as well as water, ash and protein content or colour of extracted flour after milling of wheat grains tempered with ozone-containing water. Furthermore, falling number, sedimentation volume and rheological properties of the dough were found to be similar to those of the control grains. Similar absence of changes in flour yield from ozone-treated wheat grains was reported by Desvignes et al. (2008) using gaseous ozone at the tempering step, as described in patents (Yvin et al. 2001; Coste et al. 2004). Moreover, Desvignes et al. (2008) detailed the impact of ozone on wheat grain dry fractionation. At the higher ozone dose investigated (10g of consumed ozone per kg of hard wheat grains), these authors revealed a significant reduction of coarse bran yield (around 30%), which was balanced by the increase in white shorts yield and suggested that ozone could enhance bran friability or separation between bran and the starchy endosperm. They also observed a significant reduction of the energy required for grain milling which occurs after the first breaking step.

Studies of the corresponding grain tissue mechanical properties confirmed the enhanced friability of the outer layers and of the starchy endosperm in grains treated at the higher ozone dose. These changes in mechanical properties may explain the observed reduction of milling energy required for grain fractionation. Starch damage was also found to be reduced in flours from ozone-treated grains, possibly as a consequence of changes in starchy endosperm mechanical properties and milling energy reduction. Therefore, one could expect distinct properties of flours obtained from ozone-treated grains related to water absorption during dough making and enzymatic degradation with amylases (Mariotti et al. 2006). This result, however, appears contradictory to those reported by Dubois et al. (2006), which showed a higher amount of maltose in wheat grains after ozone treatment at similar levels and conditions. Moreover, while protein content in flours from ozone-treated grains appeared similar to that from untreated grains, detailed analysis revealed an increase in the glutenin insoluble fraction (Figure 6.3), which could result from the oxidant properties of ozone, which is known to have an impact on protein crosslinking (Cataldo 2003). Therefore, contrary to the absence of wheat flour rheological changes observed by Ibanoglu (2001), dough strengthening was expected with these flours. Coste et al. (2004), using the same process and ozone dose range, demonstrated strengthening of dough with flours obtained from ozone-treated grains.

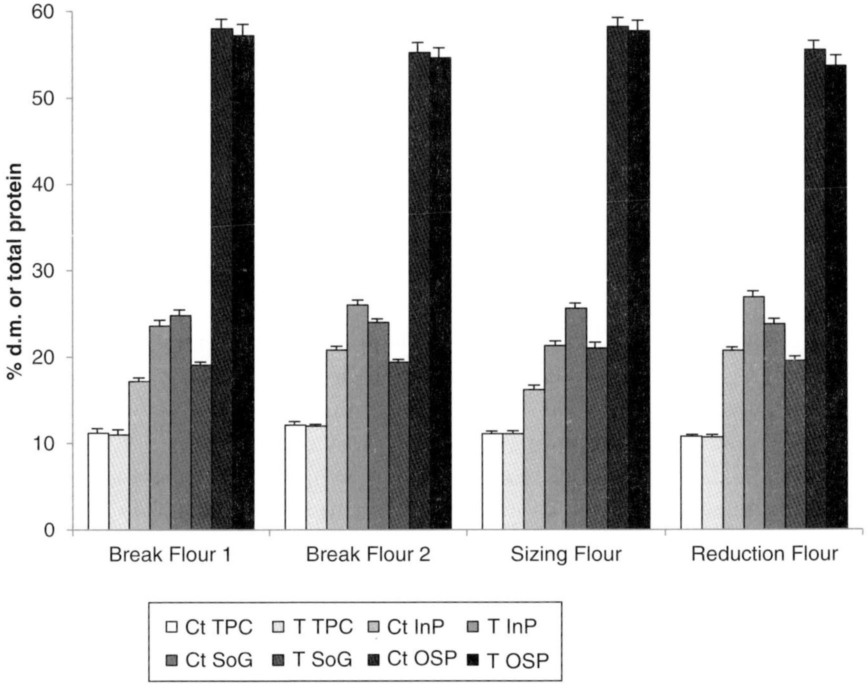

Figure 6.3 Results of protein analysis as described in Desvignes et al. (2008) after extraction with sodium phosphate buffer containing 1% sodium dodecyl sulfate and separation by size-exclusion high-performance liquid chromatography from either control hard wheat grains (Camp Remy cultivar) grains (Ct) or ozone-treated grains (T). Insoluble proteins were recovered after sonication of the pellet resuspended in the buffer (InP), total protein contents were obtained with the entire chromatograph area (TPC), glutenins were quantified in the first two peaks of the chromatograph (SoG), and other soluble proteins in the three following peaks (OSP). Results are expressed for total proteins as a percentage of the dry mass, or for other results in percentage of total proteins.

Ibanoglu (2002) also highlighted changes in dough extensibility when comparing properties of flour obtained from soft grains milled after washing with ozonated water or control (water only) for 30 minutes followed by drying. These differences, however, were not observed with flours from hard wheat grains treated similarly, suggesting a different penetration profile for ozone in hard grains.

In corn starch isolation processes, grain steeping in SO_2 at levels of 0.1–0.2% can be used in order to obtain efficient starch–protein separation, and thus a higher starch extraction yield, and to control microbial growth. However, this process could lead to problematic wastes, toxic both to the environment and to human health.

Ruan et al. (2004) showed that ozone treatment of corn grains could efficiently replace SO_2 in order to isolate high starch yields. Furthermore, ozone processing facilitated the treatment of grains at a lower temperature (20 °C versus 50 °C) and for shorter times (36 hours versus 48 hours) compared to processes using SO_2. However, depending on the ozone application timing, different results were found, suggesting that the observed effects actually depend on ozone concentration and penetration. These results are also in accordance with those of Desvignes et al. (2008), showing that ozone treatment of wheat could reduce the energy input to process cereal grains. Interestingly, ozone was also found to modify pasting properties of starch isolated from rice (An and King 2009).

6.4 Industrial applications and scale-up

A number of patents have been issued which describe equipment and methods for employing ozone to decontaminate grains and to modify their physiological status or technological properties. Yvin and Coste (1997) described equipment which facilitates an increase in seed germination or bulb growth using gaseous ozone. Barley and maize grain germination levels were found to increase after 5–20 minutes of ozone treatment with 1–28 g O_3/kg of grains (O_3 concentration in the gas vector 20–90 g/Nm^3, pressure of the gaseous atmosphere in the reactor 100–300 mbar) and with grain moisture content of around 10–20%. Murphy and Hitchens (2002) described two different devices to treat grains, in which ozone is forced under pressure through the bulk grain for a fixed duration or is treated during transit from one storage bin to another. These authors argue that these processes could be used to reduce mycotoxin level in grains. Vetter et al. (2007) also reported a system for grain sanitation with ozone in a chamber where ozone concentration is maintained between 10 and 250 ppm depending on the following parameters: type of grain, density, compaction factor, moisture content and temperature. Their design, as well as those reported by Decker et al. (2009), was clearly defined for grain storage improvement.

In the early 2000s, ozone treatment of wheat grains before processing was reported to improve flour safety and potentially its technological properties (Yvin et al. 2001; Pernot et al. 2007a,b). A specific process known as Oxygreen (Figure 6.4) was developed, which enables treatment of grain with 0.5–20 g of ozone per kg of grain (80–160 g/m^3 TPN ozone in gas vector) for a treatment time of 5–70 minutes. The wheat grain water content is increased after ozone treatment by around 4%; treated grains generally have an optimal water content (15–17%) which allows grains to be processed directly by milling after a storage time of 8–36 hours. However, modification of the flour properties is dependent on the wheat cultivar and the treatment conditions (Coste et al. 2004; Pernot et al. 2007b). Indeed, the amount and timing of water addition, as well as the duration of the storage time before milling, were found to be important parameters to vary in order to control flour properties.

Ozone treatment generally results in flours with an increased dough toughness, which is in accordance with the observed effect on the wheat storage protein network (Figure 6.3; Desvignes et al. 2008). Interestingly,

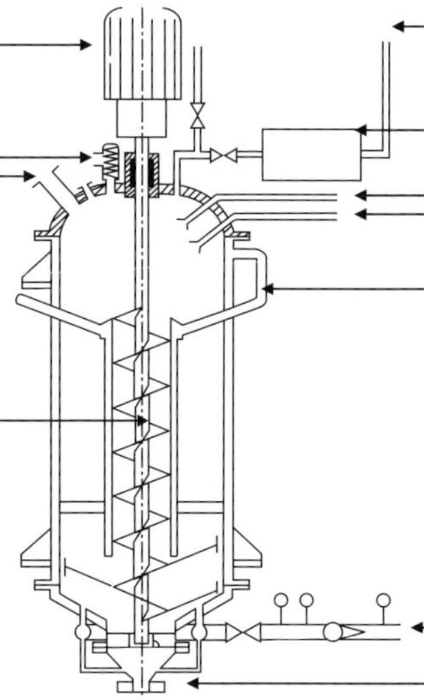

Figure 6.4 Scheme of Oxygreen reactor (Dubois et al. 2006). (From Safety of Oxygreen, an ozone treatment on wheat grains. Part 2. Is there a substantial equivalence between Oxygreen-treated wheat grains and untreated wheat grains?, M. Dubois, C. Coste, A.-G. Despres et al., *Food Additives & Contaminants*, 2006, reprinted by permission of the publisher (Taylor & Francis Group, http://www.informaworld.com)).

ozone treatment before wheat grain fractionation leads to a reduction in the required energy for grinding and the resultant flours are slightly enriched with compounds from the aleurone layer (Desvignes et al. 2008). The aleurone layer is a grain tissue which contains biochemicals that enhance flour nutritional properties. However, the observed effects on microbial mass or milling energy reduction could also be partly related to the debranning efficiency of the ozone treatment under the described conditions (Coste et al. 2007). This technology has been applied at a commercial scale in France by Paulic Minotiers and sold under the Qualista brand. A higher price is paid for flour produced by this technology. However, the actual amount of flour produced remains low as the mill production rate is limited to 1.5 T of wheat grains/hour.

Coste and Dubois (2005) reported another interesting potential application of ozone in wheat flour processing. Ozone use during dough formation was investigated, with the ozone added through the water added to flour, yeast and salt mixes, as a gas in the kneader, and through both methods. The total amount of ozone added varied between 4 and 60 mg/kg of produced dough per hour. The authors concluded that ozone addition allows a reduction in the mixing time or speed for a similar dough quality and thus could lead to an energy gain of between 15 and 23% during mixing. The resultant dough also appeared to develop faster and to be more resistant to overmixing.

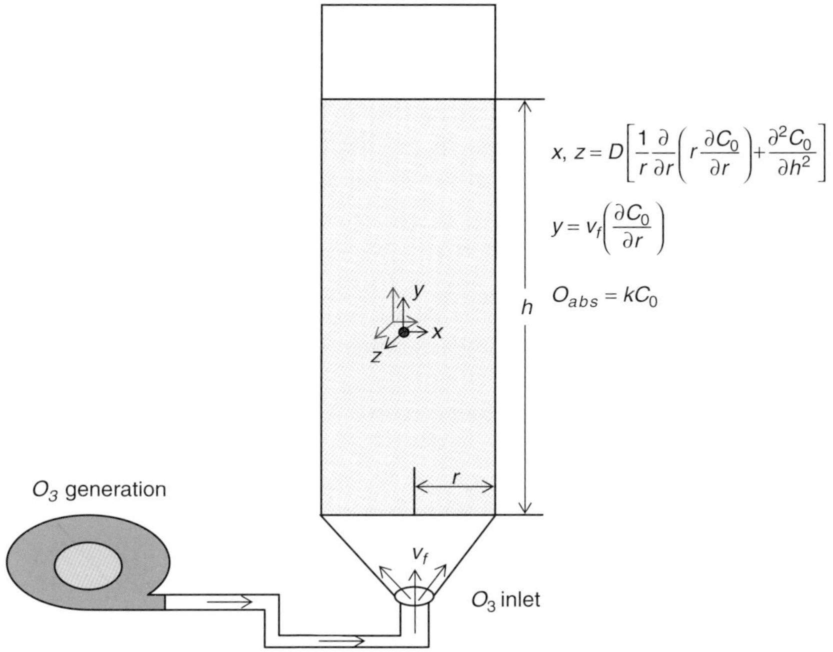

Figure 6.5 Schematic representation, reported in Tiwari et al. (2010), and equations to characterise the ozone movement in a grain column, as described in Raila et al. (2006). Ozone can move in the transverse directions (*x,z*) or in vertical direction *y*, depending on ozone gas velocity vf, adsorption of ozone (as summarised in the k factor), which is determined by grain characteristics and reactions leading to ozone degradation. Co is the applied ozone concentration, t is the exposure duration, h and r are the height and the radius of the grain column, respectively, D is the ozone diffusivity. (Reprinted from *Journal of Cereal Science*, Volume 51, Issue Number 3, B.K. Tiwari, C.S. Brennan, T. Curran, E. Gallagher, P.J. Cullen, C.P. O' Donnell, 248–255, 2010, with permission from Elsevier.)

6.5 Conclusions

Many studies have demonstrated that ozone treatment can be used to reduce levels of either biological or chemical contaminants in grains. These have been reviewed by Tiwari et al. (2010). Ozonation efficiency varies depending on the nature and state of the target, as well as the ozonation conditions (temperature, time, water content). In order to improve the control and prediction of ozone efficiency, an improved understanding of ozone diffusivity as influenced by equipment design for grain treatment, the grain tissue structure (porosity) and the ozonation parameters is needed. Raila et al. (2006) modelled ozone diffusivity in a grain column (Figure 6.5) and obtained the following equation for ozone diffusion, taking into account ozone adsorption by the grains and ozone velocity:

$$\frac{\partial C_0}{\partial t} = D\left(\frac{1}{r}\frac{\partial}{\partial r}\left[r\frac{\partial C_0}{\partial r}\right]+\frac{\partial^2 C_0}{\partial h^2}\right) - v_f\left(\frac{\partial C_0}{\partial h}\right) - kC_0$$

Similarly, the effect of ozone on grain biochemical compounds, which depend on the physiological state of the plant material, ozone penetration and treatment duration, also needs to be better characterised in order to control potential detrimental effects. Due to the strong oxidant properties of ozone, changes in the gluten protein interactions have been demonstrated to occur. However, an improved understanding of ozone action in relation to the nature of the raw matter is necessary to define optimal treatment conditions in relation to the effect of ozone on technological properties. Sensory properties (texture, taste, colour) could also be affected, depending on the ozonation conditions, and therefore a balance between benefits and drawbacks must be analysed for each potential application. Thus, a detailed evaluation of the ozone effect at each scale (grain, tissue, molecule), as influenced by the ozonation parameters, must be undertaken.

Furthermore, ozone treatment in some cases also appears to facilitate a reduction in the required processing energy. However, only a few economical evaluations of ozone treatment have been reported for grain decontamination (Pereira et al. 2008a) or for flour production. The price of flour produced with the Oxygreen process is twice that of flour for untreated grains.

Therefore, ozone treatment of grains or flours has potential industry relevance due to the necessity to control product safety and technological quality, taking into account the social and economical contexts. However, the benefits of ozonation must be compared against possible alternative techniques (as for example for grain or flour decontamination using ionisation, pulsed light, etc.) in terms of cost, efficiency and potential drawbacks. Finally, ozone treatment could also be used in combination with other type of process, as reported by Laszlo et al. (2008), who studied the efficiency of ozone and UV light in combination to reduce the microbial mass in wheat flours.

Acknowledgments

The author would like to thank the following persons: L.R. d'Antonino Faroni, Federal U. of Viçosa, Brazil; A.G. Despres-Pernot, Green Technologies, Saint-Malo, Fr.; F. Fleurat-Lessart, UR 1264, Bordeaux, Fr.; D.S. Jayas, U. Manitoba, Winnipeg, Canada; S. Naito, Food Research Institute, Nagoya, Japan; F. Violleau, EI Purpan, Toulouse, Fr., who all gave her advice or documents which were useful in the writing of this review.

References

AFSSA (2003) Opinion on the use of ozone as an auxiliary technology for the treatment of wheat grains before grinding for flours incorporated into pastries containing simple sugars, AFSSA – Saisine No. 2003-SA-0055, 24 Jul. 2003.

AFSSA (2004) Opinion on the use of ozone as an auxiliary technology for the treatment of wheat grains entering into the composition of bread and baked products, excluding traditional French bread, AFSSA – Saisine No. 2004-SA-0161, 21 Sep. 2004.

Allen, B., Wu, J.N., Doan, H. (2003) Inactivation of fungi associated with barley grain by gaseous ozone, *J Environ Sci Health*, **38**, 617–30.

An, H.J. and King, J.M. (2009) Using ozonation and amino acids to change pasting properties of rice starch, *J of Food Sci*, **74**(3): 278–83.

Cataldo, F. (2003) On the action of ozone on proteins, *Polymer Degradation and Stability*, **82**: 105–14.

Coste, C. and Dubois, M. (2005) Use of ozone for improve kneading, French Patent FR 2867658 (A1).

Coste, C., Bailli, A. and Dubois, M. (2004) Method for ozone treatment of plant materials, French Patent FR 2845015 (A1).

Coste, C., Dubois, M. and Pernot, A.-G. (2007) New process for dehusking of wheat grain, French Patent FR 3653952 (A1).

Decker, R.S., Eltiste, D.A. and Noyes, R.T. (2009) Fumigation and sanitation of biological products storage systems using ozone, US Patent 20090117016-A1.

Desvignes, C., Chaurand, M., Dubois, M., Sadoudi, A., Abecassis, J. and Lullien-Pellerin, V. (2008) Changes in common wheat grain milling behaviour and tissue mechanical properties following ozone treatment, *Journal of Cereal Science*, **47**: 245–51.

dos Santos, J.E., Martins, M.A., Faroni, L.R.D., de Andrade, M.P. and Carvalho, M.C.S. (2007) Ozonation process: saturation time, decomposition kinetics and quality of maize grains (*Zea mays* L.), in Baig. S., Pascual A., Lasalmonie A. and Serrano, C. (org) *Sustainable Agri-Food Industry Use of Ozone and Related Oxydants. Proceedings of IOA Conference and Exhibition*, Valencia, Spain, Co-chair: International Ozone Association, Scientific session 5–58, pp.1–6.

Dubois, M., Coste, C., Despres, AG., Efstathiou, T., Nio, C., Dumont, E. and Parent-Massin, D. (2006) Safety of Oxygreen, an ozone treatment on wheat grains. Part 2. Is there a substantial equivalence between Oxygreen-treated wheat grains and untreated wheat grains? *Food Additives and Contaminants*, **23**: 1–15.

Dubois, M., Canadas, D., Despres-Pernot, A.-G., Coste, C. and Pfohl-Leszkowicz, A. (2008) Oxygreen process applied on nongerminated and germinated wheat: role of hydroxamic acids, *J Agric Food Chem*, **56**: 1116–21.

Edwards, S.G. (2004) Influence of agricultural practices on *Fusarium* infection of cereals and subsequent contamination of grain by trichothecene mycotoxins, *Toxicology Letters*, **153**: 29–35.

Faroni, L.R.D., de Melo Pereira, A., de Sousa, A.H., da Silva, M.T.C. and Urruchi, W.I. (2007) Influence of corn mass temperature on ozone toxicity to *Sitophilus zeamais* (Coleoptera: *Curnulionidae*) and quality of oil extracted from ozonized grains, in Baig. S., Pascual A., Lasalmonie A. and Serrano, C. (org) *Sustainable Agri-Food Industry Use of Ozone and Related Oxydants. Proceedings of IOA Conference and Exhibition, Valencia, Spain*, Co-chair: International Ozone Association, Scientific session 5–8, pp. 1–6.

FDA (2001) Hazard analysis and critical point (HACCP): procedures for the safe and sanitary processing and importing of juice; final rule, *Federal Register*, **66**: 6137–202.

Fleurat-Lessard, F. (2004) Pest management, in Wrigley, C., Corke, H. and Walker, C. E. (eds) *Encyclopedia of Grain Science, Stored Grain*, Elsevier Science and Technology Books, pp. 244–54.

Gaou, I., Dubois, M., Pfohl-Leszkowicz, A., Coste, C., De Jouffrey, S. and Parent-Massin, D. (2005) Safety of Oxygreen, an ozone treatment on wheat grains. Part 1. A four week toxicity study in rats by dietary administration of treated wheat, *Food Additives and Contaminants*, **22**: 1113–19.

Graham, D.M. (1997) Use of ozone for food processing, *Food Technology*, **51**: 72–5.

Guzel-Seydim, Z.B., Greene, A.K. and Seydim, A.C. (2004) Use of ozone in the food industry, *Lebensm-Wiss U-Technol*, **37**: 453–60.

Homdork, S., Fehrmann, H. and Beck, R. (2000) Influence of different storage conditions on the mycotoxin production and quality of *Fusarium*-infected wheat grain, *J Phytopathology*, **148**: 7–15.

Ibanoglu, S. (2001) Influence of tempering with ozonated water on the selected properties of wheat flour, *Journal of Food Engineering*, **48**: 345–50.

Ibanoglu, S. (2002) Wheat washing with ozonated water: effects on selected flour properties, *International Journal of Food Science and Technology*, **37**: 579–84.

Işikberg, A.A. and Öztekin, S. (2009) Comparison of susceptibility of two stored-product insects, *Ephestia kuehniella Zeller* and *Tribolium confusum* du Val to gaseous ozone, *J Stored Products Res*, **45**: 149–64.

Jayas, D.S. (1999) Grain preservation: researchers aim to decrease waste by improving storage techniques, *Resource*, **7**: 7–8.

Jouany, J.-P. (2007) Methods of preventing, decontaminating and minimizing the toxicity of mycotoxins in feeds, *Animal Feed Science and Technology*, **137**: 342–62.

Kells, S.A., Mason, L.J., Maier, D.E. and Woloshuk, C.P. (2001) Efficacy and fumigation characteristics of ozone in stored maize, *J of Stored Products Research*, **37**: 371–82.

Khadre, M.A., Yousef, A.E. and Kim, J.G. (2001) Microbiological aspects of ozone applications in food: a review, *Journal of Food Science*, **66**: 1242–52.

Kim, J.G., Yousef, A.E. and Dave, S. (1999) Application of ozone for enhancing the microbiological safety and quality of foods: a review, *Journal of Food Protection*, **62**: 1071–87.

Kottapalli, B., Wolf-Hall, C.E. and Schwarz, P. (2005) Evaluation of gaseous ozone and hydrogen peroxide treatments for reducing *Fusarium* survival in malting barley, *J of Food Protection*, **68**(6): 1236–40.

Laca, A., Mousia, Z., Diaz, M., Webb, C. and Pandiella, S.S. (2006) Distribution of microbial contamination within cereal grains, *Journal of Food Engineering*, **72**: 332–8.

Laszlo, Z., Hovorka-Horvath, Z., Beszedes, S., Kertesz, S., Gyimes, E. and Hodur, C. (2008) Comparison of the effects of ozone, UV and combines ozone/UV treatment on the color and microbial counts of wheat flour, *Ozone: Science and Engineering*, **30**: 413–17.

Lauva A., Aboltins A., Palabinskis J. and Karpova-Sadigova N. 2006. Grain drying experiments with unheated air, in Dukulis, I. (ed) *5th International Sci. Conference 'Engineering for Rural Development'. Proceedings May 18–19*, Latvia U. of Agriculture, Jelgava, pp. 85–9.

Lindsay M.P. and Skerritt, J.H. (1999) The glutenin macropolymer of wheat flour doughs: structure-function perspectives, *Trends in Food Sci and Technol*, **10**: 247–53.

Lu, B., Ren Y., Du Y.-Z., Fu, Y. and Gu, J. (2009) Effect of ozone on respiration of adult *Sitophilus orizae* (L.), *Tribolium castaneum* (Herbst) and *Rhyzopertha dominica* (F.), *J of Insect Physiology*, **55**: 885–9.

Lugauskas, A. Raila, A., Zvicevičius, E., Railiene, M. and Novosinškas, H. (2007) Factors determining accumulation of mycotoxin producers in cereal grain during harvesting, *Ann Agric Environ Med*, **14**: 173–86.

Magan, N. and Aldred, D. (2007) Post-harvest control strategies: minimizing mycotoxins in the food chain, *International J of Food Microbiology*, **119**: 131–9.

Maier, D.E., Hulasare, R., Campabadal, C.A., Woloshuk, C.P. and Mason, L. (2006) Ozonation as a non-chemical stored product protection technology, in *Proceedings of the 9th International Working Conference on Stored Product Protection*, Canpinas, Brazil: Brazilian Post-harvest Association, pp. 773–7.

Mariotti, M., Pagani, M.A. and Lucisano, M. (2006) Starch damage: an interesting index for the determination of wheat flour quality, *Tecnica Molitoria*, **57**: 397–403.

McKenzie, K.S., Sarr, A.B., Mayura, K., Bailey, R.H., Miller, D.R., Rogers, T.D., Norred, W.P., Voss, K.A., Plattner, R.D., Kubena, L.F. and Phillips, T.D. (1997) Oxidative degradation and detoxification of mycotoxins using a novel source of ozone, *Food and Chemical Toxicology*, **35**: 807–20.

McKenzie, K.S, Kubena L.F., Denvir, A.J., Rogers T.D., Hitchens G.D., Bailey R.H., Harvey R.B., Buckley, S.A. and Phillips T.D. (1998) Aflatoxicosis in turkey poults is prevented by treatment of naturally contaminated corn with ozone generated by electrolys, *Poultry Sci*, **77**: 1094–102.

Mendez F., Maier, D.E., Mason, L.J. and Woloshuk, C.P. (2003) Penetration of ozone into columns of stored grains and effects on chemical composition and processing performance, *Journal of Stored Products Research*, **39**: 33–44.

Murphy, O.J. and Hitchens, G.D. (2002) Method of sterilization using ozone, US Patent 6387241 B1.

Murphy, P.A., Hendrich, S., Landgren, C. and Bryant, C.M. (2006) Food mycotoxins: an update, *J of Food Science*, **71**(5): 51–65.

Naito, S. (1989) The influence of ozone treatment on lipids contained in cereal grains, cereal grain powders, peas, beans and pulse products, *Nippon Shokuhin Kogyo Gakkaishi*, **36**(11): 878–83.

Naito, S., Okada, Y. and Sakai, T. (1988) Changes in microflora of ozone-treated cereals, grains, peas, beans, spices during storage, *Nippon Shokuhin Kogyo Gakkaishi*, **35**(2): 69–77.

Pell, E.J., Schlagnhaufer, C.D. and Arteca, R.N. (1997) Ozone-induced oxidative stress: mechanisms of action and reaction, *Physiologia Plantarum*, **100**: 264–73.

Pereira, A. de M., Faroni, L.R.D., Silva, A.G. da S. Jr, Sousa, A.H. de and Paes, J.L. (2008a) Economical viability of ozone use as fumigant of stored corn grains, *Engenharia na Agricultura*, **16**(2): 144–54.

Pereira, A. de M., Faroni, L.R.D., Sousa, A.H. de, Urruchi, W.I. and Paes, J.L. (2008b) Influence of the grain temperature on the ozone toxicity to *Tribolium castaneum*, *Rev Bras Eng Agric Ambient*, **12**(5): 493–7.
Pernot, A.-G., Dubois, M. and Coste, C. (2007a) Le procédé Oxygreen®. 1ère partie: description, objectifs et intérêts en meunerie, *Ind des Céréales*, **153**: 2–9.
Pernot, A.-G., Dubois, M. and Coste, C., (2007b) Le procédé Oxygreen®. 2ème partie: caractéristiques biochimiques et technologiques des produits finis, *Ind des Céréales*, **154**: 11–18.
Prudente, A.D. Jr and King, J.M. (2002) Efficacy and safety evaluation of ozonation to degrade aflatoxin in corn, *J Food Sci*, **67**(8): 2866–72.
Qin, Z.G., Wu, X., Deng, G., Yan, X.P., He, X.C., Xi, D.K. and Liao, X.W. (2003) Investigation of the use of ozone fumigation to control several species of stored grain insects, in *Adv in Stored Product Protection*, Massachusetts: CABI publishing, pp. 846–51.
Raila, A., Lugauskas, A., Steponavicius, D., Ralliene, M., Steponaviciene, A. and Zvicevicius, E. (2006) Application of ozone for reduction of mycological infection in wheat grain, *Ann Agric Environ Med*, **13**: 287–94.
Reddy, K.R.N., Abbas, H.K., Abel, C.A., Shier, W.T., Oliveira, C.A.F. and Raghavender, C.R. (2009) Mycotoxin contamination of commercially important agricultural commodities, *Toxin Reviews*, **28**(2–3): 154–68.
Rios, G., Pinson-Gadais, L., Abecassis, J., Zakhia-Rozis, N. and Lullien-Pellerin, V. (2009) Assessment of dehulling efficiency to reduce deoxynivalenol and *Fusarium* level in durum wheat grains, *J Cereal Sci*, **49**: 387–92.
Rozado, A.F., Faroni, L.R.A., Urruchi, W.M.I., Guedes, R.N.C. and Paes, J.L. (2008) Ozone application against *Sitophilus zeamais* and *Tribolium castaneum* on stored maize, *Rev Bras Eng Agric Ambient*, **12**(3): 282–5.
Ruan, R., Lei H., Deng, S., Lin, X., Li, Y., Wilcke, W. and Fulcher, G. (2004) Ozone aided corn steeping, *Cereal Chemistry*, **81**(2): 182–7.
Sandermann, H. Jr. (1996) Ozone and plant health, *Ann Rev Phytopathol*, **34**: 347–66.
Sousa, A.H. de, Faroni, L.R.D.A., Perreira, de M.A., Cardoso, F. da S. and Heberle, E. (2006) Influence of grain mass temperature on ozone toxicity to *Sitophilus* zeamais, *in Proceedings of the 9th International Working Conference on Stored Product Protection*, Canpinas, Brazil: Brazilian Post-harvest Association, pp. 706–10.
Sousa, A.H. de, Faroni, L.R.D.A., Guedes, R.N.C., Totola, M.R. and Urrichi, W.I. (2008) Ozone as a management alternative against phosphine-resistant insect pests of stored products, *J of Stored Products Res*, **44**: 379–85.
Tiwari, B.K., Brennan, C.S., Curran, T., Gallagher, E., Cullen, P.J. and O'Donnell, C.P. (2010) Application of ozone in grain processing, *J Cereal Sci*, **51**(3): 248–55.
Vetter, W.I., Watkins, J. and Johnson, L. (2007) Method and system for sanitization of grain products, US Patent 0134380 A1.
Violleau, F., Hadjeba, K., Albet, J., Cazalis, R. and Surel, O. (2008) Effect of oxidative treatment on corn seed germination kinetics, *Ozone, Sci and Engineering*, **30**: 418–22.
Wang, Y., King, J.M., Xu, Z.X., Losso, J. and Prudente, A. (2008) Lutein from ozone-treated corn retains antimutagenic properties, *J Agric Food Chem*, **56**: 7942–9.

White, S.D., Murphy, P.T., Bern, C.J. and van Leeuwen, J.H. (2010) Controlling deterioration of high-moisture maize with ozone treatment, *J of Stored Products Res*, **46**: 7–12.

Wu, J., Doan, H. and Cuenca, M.A. (2006) Investigation of gaseous ozone as an anti-fungal fumigant for stored wheat, *Journal of Chemical Technology and Biotechnology*, **81**: 1288–93.

Young, J.C., Zhu, H. and Zhou, T. (2006) Degradation of trichothecene mycotoxins by aqueous ozone, *Food and Chemical Toxicology*, **44**: 417–24.

Yvin, J.C. and Coste, C. (1997) Method and system for the treatment of seeds and bulbs with ozone, US Patent 5703009.

Yvin, J.C., Bailli, A., Joubert, J.M. and Bertaud, O. (2001) Method and installation for making flour from ozone-treated grains, French Patent FR2802390 (A1).

7 Ozonation of Hydrocolloids

Joan M. King, Hee-Jung An, Seung-wook Seo
and Alfredo Prudente

7.1 Introduction

Hydrocolloids such as starch, gelatin, carageenans, alginates, pectin and various gums are widely used as food ingredients. The main purpose of hydrocolloids is for thickening of products and control of textural attributes in the presence of water. The thickening ability of individual hydrocolloids is related to their structural properties and affects the mechanisms of forming gels and/or increasing viscosity. Hydrocolloids can also provide certain appearance-related attributes to foods such as opacity. Some act synergistically with others, such as kappa carageenan and locust bean gum, which work together to provide a more elastic product. Hydrocolloids can also help to stabilise products; for example, guar gum minimises heat shock in ice cream and carageenan suspends the cocoa in chocolate milk.

Various methods have been used to modify the structure of hydrocolloids to improve or alter their properties. For example, pectin has a galacturonic acid backbone with various sidechain sugars attached. Some of the carboxylic acid groups can be replaced with methyl ester groups and the number of methyl ester groups affects the way pectin gels. High-methoxyl pectins gel in the presence of acid and sugar, while low-methoxyl pectins need calcium to gel. Other hydrocolloids that need calcium to gel include carageenans and alginate. The use of potassium instead of calcium to form kappa carageenan gel results in a gel that is more brittle. An extensive amount of work has been done with starch. Starches can be hydrolysed or depolymerised with acid or enzymatic treatments, and derivatised/stabilised or crosslinked to provide a range of starches with various thickening properties, appearance qualities, fat-like properties and characteristics such as syneresis, heat, acid and shear stability. Hydrolysis of starch can result in a gel that is stronger, clearer and more transparent. Derivatisation can result in freeze–thaw stability and/or fat-like properties of starches. Crosslinking results in a starch that resists hydrolysis under low pH and maintains stability under high heat and shear. Several different chemicals are used to alter

Ozone in Food Processing, First Edition. Edited by Colm O'Donnell, B.K. Tiwari,
P.J. Cullen, and Rip G. Rice.

the structure of hydrocolloids. Chemicals used to alter the structure of starches range from simple hydrochloric acid to enzymes to compounds such as acetic anhydride, succinic anhydride, phosphoryl chloride, sodium trimetaphosphate, sodium tripolyphosphate and monosodium orthophosphate for esterification; propylene oxide for etherification; hydrogen peroxide, peracetic acid and potassium permanganate for bleaching; and sodium hypochlorite for oxidation. The use of chemicals to modify structure results in the production of chemical wastes, which must be dealt with. Also, starch ingredients that have been altered chemically must be labelled as modified starch, which causes concern among consumers.

The US Food and Drug Administration (FDA) affirmed ozone as Generally Recognised as Safe (GRAS), originally for use as a disinfectant in bottled water (FDA 1982), but an expert panel approved ozone as GRAS for broad food processing applications (Graham 1997; EPRI 1997). Ozone is a more powerful oxidant than oxygen and reacts with most substances at ambient temperatures, but creates no ozone residues. Therefore, utilising ozone to replace chemical solvents used to treat hydrocolloids for various reasons is useful. Several studies have shown that ozone can be utilised to change the properties of various hydrocolloids.

7.2 Application of ozone in hydrocolloid processing

7.2.1 Starch

Starch is utilised in many food products. Native starches often do not have the desired properties for specific uses in foods. Viscosities of starches change when they are sheared, heated for a long time or frozen and thawed many times (Kantouch and Tawfik 1998). Starches are chemically modified for several reasons, including improving cooking properties, stabilising to prevent degradation under various conditions of processing, and preventing recrystallisation, which happens during retrogradation. As mentioned previously, numerous chemical treatments are utilised to modify starches, including acid hydrolysis, oxidation, etherification and esterification. Starch modification with chemicals can leave some residues, resulting in safety concerns; for example, benzoyl peroxide- or chlorine dioxide-treated wheat flour have residues of benzoic acid or chlorine, respectively (Voraputhaporn 1996).

Oxidised starches can be utilised in a wide range of applications, including food, paper and textiles (Li and Vasanthan 2003). Kuakpetoon and Wang (2001) and Scallet and Sowell (1967) found that oxidised starches have shorter cooking times, lower viscosities, higher stability, reduced retrogradation and better clarity. Oxidised starches have been used in jelly, pudding, sauces and marmalade for better product stability (Boruch 1985; Kokini 1994). Han (2002) found that slightly oxidising starch results in a product with excellent adhesiveness, which enhances its ability to be used in batters and breadings for fried foods. Oxidised starch has been utilised as a glaze or bakery filling for high-sugar products (Kasapis 2002). Because of its increased transparency and lower gel strength, oxidised starch has also been used in baby foods (Radly 1982).

Sodium hypochlorite has been mostly used for oxidising starch (Boruch 1985), but other oxidants such as hydrogen peroxide (Parovuori et al. 1995), ammonium persulfate (Wing 1994), chlorites, oxygen and ozone have also been tested. During oxidation, starch is depolymerised and carbonyl and carboxyl groups replace its hydroxyl groups, resulting in lower viscosity (Rutenberg and Solarek 1984) and less recrystallisation/retrogradation, since these structural changes reduce three-dimensional network formation during gelation. The rate of oxidation can be increased by using a greater concentration of starch and a higher temperature, and granular damage makes the starch more susceptible to oxidation (Wing 1994). Various starches have different physical and molecular structures and will react differently during oxidation, resulting in distinct changes to viscosity properties (Kuakpetoon and Wang 2001).

Various studies have employed ozone for treatment of starch and related food products. Packaged Japanese raw noodles treated with ozone at 0.5–50 ppm for 6 hours have an increase in shelf life of up to 5 times (Naito et al. 1989). Wheat flour is typically bleached and made more elastic through treatment with chemical oxidising agents such as chlorine gas or benzoyl peroxide. Ozonation may be a good substitute for chemicals in the oxidisation of wheat flour and could result in rapid bleaching and improved functionality.

Chan et al. (2009) studied the effects of ozone on physiochemical and functional properties of various starches in dry form, including corn, sago and tapioca. They utilised from 0.4 to 1.4 mmol of ozone treatment based on time of exposure with rotation of the starch. Their results showed that starch could be treated with ozone in a dry state to alter functional properties without loss of product and without the use of chemical solvents (Chan et al. 2009).

7.2.2 Chitosan

Chitin, which is a β-(1-4)-linked N-acetyl-D-glucosamine homopolymer, is found in the shells of crustaceans such as crab, crawfish and shrimp. Chitosan is formed from deacetylated chitin. Currently, depolymerisation and decolourisation of chitosan are achieved by chemical or enzymatic methods, which are time-consuming and expensive. There are four main steps to isolate chitosan: (1) deproteinisation (DP) of the chitin; (2) demineralisation (DM) to remove calcium carbonate from the chitin; (3) decolouration (DC) of chitin pigments; and (4) deacetylation (DA) to removal acetyl groups to form chitosan. Treatment of ground shells with sodium hydroxide solution (1–10%) for 0.5 to 12 hours at elevated temperature (65~100 °C) removes the proteins present. Extended treatment under alkaline conditions at the higher end of the temperature range can cause depolymerisation and deacetylation. Dilute hydrochloric acid (up to 10%)

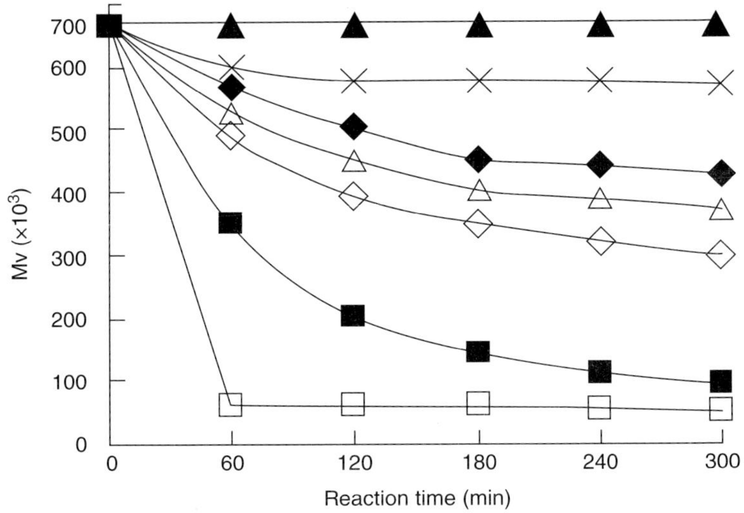

Figure 7.1 The efficiency of the ozone and ultrasonic radiation treatment compared with acid hydrolysis for the degradation of chitosan. ▲, acid hydrolysis of chitosan in 350 mL of 0.25 M hydrochloric acid solution at 20 ± 2 °C; ×, ultrasonic radiation alone in 350 mL of 0.25 M hydrochloric acid solution at 20 ± 2 °C; ◆, acid hydrolysis of chitosan in 350 mL of 0.5 M hydrochloric acid solution at 70 ± 5 °C; Δ, acid hydrolysis of chitosan in 350 mL of 0.6 M hydrochloric acid solution at 70 ± 5 °C; ◊, acid hydrolysis of chitosan in 350 mL of 0.7 M hydrochloric acid solution at 70 ± 5 °C; ■, only with ozone (35 ± 5 mg/min) at 20 ± 2 °C; □, ozone (35 ± 5 mg/min) in the presence of ultrasonic radiation at 20 ± 2 °C. (Reprinted from *Polymer Degradation and Stability*, Volume 93, Issue Number 10, Wu Yue, Pingjia Yao, Yuanan Wei and Haitao Mo, Synergetic effect of ozone and ultrasonic radiation on degradation of chitosan, 1814–1821, 2008, with permission from Elsevier.)

treatment at room temperature with agitation can dissolve calcium carbonate as calcium chloride for demineralisation. Just as with extended alkaline treatments, treatment with stronger acids can result in depolymerisation (Figure 7.1; Yu et al., 2008) and deacetylation of the native chitin (No and Meyers 1995). Deproteinisation and demineralisation treatments can cause discolouration of the chitin, which results in the need for a bleaching method to remove pigments. No et al. (1989) found that prior treatment of chitin residue with acetone was required in order to remove carotenoid pigments rapidly – 5 minutes versus 1 hour – by bleaching with 0.315% (v/v) sodium hypochlorite solution.

Deacetylation, which is removal of acetyl groups to convert chitin to chitosan, is accomplished by treatment at or above 100 °C with concentrated sodium hydroxide solution (40~50%) at a solid/solvent ratio of 1 : 15 for 30 minutes or longer (No and Meyers 1995). The molecular weight of chitosan can be decreased through depolymerisation by chemical, enzymatic or physical methods (No et al. 2003; Galed et al. 2005). Acidic hydrolysis using hydrochloric acid, nitrous acid or phosphoric acid is a common, rapid method for chemical depolymerisation to obtain chitosans of various molecular weights; however, this method produces lower yields and a large amount of monomeric D-glucosamine, is costly, and creates chemical wastes (Jeon and Kim 2000). Enzymatic treatment with chitosanase (Jeon et al. 2001) or protease (Li et al. 2005) may be preferable to chemical treatments due to milder conditions and greater controllability, but the cost of enzymes might limit their use in a large-scale commercial application. Oxidative degradation with ozone (Kabal'nova et al. 2001a; Yu et al., 2008), sodium nitrite (Mao et al. 2004) or hydrogen peroxide (Tian et al. 2004; Chang et al. 2001) can also be used to depolymerise chitosan. Wang et al. (1999) optimised an ozonolysis method to depolymerise polysaccharides containing β-D-aldosidic linkages similar to those in chitosan. Therefore, ozone treatment showed potential for replacing the time-consuming and expensive chemical and enzymatic methods that are currently used for depolymerising and decolourising chitosan (Seo et al. 2007). However, ozone treatment can also cause undesirable effects, such as development of yellowness to a similar level to that in severe chemical treatments, as ozone treatment time is increased (Seo et al. 2007). Yellowing was also observed by Kabal'nova et al. (2001a).

7.2.3 Gelatin

Gelatin is an ingredient that is used in many products. As with many hydrocolloids, the molecular weight is an important characteristic of gelatin that affects its viscosity, gel strength and other properties, such as emulsion stabilisation (Olijve et al. 2001). When gelatin is made, it must be sterilised and whitened; both procedures have typically been carried out by treatment with hydrogen peroxide (Donnelly and McGinnis 1977). Zhou and Regenstein (2005) studied the effects of pretreatment of pollock skin with alkali and acid to enhance the extraction of gelatin. They stated that issues to be concerned about when isolating gelatin from fish were removal of noncollagen proteins and degradation of the gelatin caused by natural proteases. Type A gelatin is acid-treated and Type B gelatin is alkali-treated. Zhou and Regenstein (2005) found that acid treatments resulted in loss of pollock skin collagen but alkali treatments did not. A patent by Losso et al. (2006) showed that collagen could be extracted from fish, alligator and crustacean byproducts using ozonated water as an aid to remove odours and microorganisms at levels that did not degrade the collagen. Ozonation may be an alternative method for breaking crosslinks in collagen in order to increase the yield of gelatin with varying gel strengths, without the aid of enzymes or chemicals. Ozone may also be useful in producing gelatin hydrolysates, which are used in several different products such as nutritional supplements, beverages and marshmallows as whipping agents, carriers and emulsifiers (Huang and Draget 2009). Cataldo (2003 and 2007) found that ozone had an effect on gelatin structure and thermal properties and also removed the yellow colour of the gelatin, which was destroyed due to ozone reaction with impurities.

7.2.4 Other hydrocolloids

Pectin is typically extracted from citrus peel or apple pomace and is a polymer of galacturonic acid units with varying levels of degree of methyl esterification. Pectin is used in numerous foods products, such as jellies, fruit chews and yoghurt. Polygalacturonase can be used to depolymerise pectin in order to affect its viscosity characteristics, but ozonation may be an alternative method. Guar is a galactomannan with a mannan polymer chain that has galactose branches which prevent crystalline regions from being formed. This allows water to easily hydrate the molecule; the molecular size of the guar affects its viscosity. Depolymerisation of guar has been accomplished by various methods, such as heat and acid treatment, hydrogen peroxide oxidation, enzymatic cleavage, ultrasonication and alkali treatment, as well as irradiation (Tayal and Khan 2000; Wielinga 2009). It may be possible to utilise ozone to depolymerise guar gum in order to produce different viscosities.

Cellulose has been modified by chemical methods to produce a range of products for use in foods, such as methyl cellulose, carboxymethyl cellulose and hydroxypropyl cellulose (Murray 2009). Microcrystalline cellulose is typically produced by acid hydrolysis, under pressure and heat, of wood pulp cellulose (Schaible and Sherwood 2005). Ultrasonication has also been tested for degrading cellulose, but it was found that the anomeric configuration of the cellulose affected the ability of sonication to degrade the material (Striegel 2007). Schaible and Sherwood (2005) had a patent application on using ozone and other oxidising treatments to produce microcrystalline cellulose from wood pulp. A study by Neely (1984) showed that a pretreatment of biomass with ozone enhanced the enzymatic digestion properties of feed for ruminants.

Sixta et al. (1991) studied the use of ozone for bleaching cellulose pulp and found an effect on viscosity. Tiwari et al. (2008) observed that ozone affects the viscosity properties of several hydrocolloids in solution, including guar, carboxyl methyl cellulose (CMC) and pectin. Tiwari et al. (2008) also found that ozone could cause a lightening of the colour of pectin and CMC solutions, which may facilitate the use of these ingredients in lighter-coloured food.

7.3 Effects of ozone on the physiochemical properties of hydrocolloids

7.3.1 Structural composition

The structure of a hydrocolloid affects its functional properties. For example, carrageenans vary in their ability to gel depending on the level of sulfated groups, while stronger and more opaque gels are obtained for starches with higher amylose, which has a linear structure. Hydrocolloids with more branching, for example, waxy starch or gum arabic, interact less in solution and result in lower viscosity than linear hydrocolloids (Williams and Phillips 2009).

An and King (2009) observed that high purity oxygen and 12 wt% ozone treatments increased apparent amylose content in rice starch. The highest apparent amylose content was observed in starch treated with gaseous ozone for 30 minutes. Other studies showed that amylose content decreased after oxidation, but Han and Ahn (2002) found that starch treated with sodium hypochlorite had increased levels of soluble amylose compared to untreated samples. Voraputhaporn (1996) found no change in amylose content of wheat flour and starch with ozonation at 25–100 ppm. Cataldo (2003, 2007) found ozone had a minimal effect on gelatin structure. Ozone treatment of gelatin dissolved in water with 450 mg of ozone over 3 hours resulted in loss of tertiary structure, but did not affect secondary or primary structure.

7.3.2 Swelling power

Chan et al. (2009) studied the effects of ozone on dry corn, sago and tapioca starches. Differences were observed in the effects of ozone on individual starch properties. Corn starch was not affected, whereas sago starch had increased solubility and tapioca starch had decreased solubility compared to controls. The decrease in solubility of tapioca starch was thought to be due to crosslinks formed by carbonyl formation, since tapioca starch had the greatest level of carbonyl content. Corn starch had increased swelling power and, sago and tapioca starches had decreased swelling power (Chan et al. 2009) (Table 7.1). The decrease in swelling power was thought to be due to structural breakdown of the granules. Voraputhaporn (1996) observed increased solubility and swelling in wheat starch treated with ozone, with almost no differences between samples treated with 25–100 ppm ozone.

Table 7.1 Swelling power (g/g) of ozone-oxidised starches. Results are expressed as means ± standard deviations ($n = 4$). Values in the same column with the same lowercase letters are not significantly different ($p > 0.05$). (Adapted from Chan et al. 2009.)

Ozone generation time (min)	Starch		
	Corn	Sago	Tapioca
Unmodified	9.45 ± 0.05 c	9.98 ± 0.19 a	15.27 ± 0.35 a
1	9.68 ± 0.06 ab	9.27 ± 0.52 b	12.49 ± 0.33 b
3	9.52 ± 0.16 bc	9.21 ± 0.14 b	12.03 ± 0.29 b
5	9.59 ± 0.09 bc	9.03 ± 0.45 bc	9.23 ± 0.31 d
10	9.78 ± 0.11 a	8.46 ± 0.12 c	10.51 ± 0.19 c

7.3.3 Molecular weight

Molecular weight is an important characteristic of hydrocolloids that is directly related to other properties such as viscosity. Larger molecules tend to have greater viscosity.

Ozone can degrade macromolecules and remove pigments due to its high oxidation potential. Seo et al. (2007) showed that the molecular weight of chitosan in acetic acid solution decreased to different levels depending on the ozone treatment time. Table 7.2 shows the molecular weights of ozone-treated chitosan for various treatment times (Seo et al. 2007).

Statistical analyses showed that there were no significant changes in molecular weight over time for water-based chitosan solutions treated with ozone (Seo et al. 2007). The molecular weight of chitosan in acetic acid solutions decreased by 92% (104 kDa) compared to the untreated chitosan (1333 kDa) after ozonation for 20 minutes, resulting in decreased viscosity of the chitosan solution. The ineffectiveness of ozonation at depolymerising chitosan in water solutions is probably due to the strong intermolecular hydrogen bonding and rigid molecular structure of chitosan in water. Acidic conditions cause protonation of amino groups in chitosan, which leads to chain repulsion and swelling (Yao et al. 1994) and promotes the solubility of chitosan in acidic solution (Hahn and Nam 2004). Ozonation for 5 minutes greatly enhanced the whiteness of chitosan, but further ozonation resulted in yellowing (Seo et al. 2007). When chitosan was ozonated in water for various time periods, no significant changes in colour were observed. Ozone treatment of chitosan in both water and acetic acid solutions did not result in deacetylation of chitosan (Seo et al. 2007). Yue et al. (2008a) also found that the degree of acetylation did not change with ozone treatment.

Yue et al. (2008a) found that treatment of chitosan in 2% acetic acid solution for 60 minutes with 65 mg/min of ozone resulted in a decrease in molecular weight from 246 to 6.1 kDA, which is a much lower molecular

Table 7.2 Molecular weight of ozone-treated chitosan. Means ± standard deviation. Means with different letters within a column indicate significant differences ($p < 0.05$). Numbers in parentheses are percentage molecular weight decrease. (Adapted from Seo et al. 2007.)

Ozone treatment time (min)	Molecular weight (kDa)	
	Ozone treatment in water	Ozone treatment in acetic acid
0	1362 ± 22.50 a (0)	1333 ± 29.69 a (0)
5	1363 ± 82.03 a (−0.07)	432 ± 105.98 b (67.59)
10	1356 ± 69.64 a (0.44)	201 ± 38.28 c (84.92)
15	1245 ± 17.90 a (8.59)	131 ± 24.76 c (90.17)
20	1229 ± 304.95 a (9.77)	104 ± 2.31 c (92.20)

weight than was found by Seo et al. (2007). The resulting chitosan was acid-free and water-soluble. Yue et al. (2008b, 2009) also found a synergistic effect between ozone and ultrasonic radiation in the degradation of chitosan to produce low-molecular-weight forms of chitosan without formation of carboxylic acid groups. With treatment at 35 mg/min ozone, the ozone alone resulted in a decrease in molecular weight from 667 to around 350 kDa, whereas the combination of ozone with ultrasonic radiation resulted in a decrease to 51 kDa after only 60 minutes (Figure 7.1).

Studies have shown that ozone alone or ozone with hydrogen peroxide can depolymerise cellulose or chitosan (Demin et al. 1993; Kabal'nova et al. 2001a,b). Ozone rapidly reacts with chitosan. Kabal'nova et al. (2001a) found that with a 2% ozone treatment of chitosan solution for 15 minutes, the ratio of reacted moles of ozone to moles of chitosan units was 0.13, and after 4 hours the ratio was 1.5. The number-average molecular mass of chitosan decreased during treatment of chitosan in dilute acid solution (0.33 M CH_3COOH or 0.1 M HCl) with 2% ozone. Higher temperatures can increase the initial rate of reaction and further decrease the degree of polymerisation (Kabal'nova et al. 2001a).

The antimicrobial activities of chitosan are greatly affected by its molecular weight (Jeon et al. 2001; No et al. 2002), which varies with the source of chitin and the methods used to prepare the chitosan. Several studies have shown that chitosan can inhibit bacterial growth. The molecular weight and concentration of chitosan used can affect its antimicrobial activity and the type of bacteria against which it is effective (Jeon et al. 2001; No et al. 2002; Zheng and Zhu 2003). Seo et al. (2008) found that ozone-depolymerised chitosan with molecular weights ranging from 102 to 244 kDa showed greater antimicrobial activity against *Listeria monocytogenes*, *Staphylococcus aureus* and *Pseudomonas fluorescens*. For *E.coli*, high-molecular-weight chitosan was more effective in growth inhibition than low-molecular-weight chitosan. The antimicrobial effects strengthened as the concentration of chitosan increased, regardless of molecular size (Seo et al. 2008).

7.3.4 Viscosity

Voraputhaporn (1996) found that wheat starch treated with ozone had increased peak viscosity compared to native starch. An and King (2009) treated 20% rice starch solutions in distilled water with 20 wt% ozone and found that ozone increased the peak viscosity of rice starch (indicating greater swelling), resulting in less cooking stability, as indicated by increased breakdown compared to untreated starch. There was also a decreased tendency for retrogradation, as indicated by the lower total setback due to the lower final viscosity during cooling. Peak viscosity increased by 284 mPa.s after ozone treatment for 15 minutes, but final viscosity and total setback decreased by 458 and 616 mPa.s, respectively,

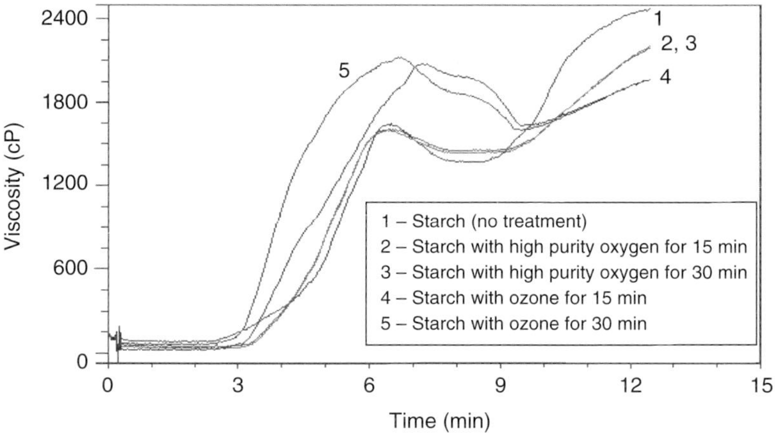

Figure 7.2 Effects of high purity oxygen and ozone treatments for 0, 15 and 30 minutes on sigma rice starch. (Adapted from An 2005.)

compared to non-oxidised starches (An and King 2009). Rice starch ozonated for 30 minutes reacted similarly, but to a greater extent (Figure 7.2). Kuakpetoon and Wang (2001) reported that starch slightly oxidised with hypochlorite showed greater swelling and peak viscosity due to the loss of granule integrity, which resulted in more water entering into the starch granules. Rutenberg and Solarek (1984) reported that oxidative scission results in structural changes that make gel formation difficult, resulting in less potential for retrogradation. Therefore, ozonated starch exhibited similar pasting properties to those from oxidised starch treated with chemical oxidising agents.

In the study by Chan et al. (2009), differences were observed in how the rapid visco analyser (RVA) viscosities were affected by ozone for tapioca, sago and corn starches. There was no trend in the effect of ozone treatment on peak, hot-paste and cold-paste viscosities and breakdown (BKD) in tapioca starch, but all viscosities and BKD decreased for corn starch and increased for sago starch (Chan et al. 2009). Setback decreased for both ozone-treated corn and tapioca starches, indicating that they became more stable to retrogradation, whereas sago starch showed the opposite effect.

Two international patents by Kesselmans and Bleeker (1997a,b) covered the use of ozone in the presence of a buffer or halogenide-containing catalyst to oxidise root, tuber or waxy cereal starches and cellulose under dry or semi-dry conditions. In the examples provided, the intrinsic viscosity of potato starch treated with ozone decreased compared to native potato starch.

Figure 7.3 shows the changes in viscosity of chitosan solution with ozone treatment in different conditions. An initial 5-minute ozone treatment in acetic acid rapidly reduced the viscosity by 91% from 331 to 29 cP, which then slowly decreased to 14, 12 and 10 cP in 10, 15 and 20 minutes,

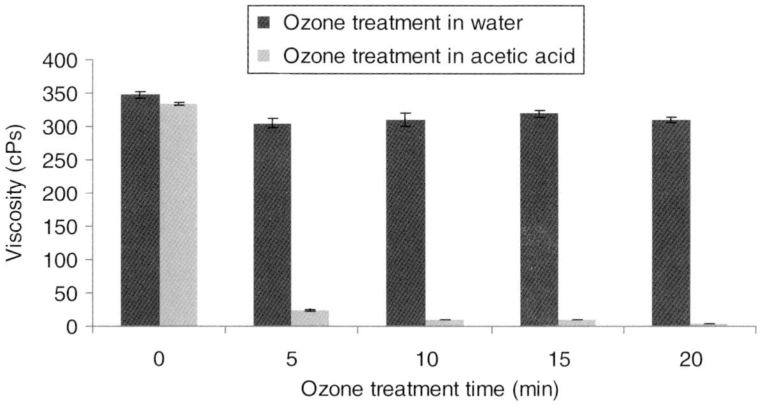

Figure 7.3 Effect of ozone treatment times on viscosity of chitosan solution. (Seo et al. 2007.)

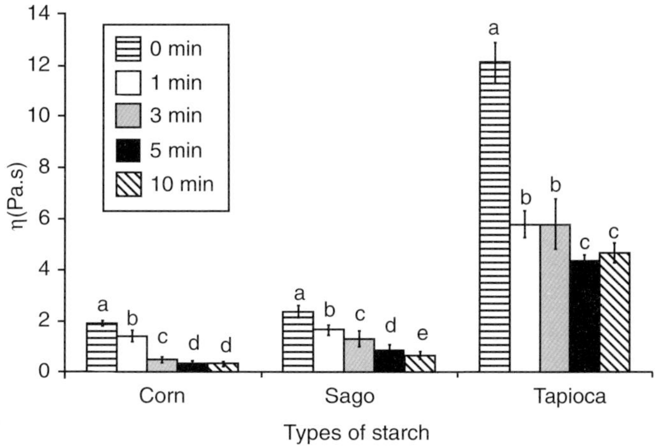

Figure 7.4 Viscosity of 5% (w/w) unmodified and ozone-oxidised corn, sago and tapioca starches. Measurements were made at 25 °C. (Reprinted from *Food Chemistry*, Volume 126, Issue Number 3, Hui-Tin Chan, Chiu Peng Leh, Rajeev Bhat, Chandra Senan, Peter A. Williams and Alias A. Karim, Molecular structure, rheological and thermal characteristics of ozone-oxidised starch, 1019–1024, 2011, with permission from Elsevier.)

respectively (Seo et al. 2007). Chan et al. (2011) observed a significant decrease in the viscosity for 5% (w/w) ozone-oxidised corn, sago and tapioca starches at 25 °C compared to unmodified starch. They observed that tapioca starch exhibited a pronounced decrease in viscosity upon oxidation by ozone compared to corn and sago (Figure 7.4). The authors further observed a reduction in the molecular weight of corn and sago starches leading to a reduction in the viscosity; however, they could not relate the viscosity reduction of ozonated tapioca starch with the molecular weight, as they observed an increase in molecular weight after ozonation

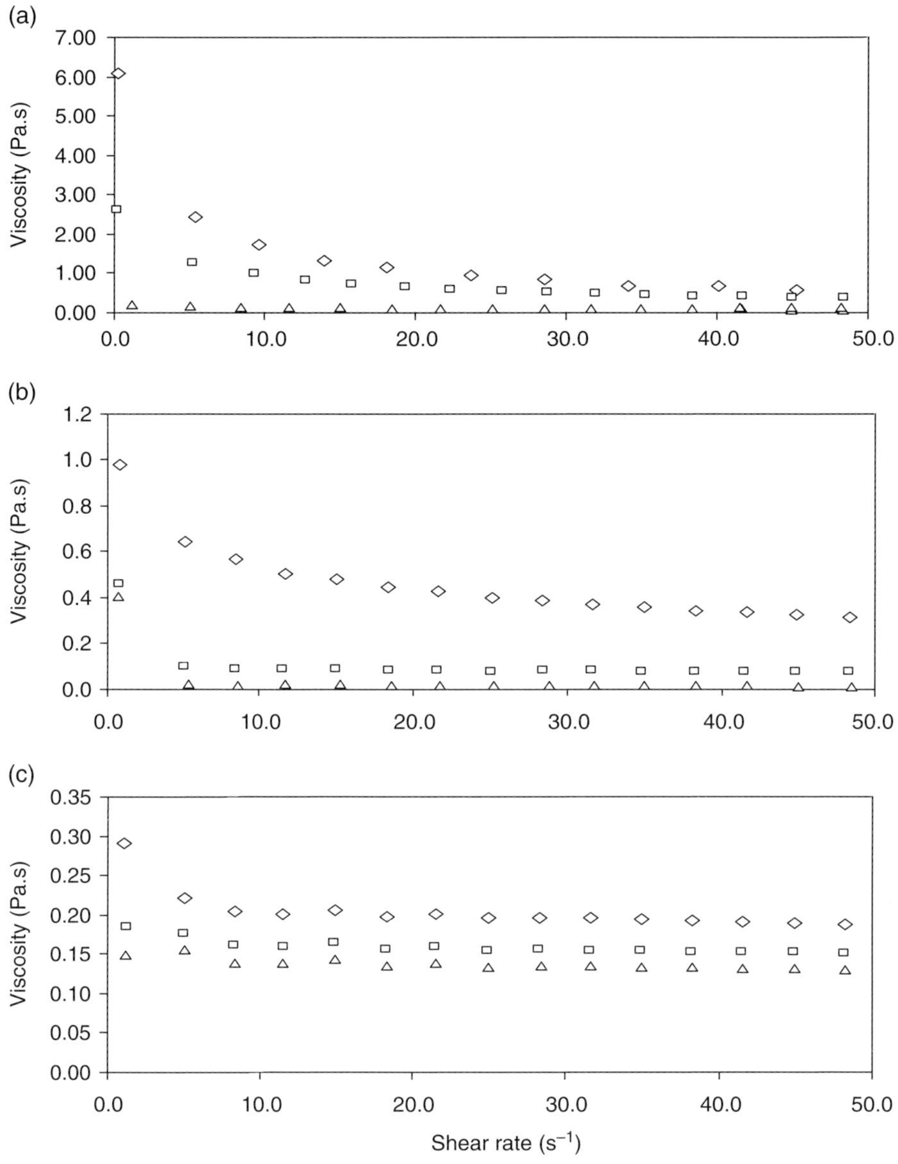

Figure 7.5 Effect of ozonation on the apparent viscosity of: (a) guar gum (1%); (b) CMC 1%; and (c) pectin (2%). Control (◊), ozone (7.8% (w/w)) for 5-minute (□) and 10-minute (Δ) processing times. (Reprinted from *Food Research International*, Volume 41, Issue Number 10, B.K. Tiwari, K. Muthukumarappan, C.P. O'Donnell, M. Chenchaiah and P.J. Cullen, Effect of ozonation on the rheological and colour characteristics of hydrocolloid dispersions, 1035–1043, 2011, with permission from Elsevier.)

for tapioca starch. The significant reduction in the viscosity of oxidised tapioca starch was reported to be related to a higher content of carbonyl groups. In another study, Chan et al. (2009) observed a greater fraction of carbonyl groups in oxidised tapioca starches compared to corn and sago starches oxidised at the same ozone generation time of 10 minutes.

Tiwari et al. (2008) found that treatment with 2.4–7.8% (w/w) ozone for 5–10 minutes affected the properties of guar, CMC and pectin in 1 or 2% solutions. Apparent viscosity of all hydrocolloids tested decreased with ozone treatment compared to control, which may have been due to depolymerisation. Guar gum and CMC were the most susceptible (Figure 7.5). Cataldo (2003) observed only a 5% change in viscosity of gelatin after 3 hours of ozonation with 450 mg of ozone bubbled into a solution of gelatin. Sixta et al. (1991) found that viscosity decreased in cellulose pulp after ozone treatment at a concentration of 90–100 g/cm^3. Simoes and Castro (2001) found that higher temperatures facilitated ozone (5 mg/L) depolymerisation of cellulose, resulting in lowered viscosity from 425 to 322 cm^3/g after 60 minutes of treatment at 2 °C, but viscosity decreased to approximately 125 cm^3/g at 33 °C.

7.3.5 *Thermal properties*

Voraputhaporn (1996) treated dry wheat flour with 25–100 ppm ozone. Voraputhaporn (1996) found a higher peak gelatinisation temperature with a lower enthalpy, but the amylose–lipid complex shifted to a lower temperature than in unoxidised starch. It was also found that the second endotherm from ozonated starch had higher enthalpy compared to the control, which may have been due to oxidation of starch lipid in the amylose–lipid complexes. Thermal analysis of the ozonated gelatin showed a higher peak temperature with a shoulder and lower enthalpy than a reference gelatin, due to structural changes (Figure 7.6) (Cataldo 2007). The treated gelatin was ozonated with 15 wt% ozone at a flow rate of 150 mg/h.

7.4 Mechanism and structural effects of ozone action on hydrocolloids

Rakovsky et al. (1996) found that starch could be cleaved by ozone at the 2C–3C and 1C–4C bonds. A study was conducted on ozone treatment of cellobiose to determine the mechanism of ozone depolymerisation of cellulose (Lemeune et al. 2000). Lemeune et al. (2000) proposed that ozone attacked either the C1 position under acidic conditions or the oxygen ether link connecting the glucoses together directly under both acidic and basic conditions (Figure 7.7). It was also proposed that ozone attacks the free anomeric carbon end to form a carboxylic acid under basic conditions (Lemeune et al. 2000).

Demin et al. (1993) and Kabal'nova et al. (2001b) stated that the basic mechanism of ozone depolymerisation of chitosan is the formation of labile hydrotrioxides at the C(1)-H bond, followed by cleavage of the C1–C4 linked β-D-glycoside bonds (Figure 7.8). They also stated that

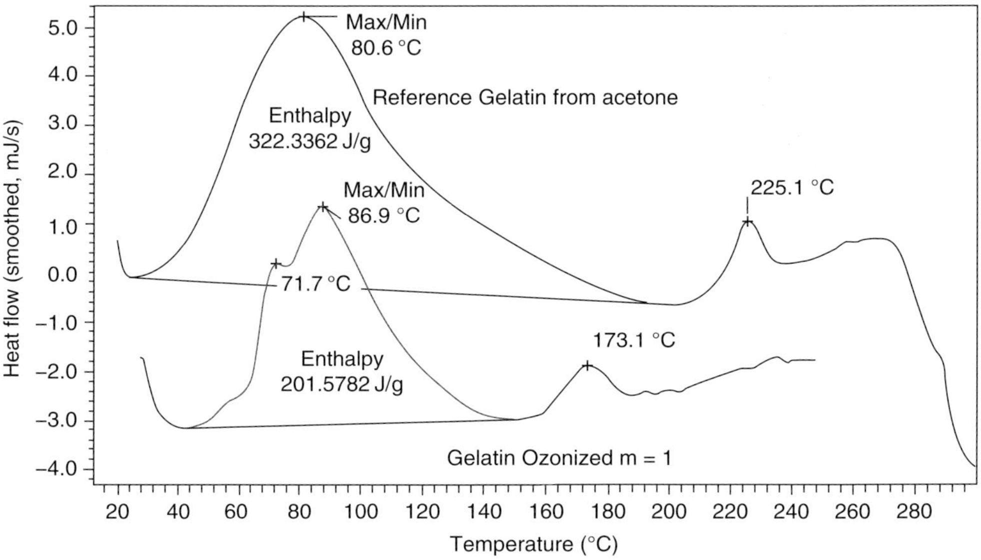

Figure 7.6 Differential scanning calorimetry (DSC) traces of samples heated at 5 °C/min in static air inside Al crucibles. The upper trace is due to reference gelatin, while the trace at the bottom is due to ozonated gelatin at a molar ratio ozone/gelatin mr = 1. (Reprinted from *International Journal of Biological Macromolecules*, Volume 41, Issue Number 2, Franco Cataldo, On the action of ozone on gelatin, 210–216, 2007, with permission from Elsevier.)

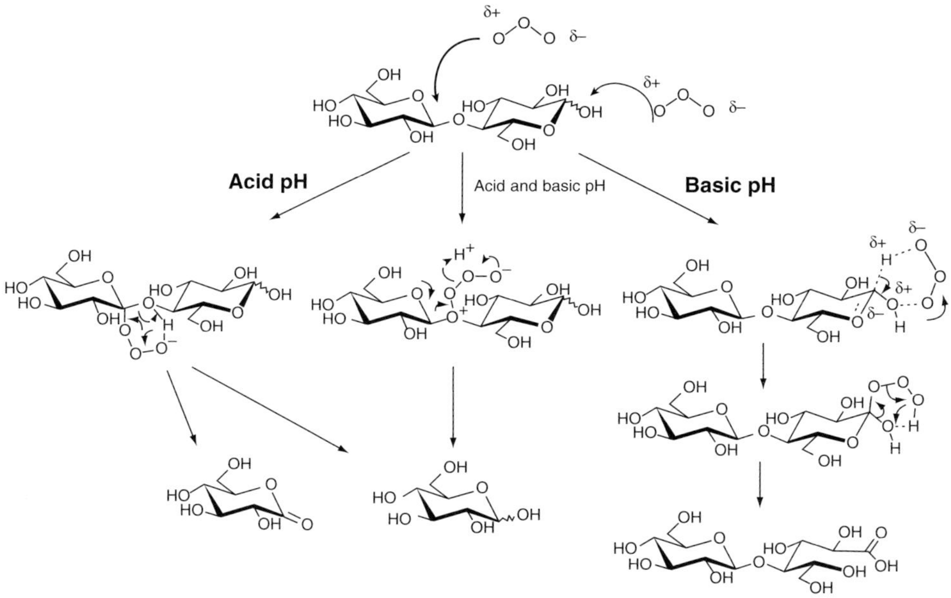

Figure 7.7 Proposed mechanism for cellobiose ozonation. (Adapted from Degradation of cellulose models during an ozone treatment. Ozonation of glucose and cellobiose with oxygen or nitrogen as carrier gas at different pH, S. Lemeune, J.M. Barbe, A. Trichet et al., *Ozone: Science & Engineering*, 2000, reprinted by permission of the publisher (Taylor & Francis Group, http://www.informaworld.com).)

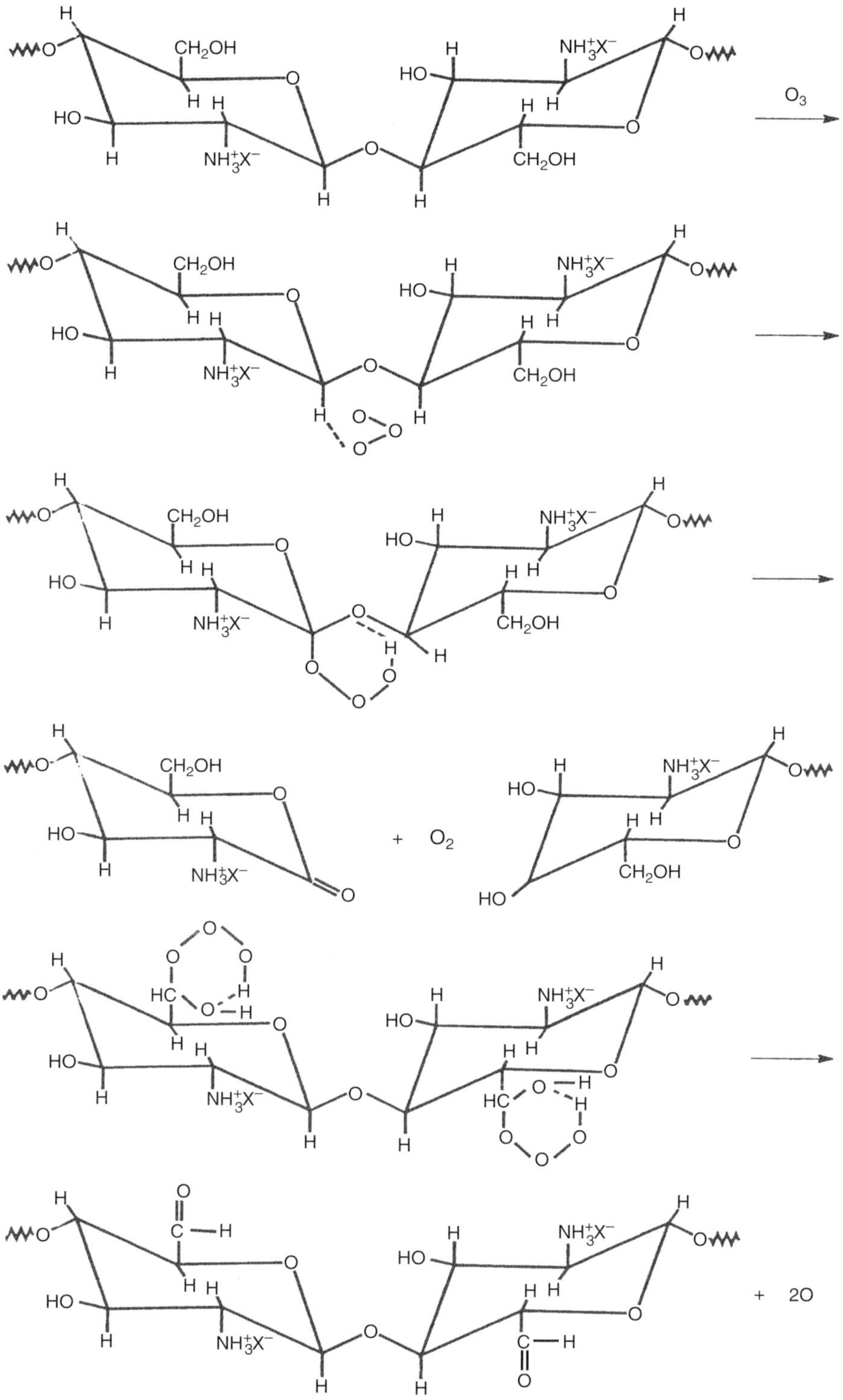

Figure 7.8 Mechanism of oxidative destruction under the action of ozone. (Adapted from Kabal'nova et al. 2001b.)

carbonyl groups could be formed at the C6 position with longer ozone treatment (6 hours) under higher temperature (70 °C). Chan et al. (2009) found higher levels of carbonyl and carboxyl contents with longer ozone exposure in tapioca, sago and corn starches tested. Kesselmans and Bleeker (1997a) observed that carbonyl content increased the most for tapioca and potato starches and the least for cellulose. One of the first reported studies using ozone on pectin reported that degradation proceeded through initial loss of hydrogen at position C5, followed by loss of the glycosidic residue at the C4 position, following a trans elimination reaction (Albersheim et al. 1960).

Wang et al. (1999) investigated the ozone depolymerisation of GBS (Group B *Streptococcus*) polysaccharides, which are important for making vaccines through conjugation with proteins. Through testing of GBS polysaccharides, cellobiose, lactose and dextran, they found that ozone degradation of carbohydrates in aqueous solution follows three different mechanisms: ozonolytic degradation of β-D-glycosidic linkages, oxidative degradation by hydroxyl radicals formed from water, and acid hydrolysis. Radical reactions and acid hydrolysis degrade carbohydrates nonselectively, whereas direct ozone attack selectively depolymerises polysaccharides.

It can be concluded that the studies discussed in this chapter demonstrate that ozonation is a powerful tool that can be employed to replace harmful chemical treatments in the depolymerisation and decolouration of various hydrocolloids to alter their physicochemical properties.

References

Albersheim, P., Neukom, H. and Deuel, H. (1960) Splitting of pectin chain molecules in neutral solutions, *Arch Biochem Biophys*, **90**: 46–51.

An, H.J. (2005) Effects of ozonation and addition of amino acids on properties of rice starches, PhD dissertation, Louisiana State University.

An, H.J. and King, J.M. (2009) Using ozonation and amino acids to change pasting properties of rice starch, *Journal of Food Science*, **74**(3): C278–C283.

Boruch, M.L. (1985) Transformation of potato starch during oxidation with hypochlorite, *Starch*, **37**(3): S91–S98.

Cataldo, F. (2003) On the action of ozone on proteins, *Polymer Degradation and Stability*, **82**: 105–14.

Cataldo, F. (2007) On the action of ozone on gelatin, *International Journal of Biological Macromolecules*, **41**: 210–16.

Chan, H.T., Bhat, R. and Karim, A.A. (2009) Physicochemical and functional properties of ozone-oxidized starch, *Journal of Agricultural and Food Chemistry*, **57**: 5965–70.

Chan, H.T., Leh, C.P., Bhat, R. Senan, C., Williams P.A. and Karim, A.A. (2011) Molecular structure, rheological and thermal characteristics of ozone-oxidized starch, *Food Chemistry*, **126**(3): 1019–24.

Chang, K.L.B., Tai, M.-C. and Cheng, F.-H. (2001) Kinetics and products of the degradation of chitosan by hydrogen peroxide, *Journal of Agricultural and Food Chemistry*, **49**: 4845–51.

Demin, V.A., Kabal'nova, N.N., Osipova, G.I. and Shereshovetz, V.V. (1993) Depolymerization of cellulose upon ozonation, *Russian Journal of Applied Chemistry*, **66**: 2562.

Donnelly, T.H. and McGinnis, R.S. (1977) Gelatin manufacture: peroxide liquefaction process, US Patent 4043996.

EPRI (1997) Ozone: GRAS affirmation for use in food, *Food Industry Currents*, **1**(1): 1–7.

FDA (1982) GRAS status of ozone, *Fed Reg*, **47**: 50 209–10.

Galed, G., Miralles, B., Panos, I., Santiago, A. and Heras, A. (2005) N-acetylation and depolymerization reactions of chitin/chitosan: influence of the source of chitin, *Carbohydrate Polymers*, **62**: 316–20.

Graham, D.M. (1997) Use of ozone for food processing, *Food Technol*, **51**: 72–5.

Hahn, H.G. and Nam, K.D. (2004) Fungicidal activities of chitosan against plant pathogens, Journal of Chitin and Chitosan, **9**(2): 73–8.

Han, J.S. (2002) Changes of dynamic viscoelastic properties of oxidized corn starch suspensions during heating and cooling, *Food Sci Biotechnol*, **11**(3): 231–7.

Han, J.S. and Ahn, S.Y. (2002) Physicochemical properties of corn starch oxidized with sodium hypochlorite, *J Korean Soc Food Sci Nutr*, **31**(2): 189–95.

Huang, I.J. and Draget, K.I. (2009) Gelatin, in Phillips, G.O. and Williams, P.A. (eds) *Handbook of Hydrocolloids*, 2 edn, Boca Raton, FL: CRC Press LLC, pp. 142–63.

Kabal'nova, N.N., Shereshovets, V.V., Murinov, K.Y., Galiaskarova, G.G., Mullagaliev, I.R., Krasnogorskaya, N.N. and Monakov Y.B. (2001a) Oxidative destruction of chitosan during ozonation, *Intern J Polymeric Mater*, **50**: 47–56.

Kabal'nova, N.N., Murinov, K.Y., Mullagaliev, I.R., Krasnogorskaya, N.N., Shereshovets, V.V., Monakov, Y.B. and Zaikov, G.E. (2001b) Oxidative destruction of chitosan under effect of ozone and hydrogen peroxide, *Journal of Applied Polymer Science*, **81**: 875–81.

Kantouch, F. and Tawfik, S. (1998) Gelatinization of hypochlorite oxidized maize starch in aqueous solutions, *Starch*, **50**(2–3), S114–S119.

Kasapis, S. (2002) Viscoelasticity of oxidized starch/low methoxy pectin mixtures in the presence of glucose syrup, *International J of Food Sci and Tech*, **37**: 403–13.

Kesselmans, R.P.W. and Bleeker, I.P. (1997a) Method for oxidizing dry starch using ozone, World Intellectual Property Organization, WO/1997/032902.

Kesselmans, R.P.W. and Bleeker, I.P. (1997b) Method for oxidizing polysaccharides using ozone in the presence of a halogenide-containing catalyst, World Intellectual Property Organization, WO/1997/032903.

Kokini, T.L. (1994) Rheological properties of food, in Heldman, B.R. and Lund, D.B. (eds), *Handbook of Food Engineering*, Marcel Dekker, pp. 1–144.

Kuakpetoon, D. and Wang, Y.J. (2001) Characterization of different starches oxidized by hypochlorite, *Starch/Starke*, **53**: 211–18.

Jeon, Y.J. and Kim, S.K. (2000) Production of oligosaccharides using an ultrafiltration membrane reactor and their antibacterial activity, *Carbohydrate Polymers*, **41**: 133–41.

Jeon, Y.J., Park, P.J. and Kim, S.K. (2001) Antimicrobial effect of chitooligosaccharides produced by bioreactor, *Carbohyrate Polymers*, **44**: 71–6.

Lemeune, S., Barbe, J.M., Trichet, A. and Guilard, R. (2000) Degradation of cellulose models during an ozone treatment: ozonation of glucose and cellobiose with oxygen or nitrogen as carrier gas at different pH, *Ozone Science and Engineering*, **22**: 447–60.

Li, J.H. and Vasanthan, T. (2003) Hypochlorite oxidation of field pea starch and its suitability for noodle making using an extrusion cooker, *Food Research International*, **36**: 381–6.

Li, J., Du, Y., Yang, J., Feng, T., Li, A. and Chen, P. (2005) Preparation and characterization of low molecular weight chitosan and chito-oligomers by a commercial enzyme, *Polymer Degradation and Stability*, **87**: 441–8.

Losso, J.N., Ogawa, M., Portier, R.J. and Schexnayder, M.A. (2006) Extraction of collagen from calcified tissues, US Patent 7109300 B2.

Mao, S., Shuai, X., Unger, F., Simon, M., Bi, D. and Kissel, T. (2004) The depolymerization of chitosan: effects on physicochemical and biological properties, *International Journal of Pharmaceutics*, **281**: 45–54.

Murray, J.C. (2009) Cellulosics, in Phillips, G.O. and Williams, P.A. (eds) *Handbook of Hydrocolloids*, 2 ed, Boca Raton, FL: CRC Press LLC, pp.710–23.

Naito, S., Sawada, Y. and Yamaguchi, N. (1989) Studies on utilization of ozone in food preservation: effect of ozone treatment on storage of packaged 'Nama Men' Japanese raw noodle, *J Antibact Antifung Agents*, **17**(11): 517–26.

Neely, W.C. (1984) Factors affecting the pretreatment of biomass with gaseous ozone, *Biotechnology and Bioengineering*, **26**: 59–65.

No, H.K. and Meyers, S.P. (1995) Preparation and characterization of chitin and chitosan: a review, *Journal of Aquatic Food Product Technology*, **4**(2): 27–52.

No, H.K., Meyers, S.P. and Lee, K.S. (1989) Isolation and characterization of chitin from crawfish shell waste, *Journal of Agricultural and Food Chemistry*, **37**(3): 575–9.

No, H.K., Park, N.Y., Lee, S.H. and Meyers, S.P. (2002) Antibacterial activity of chitosans and chitosan oligomers with different molecular weights, *International Journal of Food Microbiology*, **74**: 65–72.

No, H.K., Nah, J.W. and Meyers, S.P. (2003) Effect of time/temperature treatment parameters on depolymerization of chitosan, *Journal of Applied Polymer Science*, **87**: 1890–4.

Olijve, J., Mori, F. and Toda, Y. (2001) Influence of the molecular weight distribution of gelatin on emulsion stability, *Journal of Colloid and Interface Science*, **243**:476–82.

Parovuori, P., Hamunen, A., Forssell, P., Autio, K. and Poutanen, K. (1995) Oxidation of potato starch by hydrogen peroxide, *Starch/Starke*, **47**: 19–23.

Radly, J.A. (1982) *Starch Production Technology*, New York, NY: Starch Technology Applied Sci. Pub. Ltd.,. pp. 457–69.

Rakovsky, S.K., Sheldon, R.A. and Rantwijk, F.V. (1996) Carbohydrate ozonation: selective cleavage of the 2,3 C-C bond in carbohydrates, *Oxidation Communications*, **19**(4): 482–98.

Rutenberg, M.W. and Solarek, D. (1984) Starch derivates: production and uses, in Whistler, R.L., BeMiller, J.N. and Paschall, E.F. (eds) *Starch Chemistry and Technology*, Orlando, FL: Academic Press, pp. 311–88.

Scallet, B.L. and Sowell, E.A. (1967) Production and uses of hypochlorite-oxidized starch, in Whistler R.L. and Paschall E.F. (eds) *Starch: Chemistry and Technology*, New York, NY: Academic Press, pp. 237–51.

Schaible, D. and Sherwood, B. (2005) Treatment of pulp to produce microcrystalline cellulose, US Patent Application 2005/0145351 A1.

Seo, S., King, J.M. and Prinyawiwatkul, W. (2007) Simultaneous depolymerization and decoloration of chitosan by ozone treatment, *Journal of Food Science*, **72**(9): C522–C526.

Seo, S., King, J.M., Prinyawiwatkul, W. and Janes, M. (2008) Antibacterial activity of ozone-depolymerized crawfish chitosan, *Journal of Food Science*, **73**(8): M400–M404.

Simoes, R. and Castro, J. (2001) Ozone depolymerization of polysaccharides in different materials, *Journal of Pulp and Paper Science*, **27**(3): 82–7.

Sixta, H. Gotzinger, G., Schrittwieser, A. and Hendel, P. (1991) Medieum consistency ozone bleaching: laboratory and mill experience, *Das Papier*, **45**(10): 610–25.

Striegel, A.M. (2007) Influence of the anomeric configuration on mechanochemical degradation of polysaccharides: cellulose versus amylose, *Biomacromolecules*, **8**: 3944–9.

Tayal, A. and Khan, S.A. (2000) Degradation of a water soluble polymer: molecular weight changes and chain scission characteristics, *Macromolecules*, **33**: 9488–93.

Tian, F., Liu, Y., Hu, K. and Zhao, B. (2004) Study of the depolymerization behavior of chitosan by hydrogen peroxide, *Carbohydrate Polymers*, **57**: 31–7.

Tiwari, B.K., Muthukumarappan, K., O'Donnell, C.P., Chenchaiah, M. and Cullen, P.J. (2008) Effect of ozonation on the rheological and color characteristics of hydrocolloid dispersions, *Food Research International*, **41**: 1035–43.

Voraputhaporn, W. (1996) Effect of ozone on the chemical and functional properties of wheat flour, The University of New South Wales, Department of Food Science and Technology.

Wang, Y., Hollingsworth, R.I., Kasper, D.L. (1999) Ozonolytic depolymerization of polysaccharides in aqueous solution, *Carbohydrate Research*, **319**: 141–7.

Wielinga, W.C. (2009) Galactomannans, in Phillips, G.O. and Williams, P.A. (eds) *Handbook of Hydrocolloids*, 2 edn, Boca Raton, FL: CRC Press LLC, pp. 228–51.

Williams, P.A. and Phillips, G.O. (2009) Introduction to food hydrocolloids, in Phillips, G.O. and Williams, P.A. (eds) *Handbook of Hydrocolloids*, 2 edn, Boca Raton, FL: CRC Press LLC, pp. 1–22.

Wing, R.E. (1994) Oxidation of starch by thermochemical processing, *Starch/Starke*, **46**(11): S414–S418.

Yao, K.D., Peng, T., Feng, H.B. and Feng, Y.Y. (1994) Swelling kinetics and release characteristic of crosslinked chitosan-polyether polymer network (semi-IPN) hydrogels, *Journal of Polymer Science Part A Polymer Chemistry*, **32**: 1213–23.

Yue, W., Yao, P., Wei, Y., Li, S., Lai, F. and Liu, X. (2008a) An innovative method for preparation of acid-free-water-soluble low molecular weight chitosan, *Food Chemistry*, **108**: 1082–7.

Yue, W., Yao, P., Wei, Y. and Mo, H. (2008b) Synergetic effect of ozone and ultrasonic radiation on degradation of chitosan, *Polymer Degradation and Stability*, **93**: 1814–21.

Yue, W., He, R., Yao, P. and Wei, Y. (2009) Ultraviolet radiation-induced accelerated degradation of chitosan by ozone treatment, *Carbohydrate Polymers*, **77**: 639–42.

Zheng, L.Y. and Zhu, J.F. (2003) Study on antimicrobial activity of chitosan with different molecular weights, *Carbohydrate Polymers*, **54**: 527–30.

Zhou, P. and Regenstein, J.M. (2005) Effects of alkaline and acid pretreatments on Alaska pollock skin gelatin extraction, *J Food Sci*, **70**(6): C392–C396.

8 Ozone in Meat Processing

Fred W. Pohlman

8.1 Introduction

Meat safety continues to be a global concern. Illnesses related to *E. coli*, *Salmonella*, *Campylobacter*, *Listeria* and other pathogens remains a serious issue. Because of these concerns, the meat industry continues to research and adopt new strategies to combat pathogens. Technologies have been researched from farm to fork. These have included animal feeding strategies to reduce the incidence of pathogen shedding, animal sanitation technologies for improving the pathogen loads entering meat plants, carcass washing and interventions to remove extraneous materials and contamination that may host pathogens on carcass surfaces, as well as primal, subprimal retail and product interventions to provide reduced likelihood of pathogen survival. Many industries have adopted multiple intervention technologies in a 'hurdle' technology approach. This approach provides multiple obstacles or hurdles to destroy pathogens if present or to decrease their likelihood of survival and/or proliferation along the processing chain from live animal to finished product. Many regulatory agencies have adopted such an approach for safe meat production. For instance, in the USA, processing is conducted through the Hazard Analysis Critical Control Point system, whereby hazards are monitored at various stages through the processing chain (Critical Control Points), and interventions and validations have been encouraged through the regulatory system. To this end, in the USA, should contamination be identified, as in the case of *E. coli*, the product may be considered adulterated.

To combat potential pathogens, research has been conducted as interventions through the processing chain. Animal sanitation and chemical dehairing (Castillo et al. 1998; Xiangwu et al. 2003) have been evaluated for their effect on reducing animal hide contamination. Interventions for carcasses that have been evaluated include carcass washing (Algino et al. 2009; Gill et al. 2000; Bolton et al. 2002), lactic acid application (Gill 2009; Bosilevac et al. 2006; Gill and Badoni 2004), acetic acid application (Algino et al. 2007; Berry and Cutter 2000; Cutter 1999), steam exposure (Retzlaff et al. 2005; Minihan

Ozone in Food Processing, First Edition. Edited by Colm O'Donnell, B.K. Tiwari, P.J. Cullen, and Rip G. Rice.

et al. 2003; Nutsch et al. 1998) and others. Other interventions that have been evaluated through the processing chain up to and including ground beef production include the use of organic acids such as gluconic acid (Stivarius et al. 2002b), trisodium phosphate (Riedel et al. 2009; Rio et al. 2007; Morshedy and Sallam 2009), hot water (Tompkins et al. 2008), cetylpyridinium chloride (Riedel et al. 2009; Singh et al. 2005), sodium metasilicate (Byelashov et al. 2010; Pohlman et al. 2009; Weber et al. 2004) and ozone (Yang and Chen 1979a; Novak and Yuan 2004a,b; Reagan et al. 1996).

The effectiveness of antimicrobial interventions is a complex issue. Not only should the interventions be effective for reducing or eliminating pathogens of target, but they must also preserve meat quality characteristics. It is important that safety be improved without deterioration to meat quality attributes such as colour or appearance, aroma, sensory taste and textural properties. Therefore, attention must be given to the class of antimicrobial being used. Organic acids have been popular for evaluation as interventions because of the pH effect they provide against microorganisms. Some organic acids, such as lactic acid, naturally occur in beef; others do not, but are generally considered safe for consumption at levels applied. Therefore, they are attractive additions as intervention technologies. However, because they buffer the pH downward, they can have negative impacts on meat quality attributes such as colour and lipid oxidation. The reduced pH of meat due to organic acid application can modify the molecular structure of myoglobin, the principal pigment in meat, causing a darkening of colour. Likewise, care must be taken with organic acids with regard to the lipid fraction of meat, in order to not promote oxidation, which leads to off-flavour development.

Other antimicrobials may target pH shifts upward toward neutrality. Antimicrobials such as trisodium phosphate and sodium metasilacte are examples of antimicrobials that shift pH in this direction. Because they shift pH upward toward neutrality, this provides myoglobin with an environment more closely associated with *in vivo* conditions and favours the stability of the molecule and the stability of meat colour. Other antimicrobials such as cetylpyridinium chloride are classified as quaternary ammonium compounds. In the case of cetylpyridinium chloride, its efficacy is not found to depend upon pH shifts, therefore it has little impact on meat colour or lipid oxidation.

Another concern for antimicrobial treatments is whether they impart lasting effects or residues. Organic acids leave lasting pH effects and if not buffered, may remain. However, most of these are generally safe for consumption or Generally Recognised as Safe (GRAS) in meat products. With other classes of antimicrobials, trace amounts or residues may remain. Therefore, care should be taken to use approved amounts to avoid residue issues, and note must also be taken of proper labelling requirements where necessary.

8.2 Application of ozone in meat processing

The use of ozone offers unique features with regard to antimicrobial properties, quality and residues when used in meat systems. Because it is a gas, it can be used in a gaseous phase or in aqueous phases, where liquid mediums deliver the gaseous ozone to the target food, such as meat. Ozone decomposes rapidly in air; consequently there are both opportunities and challenges for its use in the meat industry. Residues remaining after meat product treatment are negligible. However, because ozone is the effective medium in impeding microorganisms, it becomes important to keep the ozone level high enough to illicit its positive effect. A number of studies have looked at decontaminating red meat and poultry using ozone applications. These studies have examined carcass as well as meat decontamination against endogenous bacteria, with varying results.

8.2.1 Surface decontamination of red meat

Castillo et al. (2003) evaluated the impact of an ozone treatment spray application to beef carcasses versus water washing alone using a carcass spray cabinet. Carcasses were inoculated with faeces that contained *E. coli* 0157:H7 and *Salmonella typhimurium*. The water spray treatment used a gradual spray pressure increase up to 400 lb/in^2, whereas the ozone spray treatment with ozonated water (95 mg/L ozone concentration) was applied at 80 lb/in^2. Results indicated that ozonated water treatment was similar in pathogen reduction to the water spray treatment. However, ozone was used at a much lower spray pressure, so it would appear that the use of ozone in the water spray allowed for lower pressures to be used for application of the carcass wash while still achieving similar pathogen reductions to water spray alone.

Gorman et al. (1995) compared 0.5% ozonated water with 5% hydrogen peroxide, 12% trisodium phosphate, 2% acetic acid and 0.3% commercial sanitiser for their antimicrobial efficacy against *E. coli* on beef brisket. Briskets were inoculated with a faecal paste then treated with the antimicrobials through spray application. Results indicated that water temperature played a role in bacterial count reductions, as well as the type of antimicrobial chemical intervention used. The use of hydrogen peroxide and ozonated water was more effective for reducing bacterial counts than trisodium phosphate, acetic acid or the commercial sanitiser when the antimicrobial was applied after first washing with water. The authors reported total plate count reductions using hydrogen peroxide and ozone treatments of 2.60–2.86 colony forming units (CFU)/cm^2.

Beef carcasses have the potential to become contaminated with faecal material through the normal dressing procedure. Because these bacteria may include pathogens, it is of interest to apply antimicrobial treatments to beef carcasses in order to remove any faecal contamination and pathogens that might be present. Reagan et al. (1996) evaluated trimming techniques and beef carcass washing techniques to improve the microbial quality of meat. This study evaluated intervention technologies in six high-volume commercial beef processing plants in five states in the USA. For the study, beef carcasses were intentionally inoculated with faecal material (>4.0 CFU/cm^2). Intervention treatments included knife trimming, washing with water and rinsing with ozone (0.3–2.3 ppm) or hydrogen peroxide (5%). Ozone treatment reduced carcass surface contamination by 1.30 CFU/cm^2, whereas hydrogen peroxide reduced aerobic plate counts by 1.14 CFU/cm^2. The authors also noted that trimming and washing of beef carcasses consistently reduced the bacterial load on the inoculated beef carcasses.

In other beef carcass studies, Greer and Jones (1989) evaluated the impact of gaseous ozone treatment on beef carcass bacterial spoilage profiles and on meat quality and carcass shrinkage. To do this, they placed one beef carcass paired side into a cooler supplied with continuous ozone generation and compared the findings to conventional chilling. They found that psychrotrophic bacterial growth was retarded on carcass surfaces while under ozone atmosphere. However, carcass treatment with ozone atmosphere did not reduce bacterial growth on steaks fabricated from these carcasses.

The reduction of microbial pathogens is important on carcass and tissue surfaces in order to improve the safety of intact, whole muscle cuts. However, the reduction of microbial pathogens on beef trimmings destined for ground beef is also important and is very challenging. This is because any contamination that might occur on beef tissue surfaces will become inoculated into the interior of ground beef when it is ground. Therefore, technologies that reduce microbial load on beef trimmings destined for ground beef are important to improving the safety of the ground beef product made from these trimmings.

Stivarius et al. (2002a) evaluated the impact of ozone at differing exposure levels and of chlorine dioxide as beef trimming treatments for their effect on ground beef microbial characteristics (Table 8.1). For the study, they inoculated beef trimmings with *E. coli* and *S. typhimurium*. The beef trimmings were then treated with 1% ozonated water for 7 minutes or 15 minutes, or with 200 ppm chlorine dioxide, and compared with a control. Ground beef was then produced from the beef trimmings, packaged and placed into simulated retail displays for up to 7 days. *E. coli*, *S. typhimurium*, coliform and aerobic plate counts were carried out on the ground beef during display. Results showed that the 15-minute ozone treatment of beef trimmings was effective for reducing all microorganism counts measured in ground beef, while the 7-minute ozone treatment reduced aerobic plate counts and *S. typhimurium* counts significantly.

In another study, Pohlman et al. (2002) determined the impact of ozone, chlorine dioxide, cetylpyridinium chloride and trisodium phosphate as

Table 8.1 Effect of chlorine dioxide and ozone treatments of beef trimmings before grinding on *E. coli*, coliform, *S. typhimurium*, CFU log reductions and aerobic plate count during simulated retail display. Least square means within a row bearing different letters are different ($p < 0.05$). (Reprinted from *Meat Science*, Volume 60, Issue Number 3, M.R Stivarius, F.W Pohlman, K.S McElyea, J.K Apple, Microbial, instrumental color and sensory color and odor characteristics of ground beef produced from beef trimmings treated with ozone or chlorine dioxide, 299–305, 2002, with permission from Elsevier.)

	Treatment			
Microorganism	**Control**	**Chlorine dioxide (200 ppm)**	**Ozonated water bath (1%; 7.2 °C; 7 min)**	**Ozonated water bath (1%; 7.2 °C; 15 min)**
E. coli	6.51 ± 0.08 z	5.80 ± 0.09 y	6.39 ± 0.08 z	6.37 ± 0.08 z
Coliform	5.89 ± 0.12 z	5.32 ± 0.12 x	5.74 ± 0.13 yz	5.45 ± 0.09 xy
S. typhimurium	5.70 ± 0.09 z	5.09 ± 0.09 xy	5.25 ± 0.09 y	4.92 ± 0.09 x
APC	7.20 ± 0.10 z	6.48 ± 0.10 x	6.88 ± 0.11 y	6.63 ± 0.09 xy

multiple interventions on beef trimmings, with regard to the impact on ground beef safety. For the study, beef trimmings were inoculated with *E. coli* and *S. typhimurium* and subjected to multiple interventions of: (1) 1% ozonated water followed by 5% acetic acid; (2) 1% ozonated water followed by 0.5% cetylpyridinium chloride; (3) 200 ppm chlorine dioxide followed by 10% trisodium phosphate; and (4) a control. The trimmings were then ground to produce ground beef and placed in a simulated retail display for 7 days. In general, the multiple interventions were more effective than the single antimicrobial interventions for reducing bacterial loads. Additionally, both ozone-containing treatments were effective for reducing levels of *E. coli*, *S. typhimurium*, coliforms and aerobic bacteria counts.

Ozone has also been used as a pretreatment before cooking to determine any synergistic activity on reducing microorganisms. Novak and Yuan (2004a) treated beef surfaces with ozone then cooked the treated beef at temperatures of 45–75 °C to determine the impact on enterotoxin-producing strains of *Clostridium perfringens*. The authors reported a 1–2 log CFU/g reduction in *C. perfringens* as a result of aqueous ozone treatment and heating at 45–75 °C. Additionally, they reported a reduction in spore count with the same treatments, but the magnitude of reduction was very small, indicating that the spores were much more resistant to ozone and thermal treatments. The authors concluded that ozone treatment followed by heat treatment allowed reductions at cooking temperatures that normally would not impart the reduction by themselves.

In a related study, Novak and Yuan (2004b) also looked at the impact of ozone and mild heat pretreatment on *C. perfringens* spores for beef surfaces which were subsequently packaged under modified atmosphere (MAP). *C. perfringens* spores remained dormant in beef through a 10-day storage period at 25 °C and inhibited spore germination with increasing concentrations of CO_2 in the modified atmosphere and with refrigeration. Novak and Yuan (2003) also evaluated the impact of aqueous ozone or mild heat treatment of beef then treated with heat, alkali or salt on *C. perfringens, E. coli and Listeria monocytogenes* on beef. Reductions for each of these microorganisms were greater using 3 ppm ozone for 5 minutes than for the mild heat treatment (55 °C for 30 minutes). The study also found that cells surviving ozone treatment were no more resistant to pH shifts than cells that were not exposed to ozone. The authors concluded that the pathogens evaluated were less likely to produce foodborne illnesses than those that survive sublethal heat treatments.

Ozone has also been evaluated for its effectiveness against *L. monocytogenes* in ready-to-eat products such as ham. Jhala et al. (2002) reported on the impact of ozone and a bacteriocin from *Propionibacterium shermanii* on *L. monocytogenes* in ready-to-eat cooked and cured ham. They reported a synergistic activity between ozone (0.2–1.0 ppm) and the bacteriocin, causing an inactivation of up to 3 log reductions of *L. monocytogenes*.

8.2.2 Surface decontamination of poultry

In addition to red meat, ozone has also been evaluated for poultry. While poultry scientists have evaluated the efficacy of ozone for chill and wastewater decontamination, this section will focus on poultry meat surface decontamination. Like red meat, poultry offers a medium suitable for pathogen colonisation and growth.

Yang and Chen (1979b) evaluated the effects of ozone on poultry meat microflora. To prepare for treatment, broiler carcasses were divided into thigh and breast pieces. Natural poultry microflora were incubated to produce an inoculum. The inoculated poultry was washed with a bottle dispenser with 3.88 mg/L ozone and a flow of 2050 mL/min for 20 minutes. Ozone washing reduced microbial counts with refrigerated storage. The authors estimated that ozone treatment extended poultry shelf life by 2.4 days. Furthermore, it appeared that ozone treatment of poultry was most effective for reducing Gram-negative rods.

Sheldon and Brown (1986b) also studied the impact of ozone treatment as a disinfectant for poultry carcasses. Using the carcass chill water as the delivery medium, carcasses chilled with ozonated water had lower microbial counts than those chilled with conventional water. In another study, Fabrizio et al. (2002) studied the impact of various antimicrobials, including ozone, on *S. typhimurium* levels on poultry. For this study, broiler carcasses were inoculated with *S. typhimurium* and subjected to antimicrobial treatments. Aqueous ozone (10 ppm) was effective at reducing *S. typhimurium* levels over 7 days of storage. Levels of *S. typhimurium* were only able to be detected after selective enrichment.

Evaluating ozone and other antimicrobials, Vadhanasin et al. (2004) investigated the use of ozone as a critical control point to reduce the

Table 8.2 Microbial counts on chicken stored under normal and ozone-rich atmospheres. (Reprinted from *International Journal of Refrigeration*, Volume 7, Issue Number 6, J.C Nieto, F. Jiménez-Colmenero, Ma.C. Peláez, Effect of ozone on bacterial flora in poultry during refrigerated storage, 389–392, 1984, with permission from Elsevier.)

		Days in storage					
Micro-organisms	**Samples****	**0**	**2**	**4**	**7**	**10**	**13**
Viable aerobes	NA	3.7 × 10	4.3 × 10^3	7.7 × 10^5	2.0 × 10^7	1.0 × 10^8	2.5 × 10^8
	OA		5.7 × 10^2	2.1 × 10^4	1.5 × 10^6	1.1 × 10^7	1.3 × 10^7
Psychrotrophes	NA	1.0 × 10	8.4 × 10^2	1.1 × 10^5	7.7 × 10^6	1.3 × 10^7	1.3 × 10^8
	OA		5.4 × 10^2	5.7 × 10^4	1.3 × 10^5	5.5 × 10^6	1.2 × 10^7
Coliforms	NA	0.5	0.7	1.4	2.5	3.1	3.1
	OA		0.1	0	0	0	0
Escherichia coli	NA	0.5	0.1	1.4	1.4	1.4	1.4
	OA		0	0	0	0	0
Salmonella	NA		+	+	+	+	+
(if present in 25 g sample)	OA	+	–	–	–	–	–

* Each result is the mean of two replicates
** NA = Normal atmosphere; OA = ozone atmosphere

incidence of *Salmonella* on broiler carcasses. Ozone and hydrogen peroxide were similarly effective at reducing *Salmonella* incidences, while peracetic acid was found to be more effective and chlorination was reported to be less effective than either ozone or hydrogen peroxide.

Nieto et al. (1984) studied the effect of ozone during refrigerated storage of poultry on several microorganisms. Chicken carcasses were stored under normal and ozone-rich atmospheres (40 mg/h) for 13 days at a temperature of 2 ± 1 °C and a relative humidity (RH) of 93 ± 2%. The authors reported that ozone not only inhibited the growth of surface flora but destroyed the pathogens responsible for toxic infection and food poisoning (Table 8.2), without any adverse effect on the sensory quality of the product.

8.2.3 Other meat applications

Steffen and Rice (2010) demonstrated the use of ozone (in gas and aqueous phases) in a new Swiss Ventafresh technology for the preparation of complete meals in a central food processing plant; the meals were then packaged and sealed in a sterilised manner (Figure 8.1). Ozone is reported to be effective against contamination of natural casings, which are generally contaminated with several microorganisms. Benli et al. (2008) employed a hog casing wash system (Figure 8.2). They observed a decrease in the bursting strength of natural hog casings after 4 hours of ozone treatment at ~7 mg/L with improved colour by a decrease in the a* and b* values. Benli et al. (2008) suggested that a combination of treatments, such as washing with ozonated water to whiten casings with another treatment, such as

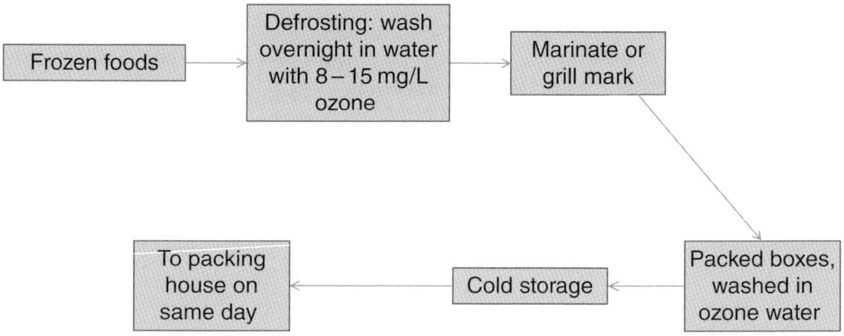

Figure 8.1 Ventafresh technology for the preparation of complete meals. (Reproduced from *User Experiences with Ozone, Electrolytic Water (Active Water) and UV-C Light (Ventafresh, Technology) in Production Processes and for Hygiene Maintenance in a Swiss Sushi Factory*, Hanspeter Steffen, Marc Duerst, Rip G. Rice, 2010, reprinted by permission of the publisher (Taylor & Francis Group, http://www.informaworld.com).)

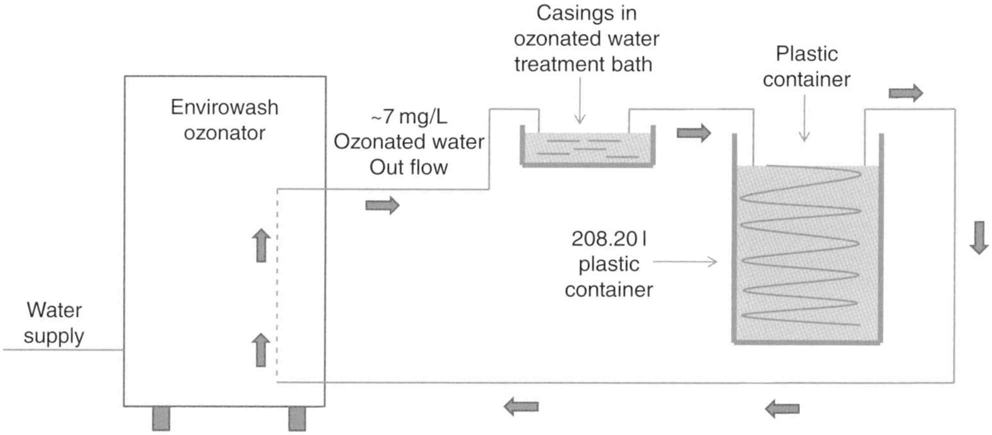

Figure 8.2 Ozonator set-up and flow diagram for ozonation of natural casings in recirculation mode and chilling coil in a tank filled with iced water. (Reprinted from *Meat Science*, Volume 79, Issue Number 1, H. Benli, B.S. Hafley, J.T. Keeton, L.M. Lucia, E. Cabrera-Diaz, G.R. Acuff, Biomechanical and microbiological changes in natural hog casings treated with ozone, 155–162, 2008, with permission from Elsevier.)

irradiation, might prove effective for enhancing the casing value while ensuring the destruction of potential foodborne pathogens.

8.3 Effect on meat quality

Ozone has been shown to reduce microorganism levels on red meat and poultry carcasses and cuts. Although it is likely there are a number of mechanisms involved in the mode of action of ozone against microflora, one is the oxidation potential of ozone. Ozone has a high oxidation potential (−2.07 V) and consequently poses some unique challenges to retaining meat quality. Specifically, oxidation in meat systems can influence meat colour and lipid stability characteristics. With regard to meat colour, oxymyoglobin is the pigment principally responsible for the redness of meat. When subjected to oxidation potentials, this pigment can revert to the oxidised pigment form, metmyoglobin, which is brown in colour. Because consumers often associate freshness with red colouration, the development of metmyoglobin or brown colour due to pigment oxidation may affect consumer acceptance of the ozone-treated product(s).

In addition to causing negative effects on meat pigment chemistry, oxidation can also have an impact on lipid stability. Meat contains a lipid or fatty acid profile. Generally, both saturated and unsaturated lipids are present. Because unsaturated fatty acids contain double bonds, they are susceptible to oxidation. Therefore, an oxidising environment can induce lipid oxidation and development of oxidation byproducts. The oxidation of lipids and generation of byproducts can cause aroma and flavour profile changes in meat. Often, these sensory changes are perceived by consumers as negative, as in the case of rancidity development in meat due to the oxidation of lipids and the generation of undesirable aroma and flavour byproducts. Therefore, like meat pigments, oxidation of lipids can lead to adverse quality attributes in meat. And because ozone is a strong oxidiser, this can pose concern for meat quality in products treated with ozone if care is not taken.

While there have been a number of studies that have evaluated the impact of ozone on meat microflora, fewer studies have investigated the impact of ozone treatments on meat colour, processing or quality characteristics. Greer and Jones (1989) evaluated the impact of gaseous ozone treatment on beef carcass bacterial spoilage profiles and on meat quality and carcass shrinkage. Using gaseous ozone to treat beef carcasses, they found ozone-treated carcasses had more cooler shrinkage and required more trimming of discoloured and dry muscle tissue compared with conventionally treated beef carcass sides. Using instrumental colour measurements, they found that the loin eyes from ozone-treated sides were darker (L^*), less red (a^*) and less yellow (b^*) than loin eyes from control sides. However, they detected no difference in fat colour due to ozone treatment. When steaks were evaluated in a retail case, sensory panellists did not detect any differences in odour or appearance between steaks from ozone-treated carcasses and those from carcasses that were conventionally treated.

Stivarius et al. (2002a) evaluated the impact of ozone and chlorine dioxide on microbial, colour and sensory attributes of ground beef made from treated beef trimmings. Ozone treatment of beef trimmings caused ground beef made from the trimmings to be lighter (L^*) in colour than control ground beef. However, ground beef made from trimmings that had been ozonated for 15 minutes was similar in redness (a^*) and colour saturation.

Sensory panellists did find a slightly browner colour with the ozone treatment, but indicated no difference in percentage discolouration, beef odour or off-odour characteristics between ground beef made from ozone-treated trimmings and that from the control. Sensory panellists also found no differences in overall colour between ground beef produced from trimmings that had been treated for 7 minutes in ozonated water and control ground beef through 7 days of simulated retail display.

Pohlman et al. (2002) evaluated ozone as a multiple intervention with other antimicrobials to improve the safety of ground beef manufactured from treated beef trimmings. As with Stivarius et al. (2002b), Pohlman et al. (2002) reported that ground beef made from beef trimmings that were treated with ozone followed by 0.5% cetylpyridinium chloride or 5% acetic acid was lighter in colour (L*) than the control ground beef. Instrumental redness (a*) was lower for treatments that used ozone as a decontaminant initially during display, but by day 1 of display the ozone/cetylpyridinium treatment was similar in redness (a*) to the control, whereas the ozone/acetic acid treatment was always less red than the control throughout the display period. This was not surprising, since oxidation is not the only thing to have an adverse impact on meat colour; so too does reduced pH due to organic acids. Sensory panellists found a difference in beef odour and off-odour characteristics between the control and the ozone/acetic acid-treated ground beef. However, sensory panellists found no difference in beef odour or off-odour characteristics between the control and ozone/cetylpyridinium-treated ground beef. Therefore, if care is taken, it is possible to minimise or reduce the impact of ozone on meat colour and lipid stability.

Fewer studies have been conducted with regard to the impact of ozone on poultry quality characteristics. Sheldon and Brown (1986b) evaluated the impact of ozone as a means for disinfecting poultry carcasses and chill water. Poultry carcass colour, lipid oxidation and sensory characteristics were monitored for the ozone decontamination treatments. The authors

Table 8.3 Mean and standard deviation of 2-thiobarbituric acid numbers of the breast, thigh, drumstick and back skin of water- and ozonated water-chilled broiler carcasses. mg TBA/kg tissue, two replicates. Means with different letter superscripts within rows are significantly different ($p < 0.05$). (Adapted from Sheldon and Brown 1986.)

	Treatment	
Broiler part	**Water**	**Ozone**
Breast	0.314 (0.158) b	0.162 (0.047)
Thigh	0.225 (0.013) b	0.146 (0.035) c
Drumstick	0.225 (0.010) c	0.274 (0.006) b
Skin (back)	0.650 (0.001) b	0.170 (~0.059) c

reported that no significant carcass colour changes occurred as the result of ozone treatment. Because they were concerned about the oxidation potential of the unsaturated fatty acids in poultry, they measured thiobarbituric acid reactive substances. They observed that 2-thiobarbituric acid numbers in broiler tissues exposed to either ozonated chill water or chill water alone showed significant treatment differences (Table 8.3), with an exception for drumsticks. Further, they observed no significant sensorial differences between the two treatments and panellists did not detect any flavour or off-flavour differences between the two treatments.

Ozone treatment did not cause any significant lipid oxidation or sensory evaluated off-flavour production. Therefore, as with red meat, with proper care it may be possible to use ozone as a poultry antimicrobial without adversely affecting appearance and lipid stability.

References

Algino, R.J., Ingham, S.C. and Zhu, J. (2007) Survey of antimicrobial effects of beef carcass intervention treatments in very small state-inspected slaughter plants, *J Food Sci*, **72**(5): M173–M179.

Algino, R.J., Badtram, G.A., Ingram, B.H. and Ingham, S.C. (2009) Factors associated with *Salmonella* prevalence on pork carcasses in very small abattoirs in Wisconsin, *J of Food Protection*, **72**(4): 714–21.

Benli, H., Hafley, B.S., Keeton, J.T., Lucia, L.M., Cabrera-Diaz, E. and Acuff, G.R. (2008) Biomechanical and microbiological changes in natural hog casings treated with ozone, *Meat Science*, **79**, 155–62.

Berry, E.D. and Cutter, C.N. (2000) Effects of acid adaptation of *Escherichia coli* 0157:H7 on efficacy of acetic acid spray washes to decontaminate beef carcass tissue, *Appl Environ Microbiology*, **66**(4): 1493–8.

Bolton, D.J., Pearce, R.A., Sheridan, J.J., Blair, I.S., McDowell, D.A. and Harrington, D. (2002) Washing and chilling as critical control points in pork slaughter hazard analysis and critical control point (HACCP) systems, *J Applied Microbiology*, **92**(5): 893–902.

Bosilevac, J.M., Nou, X., Barkocy-Gallagher, G.A., Arthur, T.M. and Koohmaraie, M. (2006) Treatments using hot water instead of lactic acid reduce levels of aerobic bacteria and *Enterobacteriaceae* and reduce the prevalence of *Escherichia coli* 0157:H7 on preevisceration beef carcasses, *J Food Protection*, **69**(8): 1808–1813.

Byelashov, O.A., Belk, K.E., Smith, G.C., Sofos, J.N., Adler, J.M., Geornaras, I. and Ko, K.Y. (2010) Evaluation of brining ingredients and antimicrobials for effects on thermal destruction of *Escherichia coli* 0157:H7 in a meat model system, *J Food Science*, **75**(4): 209–17.

Castillo, A., Dickson, J.S., Clayton, R.P., Lucia, L.M. and Acuff, G.R. (1998) Chemical dehairing of bovine skin to reduce pathogenic bacteria and bacteria of fecal origin, *J Food Protection*, **61**(5): 623–5.

Castillo, A., McKenzie, K.S., Lucia, L.M. and Acuff, G.R. (2003) Ozone treatment for reduction of *Escherichia coli* 0157:H7 and *Salmonella* serotype typhimurium on beef carcass surfaces, *J Food Protection*, **66**(5): 775–9.

Cutter, C.N. (1999) Combination spray washes of saponin with water or acetic acid to reduce aerobic and pathogenic bacteria on lean beef surfaces, *J Food Protection*, **62**(3): 280–3.

Fabrizio, K.A., Sharma, R.R., Demirci, A. and Cutter, C.N. (2002) Comparison of electrolyzed oxidizing water with various antimicrobial interventions to reduce *Salmonella* species on poultry, *J Poultry Science*, **81**: 1598–605.

Gill, C.O. (2009) Effects on the microbiological condition of product of decontaminating treatments routinely applied to carcasses at beef packing plants, *J Food Protection*, **72**(8): 1790–801.

Gill, C.O. and Badnoni, M. (2004) Effects of peroxyacetic acid, acidified sodium chlorite or lactic acid solutions on the microflora of chilled beef carcasses, *Int J Food Microbiology*, **91**(1): 43–50.

Gill, C., Bryant, J. and McGinnis, C. (2000) Microbial effects of the carcass washing operations at three beef packing plants, *Fleischwirtschaft*, **80**(11): 121–3.

Gorman, B.M., Sofos, J.N., Morgan, J.B., Schmidt, G.R. and Smith, G.C. (1995) Evaluation of hand-trimming, various sanitizing agents, and hot water spray-washing as decontamination interventions for beef brisket adipose tissue, *J Food Protection*, **58**: 899–907.

Greer, G.G. and Jones, S.D.M. (1989) Effects of ozone on beef carcass shrinkage, muscle quality and bacterial spoilage, *Can Inst Food Sci Technol J*, **22**: 156–60.

Jhala, R., Muthukumarappan, K., Julson, J.L., Dave, R.I. and Mahapatra, A.K. (2002) Synergistic effect of ozone and microgard 300 for controlling *Listeria monocytogenes* in ready-to-eat cooked and cured ham, *ASAE Annual International Meeting/CIGR XVth World Congress*, paper number: 026143, pp. 1–7.

Minihan, D., Whyte, P., O'Mahony, M. and Collins, J.D. (2003) The effect of commercial steam pasteurization on the levels of *Enterobacteriaceae* and *Escherichia coli* on naturally contaminated beef carcasses, *J Vet Med B Infect Dis Vet Public Health*, **50**(7): 352–6.

Morshedy, A.E.M.A. and Sallam, J.I. (2009) Improving the microbial quality and shelf life of chicken carcasses by trisodium phosphate and lactic acid dipping, *Int J Poultry Science*, **8**(7): 645–50.

Nieto, J.C., Jiménez-Colmenero, F. and Peláez, Ma.C. (1984) Effect of ozone on bacterial flora in poultry during refrigerated storage, *International Journal of Refrigeration*, **7**(6): 389–92.

Novak, J.S. and Yuan, J.T.C. (2003) Viability of *Clostridium perfringens*, *Escherichia coli*, and *Listeria monocytogenes* surviving mild heat or aqueous ozone treatment on beef followed by heat, alkali, or salt stress, *J Food Protection*, **66**(3): 382–9.

Novak, J.S. and Yuan, J.T.C. (2004a) Increased inactivation of ozone-treated *Clostridium perfringens* vegetative cells and spores on fabricated beef surfaces using mild heat, *J Food Protection*, **67**(2): 342–6.

Novak, J.S. and Yuan, J.T.C. (2004b) The fate of *Clostridium perfringens* spores exposed to ozone and/or mild heat pretreatment on beef surfaces followed by modified atmosphere packaging, *Food Microbiology*, **21**: 667–73.

Nutsch, A.L., Phebus, R.K., Riemann, M.J., Kotrola, J.S., Wilson, R.C., Boyer J.E. Jr and Brown, T.L. (1998) Steam pasteurization of commercially slaughtered beef carcasses; evaluation of bacterial populations at five anatomical locations, *J Food Protection*, **61**(5): 571–7.

Pohlman, F.W., Stivarius, M.R., McElyea, K.S., Johnson, Z.B. and Johnson, M.G. (2002) The effects of ozone, chlorine dioxide, cetylpyridinium chloride and trisodium phosphate as multiple antimicrobial interventions on microbiological, instrumental color, and sensory color and odor characteristics of ground beef, *Meat Science*, **61**: 307–13.

Pohlman, F.W., Story, R.P., Baublits, R.T., Rajaratnam, G., Bokina, C., Quilo, S.A., Dias-Morse, P.N., Crandall, P.G. and Brown A.H. Jr (2009) Microbial, instrumental color and sensory characteristics of ground beef processed from beef trimmings treated with potassium lactate, sodium metasilicate, peroxyacetic acid or acidified sodium chlorite as single antimicrobial interventions, *J Muscle Foods*, **20**(1): 54–69.

Reagan, J.O., Morgan, J.B., Marsden, J.L., Smith, G.C., Nickelson R. II, Kastner, C.L., Buege, D.R., Acuff, G.R., Dickson, J.S. and Buyck, M.J. (1996) Trimming and washing of beef carcasses as a method of improving the microbiological quality of meat, *J Food Protection*, **59**(7): 751–6.

Retzlaff, D., Phebus, R., Kastner, C. and Marsden, J. (2005) Establishment of minimum operational parameters for a high-volume static chamber steam pasteurization system (SPS 400-SC) for beef carcasses to support HACCP programs, *Foodborne Pathog Dis*, **2**(2): 146–51.

Riedel, C.T., Haxgart, S.N., Christensen, B.B., Brindsted, L. and Rosenquist, H. (2009) Chemical decontamination of *Campylobacter jejuni* on chicken skin and meat., *J Food Protection*, **72**(6): 1173–80.

Rio, E. del, Alonso-Calleja, C., Capita, R., Muriente, R. and Prieto, M. (2007) Effectiveness of trisodium phosphate, acidified sodium chlorite, citric acid, and peroxyacids against pathogenic bacteria on poultry during refrigerated storage, *J Food Protection*, **70**(9): 2063–71.

Sheldon, B.W. and Brown, A.L. (1986a) Bactericidal activity, egg-shell conductance, and hatchability effects of ozone versus formaldehyde disinfection, *Poult Sci*, **68**: 1074–7.

Sheldon, B.W. and Brown, A.L. (1986b) Efficacy of ozone as a disinfectant for poultry carcasses and chill water, *J Food Sci*, **51**: 305–9.

Singh, M., Herald, T.J., Nutsch, A.L., Marsden, J.L., Thippareddi, H. and Phebus, R.K. (2005) Efficacy of cetylpyridinium chloride against Listeria monocytogenes and its influence on color and texture of cooked roast beef, *J Food Protection*, **68**(11): 2349–55.

Steffen, H. and Rice, R.G. (2010) New restaurant concept relies on ozone, UV radiation, ultrasound and modified air packaging, *Ozone: Science & Engineering*, **32**(2): 137–43.

Stivarius, M.R., Pohlman, F.W., McElyea, K.S. and Apple, J.K. (2002a) Microbial, instrumental color and sensory color and odor characteristics of ground beef produced from beef trimmings treated with ozone or chlorine dioxide, *Meat Science*, **60**: 299–305.

Stivarius, M.R., Pohlman, F.W., McElyea, K.S. and Apple, J.K. (2002b) The effects of acetic acid, gluconic acid and trisodium citrate treatment of beef trimmings on microbial, color and odor characteristics of ground beef through simulated retail display, *Meat Science*, **60**: 245–52.

Tompkins, N.M., Salman, M.D., Kendall, P.A. and Avens, J.S. (2008) Effect of boiling water carcass immersion on aerobic bacteria counts of poultry skin and processed ground poultry meat, *Zoonoses and Public Health*, **55**(5), 235–41 [electronic resource].

Vadhanasin, S., Bangtrakulnonth, A. and Chidkrau, T. (2004) Critical control points for monitoring *salmonellae* reduction in Thai commercial frozen broiler processing, *J Food Protection*, **67**(7): 1480–3.

Weber, G.H., Bender, F.G. and O'Brien, J.K. (2004) Control of *Escherichia coli* 0157:H7 with sodium metasilicate, *J Food Protection*, **67**(7): 1501–6.

Xiangwu, N., Rivera-Betancourt, M., Bosilevac, J.M., Wheeler, T.L., Shackelford, S.D., Gwartney, B.L., Reagan, J.O. and Koohmaraie, M. (2003) Effect of chemical dehairing on the prevalence of *Escherichia coli* 0157:H7 and the levels of aerobic bacteria and *Enterobacteriaceae* on carcasses in a commercial beef processing plant, *J Food Protection*, **66**(11): 2005–9.

Yang, P.P.W. and Chen, T.C. (1979a) Stability of ozone and its germicidal properties on poultry meat microorganisms in liquid phase, *J Food Sci*, **44**: 501–4.

Yang, P.P.W. and Chen, T.C. (1979b) Effects of ozone treatment on microflora of poultry meat, *J Food Processing and Preservation*, **3**(2): 177–85.

9 Ozone in Seafood Processing

Shigezou Naito

9.1 Introduction

In Japan, fish and seafood products are the principal sources of protein. Fish preservation by applying the principles of chemistry, engineering and other branches of science extends the shelf life of seafood products. Although dried and salted fish have been traded internationally, consumption of fish has tended to be localised close to the areas of their capture. Developments in seafood technology, particularly freezing and ozone treatment, have provided means to overcome the problem of perishability of fish and seafood products, so that these products are now common items of international trade and are consumed in inland areas remote from seacoasts. Utility and the maintenance of quality are the principal microbiological problems associated with fish and seafood products. Nevertheless, fish and seafood products have in common with other products traded internationally the potential to act as transporters of putrefactive microorganisms. In particular, shrimp and tuna, which are harvested in almost every region of the world, are subjected to primary handling and processing operations that range from highly sophisticated to quite primitive, thus providing qualities ranging from impeccably hygienically clean to dirty and potentially dangerous. Health risks are related to broader environmental conditions. Generally, in regions where water and air temperatures are low, public health risks to the ultimate consumer are less than in higher temperature regions such as the tropics.

Molluscs, as sessile filter feeders, concentrate bacteria and viruses from water and can be dangerous carriers of enteric pathogens. They can be doubly dangerous because many are eaten raw or only lightly cooked. The frequent disposal of human wastes directly into estuarine and inshore marine waters, rivers and lakes, and the continuing increase in human populations in cities, raises the level of concern about these hazards.

Ozone is both a cleaning and a sanitising agent. Ozone-based systems have been marketed to the food industry for some years but have not gained wide acceptance until recently. The reasons for this lag period

Ozone in Food Processing, First Edition. Edited by Colm O'Donnell, B.K. Tiwari, P.J. Cullen, and Rip G. Rice.

include inadequate science, ineffective validations, lack of service after purchase and general scepticism about new technologies.

Recently, seafood processors have begun using interventional ozone cleansing of indirect surfaces at breaks and shift changes during processing. Ozone's use during production yields a cleaner plant and decreases the labour time needed for full plant sanitation. Ancillary benefits include reduced energy costs resulting from a large reduction in hot water consumption and chemical cost reductions resulting from decreased chemical usage.

9.2 Application of ozone in fish and storage of processed seafood products

9.2.1 *Fresh fish and seafood*

The subsurface flesh of live, healthy fish is not considered bacteriologically sterile. The largest concentrations of microorganisms are found in the intestine, gills and slime. The numbers and types of microorganisms found on fresh-caught fish are influenced by the geographic location of catch, season and method of harvest (Powell et al. 1979). Bacteria which attack seafood proteins convert them into amino acids, which subsequently are broken down into foul-smelling end products. Degradation of cysteine, for example, yields hydrogen sulfide, imparting a rotten-egg smell to seafood. Decomposition of tryptophan yields indole and ketones, which give seafood a faecal odour. In fresh fish and bivalve molluscs, ozone application suppresses the smell characteristics, which can sometimes be disagreeable, improving the sensory attributes of these products. It is worth noting that ozone, in this case, is not used to mask low quality, thus avoiding economic fraud. Chilled tilapias were stored at 0 and 5 °C after short ozone (6 ppm) pretreatment of live fish. Sensory analysis showed that ozone pretreatment prolonged their quality characteristics during 1 month's storage at 0 °C. The combination of ozone pretreatment with storage at 0 °C appears to be a feasible means of prolonging the storage life of fish, and extending their marketability and exportation potential (Nash 2002; Gelman et al. 2005).

The flora of fish is the product of the environments from which the fish are harvested (Shewan 1971). Differences to be expected include variation in salinity, temperature, humidity and organic matter found in the catch harvest area. The average number of bacteria found on imported frozen fresh-caught Vietnam shrimp was determined to range from 103 to 104 per gram, while the water above the defrosted shrimp after water washing had counts from 104 to 105 per gram. The flora of shrimp harvested from Vietnam was predominantly *Pseudomonas*, *Moraxella*, *Bacillus* and *Micrococcus*. Most common spoilage microorganisms were isolated from the manufacturing process. The use of ozone treatment alternating with ultraviolet (UV) irradiation during the manufacturing process resulted in a synergistic disinfection effect owing to different mechanisms (Naito 1992a). The use of ozone-containing water for dipping and washing fish or fish fillets resulted in an effective reduction of microbiological flora and simultaneously had no effect on the product (Ravesi et al. 1988; Gelman et al. 2005). The catfish study showed highly statistically significant reductions in plate counts when live fish and fillets were washed in ozone-containing water. It has been claimed that ozone gassing can be used as a powerful surface disinfectant (Campos et al. 2005).

Soaking peeled shrimp meat in ozone-containing water was found to be more effective than spraying shrimp with ozone-containing water. The higher ozone concentrations and longer treatment times studied were more effective for reducing levels of spoilage bacteria on the shrimp. The application of ozone-containing water did not increase lipid oxidation in the shrimp immediately after treatment (Chawla et al. 2007). These researchers found that soaking shrimps in 3 ppm dissolved ozone for 40 and 60 seconds caused the greatest reduction of total aerobic counts on the shrimp meat.

In a shrimp processing factory in Japan, frozen shrimp were washed with water on a massive scale in order to remove bubble foam. It is recognised that the predominant shrimp spoilage bacteria belong to the genus *Pseudomonas* and *Bacillus*. Since *Pseudomonas* is normally found in the shrimp water of a washing line, its presence raises the possibility of shrimp spoilage. The recommended pasteurisation process of 0.5~1.0 ppm of ozone-containing water at 10~20 °C for the treatment line and floor is adequate to kill *Pseudomonas* that might have been present on the shrimp. Since the most common contamination places are related to the shrimp surface and the manufacturing floor, the ozone-containing water rinse method offers a rapid, reliable and nondestructive means of shrimp processing. However, *Bacillus* spores are not sensitive to ozone. The synergistic sporicidal activity of ozone with ascorbic acid and isoascorbic acid was effective on *Bacillus subtilis* spores (Naito 1992b,c).

The conditions under which the sporicidal effect was evaluated were:

- ozone gas concentration: 5~50 ppm (v/v);
- time: 1~6 hours;
- relative humidity (RH): 20%, 95%;
- temperature: 10~50 °C.

The sporicidal effect of ozone was enhanced by increasing the treatment temperature and ascorbic acid or isoascorbic acid concentration. These findings suggest that oxygen radicals are involved in the sporicidal activity.

Paranjpye et al. (2008) investigated the effect of washing and soaking of naturally contaminated *L. monocytogenes* on individually quick-frozen (IQF) shrimp in 5 ppm ozone-containing water and ozone gas. They found that regardless of whether shrimp were washed or soaked with ozone-containing water for 20 or 60 minutes, or exposed to ozone gas for similar durations, the treatment was ineffective for inactivating *L. monocytogenes*.

After ozonation treatments or when ozone autodecomposes, the stable end product from ozone is oxygen, which reacts with a large variety of organic compounds, although at varying rates. Soaking or washing peeled shrimp meat in ozone-containing water was found to cause protein to flow from the shrimp into the bubble foam. Before soaking or washing peeled shrimp meat in ozone-containing water, shrimp should be washed with water to remove surface protein.

The browning of cured surimi (kamaboko) caused by bacteria is typical of seafood spoilage. This spoilage is caused by *Pseudomonas* bacteria (Mori et al. 1974). Kamaboko is a Japanese processed seafood product, in which various white fish are puréed, combined with additives such as MSG (monosodium glutamate), formed into distinctive loaves and then steamed until fully cooked and firm. It was confirmed that the *Pseudomonas* isolated caused the browning of kamaboko containing either glucose or sucrose within 1 or 2 days of incubation. It was found that glucose or sucrose was responsible for the browning of kamaboko (Mori et al. 1974). The bacterial spoilage of film-packaged kamaboko was caused by *Bacillus licheniformis* and this strain contaminated the kamaboko manufacturing process (Mori et al. 1973). The numbers and types of microorganisms are determined by the natural flora and the manner in which the seafood is handled between the manufacturing process and storage. The actual storage temperature profile included not only the final market box temperature but the ambient temperature during the manufacturing process, delays in refrigerating storage and fluctuations in storage temperature.

Ozone strongly oxidises the cell walls and cytoplasmic membranes of bacteria directly. The microbiocidal effect of ozone-containing water took place within the first 5 seconds of treatment (Yamayoshi and Tatsumi 1993). The most sensitive strain to ozone-containing water was *Pseudomonas aeruginosa*, while the most resistant strain was *Staphyrococcus aureus* (Yamayoshi and Tatsumi 1993). When *P. aeruginosa* was ozonated in water at a concentration of 1 ppm for 5 seconds, there was a threefold greater reduction in the number of surviving bacteria. When the most resistant strain of *S. aureus* was ozonated at a concentration of 5 ppm for 5 seconds, there was a threefold reduction in the number of surviving bacteria. The reaction rate constant at 20 °C, which was the reaction temperature selected for the experiment, was then calculated from the half-time (Yamayoshi and Tatsumi 1993).

Ozone-containing water should be useful in reducing the number of bacterial infections caused by inadequate disinfection of fresh and frozen fish and crustaceans. The preserving effect of ozone was reported for fresh jack mackerel (*Trachurus trachurus*) and shimaaji (*Garanx mertensi*) (Haraguchi et al. 1969). In the preliminary experiment, moulds, yeasts and aerobic bacteria which had been streaked on agar plates were killed after exposure to an ozone-containing atmosphere (0.6 ppm (v/v)) for 30–60 minutes. Viable bacterial counts on skin surfaces of gutted fish, soaked in 3% NaCl solution containing 0.6 ppm of ozone for 30–60 minutes, decreased to 1/100–1/1000 of those of the control fish. The storage life of the fish was lengthened by 1.2–1.6 times as a result of the ozone treatment once every 2 days.

Ozone has certain characteristics that make it attractive for use as a sanitiser in food processing, and it is probably safer than other sanitiser systems. It has been shown to be a more powerful disinfectant than the most commonly used disinfectant, chlorine, for deactivation of a very large number of organisms, including the most resistant.

In order to study the seafood preservative effects of ozone, microbiocidal effects of ozone in aqueous solution were investigated employing *Aspergillus* spores, *Penicillium* spores, yeasts, yeast spores, lactic acid bacteria and *Bacillus* spores (Naito and Shiga 1982a). The conditions under which the microbiocidal effects on various microbial suspensions of 10^{6-9}/mL were evaluated included:

- ozone gas concentration: 0.3–0.5 mg/L;
- pH: 6~7;
- time: 5 seconds to 4 hours;
- flow rate: 1455 mL/min;
- temperature: 5 and 20 °C.

The microbiocidal times of the microbial suspensions were found to be:

- *Aspergillus* spores: 90–180 minutes;
- *Penicillium* spores: 45–60 minutes;
- yeasts: 5–10 minutes;
- yeast spores: 8–10 minutes;
- lactic acid bacteria: 15–120 seconds;
- *Bacillus* spores: 180–240 minutes.

Lower pH value and lower temperature (between 5 and 10 °C) resulted in greater microbiocidal effects. The microbiocidal effects of ozone against all tested microorganisms were elevated by the addition of 1–5% sodium chloride and lowered by the addition of 1–10% sucrose. Each seafood product must be evaluated on the basis of its own characteristics and ozone treatment must be established for the treatment of both the desired final product(s) as well as the secondary, contaminating, materials that accompany the desired product(s). The hygienic quality of fish and marine products declines rapidly due to microbial cross-contamination from various sources, ultimately leading to spoilage (Gram and Huss 1996).

An accelerating effect of garlic powder on ozone treatment for microorganism control was seen only for *Bacillus* spores at concentrations of 0.2–1.0% (Naito and Shiga 1982b). When *B. coagulans* spores were exposed to ozone, the germicidal effect was significantly enhanced by the presence of DADS (diallyl disulfide) or DAS (diallyl sulfide) (Naito and Shiga 1982b). However, these effects were not demonstrated for *B. stearothermophilus* spores.

DADS- or DAS-initiated spore germination

It seems that the increased effect of ozone for disinfection in the presence of DADS or DAS depends on spore germination. The claimed odour components of retorted kamaboko were hydrogen sulfide and several other sulfides. Some of the microorganisms survive in retorted kamaboko as spore-forming bacteria, and subsequently cause spoilage of the kamaboko.

Germicidal activity of ozone decreased in the presence of 20 mM hydrogen sulfide, but increased in the presence of 2 or 0.2 mM hydrogen sulfide. This result was considered to have been caused by the difference in residual ozone concentration and final pH (Naito and Shiga 1982b). Ozone at 0.3~0.6 mM reacts well with 0.2~2.0 mM hydrogen sulfide to form sulfur compounds, and as this reaction was performed rapidly, residual ozone concentration increased and pH decreased. However, 0.3~0.6 mM ozone did not react completely with 20 mM hydrogen sulfide to form sulfur compounds, yet residual ozone was not detected (Naito and Shiga 1982b).

A microorganism in juten-tofu with yellow spots was isolated and identified as *Leuconostoc mesenteroides*. As this strain was detected in airborne microorganisms, it was estimated that contamination by it took place during the process of manufacturing juten-tofu. This strain is usually detected in spoiled fish and seafood products. Inactivation of *L. mesenteroides* by ozone-containing water was studied using an ozone-containing water manufacturing machine (Figure 9.1) (Naito et al. 2001a). Survival curves for the inactivation of *L. mesenteroides* isolated from juten-tofu with ozone-containing water were determined. Its survival ratio was significantly decreased over 120 seconds by increasing the dissolved ozone concentration from 0.7 to 3 mg/L (Naito et al. 2001a). Effects of pH on the survival curves for the inactivation of *L. mesenteroides* isolated from juten-tofu with ozone-containing water were determined. Treatment at pH 7 and 8 was usually more effective for ozone inactivation of this strain than at pH 3 and 4 (Figure 9.2) (Naito et al. 2001a).

In ozone-containing water, ozone may react directly with dissolved substances, or it may decompose to form secondary oxidants such as •OH radicals, which then immediately react with solutes. Therefore, their relative importance must be known when the oxidation effects of ozone and the rate of ozone consumption are to be predicted or generalised. Early observations of the lifetime of ozone-containing water indicated that decomposition of the ozone is accelerated in the presence of ethyl alcohol or by increasing the pH. It became evident that the decomposition of ozone at a given pH or in the presence of ethyl alcohol is often accelerated by a radical-type chain reaction, which in impure solutions can be initiated, promoted or inhibited by various solutes. The decomposition of aqueous ozone is generally due to

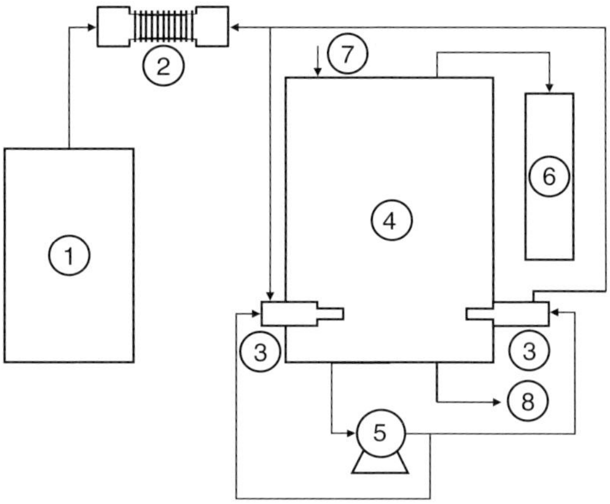

Figure 9.1 Diagram of ozone-containing water manufacturing machine (Naito and Shiga 1982a). (1) oxygen, (2) ozone generator, (3) ejector (gas nozzle), (4) dissolution vessel, (5) pump, (6) ozone decomposition catalyst, (7) water, (8) ozone-containing water.

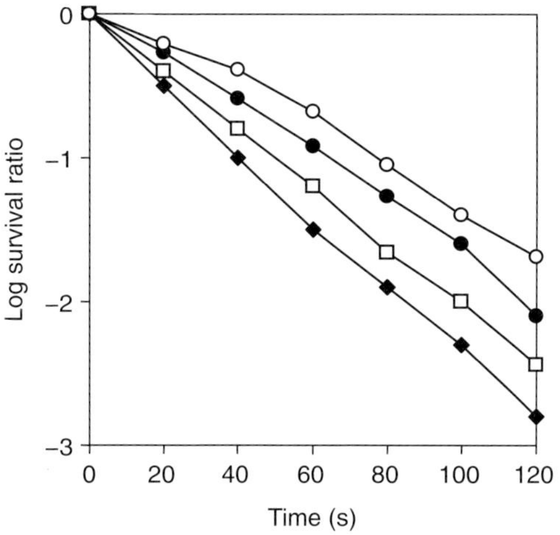

Figure 9.2 Effect of pH on the survival curves for the inactivation of *L. mesenteroides* isolated from juten-tofu with ozone-containing water (Naito and Shiga 1982a). The bacterial cells (4.8×105 cells/mL) were treated with ozone-containing water containing 1 mg/L dissolved ozone in water at an indicated period of up to 120 seconds at 20 °C and pH 3.0 (○), 5.0 (●), 7.0 (□) and 8(♦).

a chain reaction involving •OH radicals (Staehelin and Hoigné 1985). The effects of *tert*-butyl alcohol on the survival curves for the inactivation of *L. mesenteroides* isolated from juten-tofu with ozone-containing water were determined. The survival ratio increased with addition of 0.1~1 mM tert-butyl alcohol (Figure 9.3) (Naito et al. 2001a).

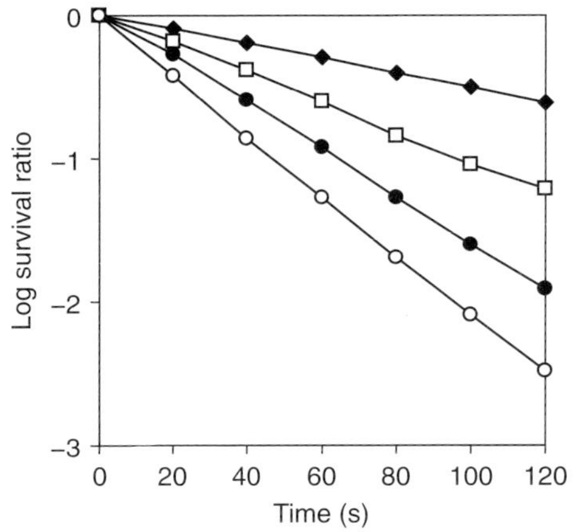

Figure 9.3 Effect of *tert*-butyl alcohol on the survival curves for the inactivation of *L. mesenteroides* isolated from juten-tofu with ozone-containing water (Naito and Shiga 1982a). The bacterial cells (4.8×105 cells/mL) were treated with ozone-containing water containing 1 mg/L dissolved ozone in water at an indicated period of up to 120 seconds at 20 °C and pH 7.0 with 0 (○), 0.1 (●), 0.5 (□) and 1 (♦) mM tert-butyl alcohol.

In the fishery industry, fish were treated with ozone-containing water to disinfect and to improve sensory qualities. Treatment of frozen or fresh shrimp, squid, octopus, mackerel, tuna, yellowtail and salmon with 1.5% NaCl solution containing 2.0 mg/L of ozone-containing water for 5–10 minutes decreased the viable bacterial count by 2–3 logs. The storage life of the above fish increased by 50–80% when the ozone-containing water treatment was applied.

In-plant sterilisation in the fishery industry was effective with ozone-containing water for the inactivation of spoilage bacteria such as *Vibrio, Salmonella, Escherichia coli* and *Staphylococcus aureus* (Chen et al. 1987). Chen et al. (1987) studied ozone for the in-plant sterilisation of frozen fishery products. They found that ozone was effective in distilled water and 3% NaCl solution for the inactivation of microorganisms such as *Vibrio cho*lera, *E. coli, Salmonella, Typhimurium, V. parahaemolyticus* and *S. aureus*). Ozone treatment of shrimp decreased *E. coli* counts by 98.5%.

Dondo et al. (1992) reported that ozone decreased surface contaminants of fish during several days of refrigerated storage. Ozone-containing water treatment improved the sensory quality of fish by decreasing the formation of trimethylamine. A beneficial decolourisation effect of horse mackerel (*T. japonicus*) mince resulted from washing with ozone-containing water for 10–20 minutes (Chen et al. 1997). However, a marked decrease in pH and undesirable mince gel strength, as well as oxidation of fish oil, occurred during this ozone treatment. Ozone promoted detachment of the surface slime of redfish aboard fishing vessels, and ozonation during transport reduced bacterial counts and extended the shelf life of the fish by ~1.5 days

(Koetters et al. 1997). Ozone was tested for improvement of the washing process that is employed during manufacture of dark-fleshed fish surimi (Chen et al. 1997). The authors found that ozone washing treatment minimised the washing time and improved colour; however, an undesirable gel strength and a decrease in the pH of the mince were observed.

To improve the effectiveness of treatments using ozone-containing water, fish should be treated with ozone-containing water when it is fresh, because bacteria invade the fish body from its skin over time. Ozone-containing water washing treatment also minimises the washing time and improves colour (Naito and Takahara 2006).

9.2.2 Dried and smoked products

Smoking of seafood products serves a number of different functions. Although traditionally used as a method of preservation, smoking is used to give seafood products a distinct flavour. In addition, smoking contributes to an increase in the shelf life of the product. Seafood smoking is conducted mostly at temperatures of around 50–60 °C and results in a cooked product. Smoked seafood products are dried to varying degrees, depending on the smoking procedures employed. This gives a wide range of free water content (Aw), resulting in a great variety of microbial distributions. The pathogens from seafood have been isolated consistently from production lines of fresh to cold-smoked fish (Ben Embarek 1994). Extension of the shelf life of perishable foods using ozone to reduce levels of microbial activity has also been reported by Rice et al. (1982). The remarkable effect of ozone on pure cultures of *Listeria* spp. stimulated challenge studies with cold-smoked fish (Vaz-Velho et al. 2006). The purpose of Vaz-Velho et al. (2006) study was to ascertain if the application of gaseous ozone to raw fish has any effect on *Listeria* numbers during cold-smoking processing and a further chilled storage period of 3 weeks at 5 °C in vacuum packs. The applicability of ozone treatments in the cold-smoked fish industry made it necessary to choose the best practicable time to apply the treatment to the fish. It was assumed for a smoking plan that the best moment would be after filleting and washing and just before salting.

Swelling spoilage of smoked squid products is occasionally caused by gas production of heterofermentative lactic acid bacteria. A microorganism in a smoked squid product affected by swelling spoilage was isolated and identified (Naito et al. 2001b). This microorganism was isolated on an MRS agar plate and BCP (plate count agar with bromcresol purple) plated count (agar plate) at 30 °C under anaerobic conditions. The microorganism, a type of heterofermentative lactic acid bacteria, was identified as *Lactobacillus fructivorans*. It has been presumed that the swelling spoilage with gas production by *L. fructivorans* is related to airborne microorganisms, and the contamination by this strain took place in the process of seasoning the squid for the second time. The inactivation of this strain by ozone-containing

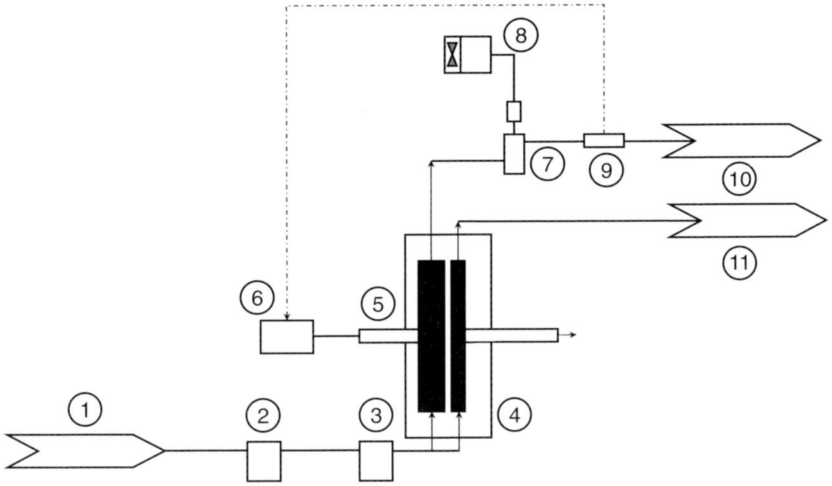

Figure 9.4 Diagram of the ozone-containing water manufacturing machine (Naito et al. 2001). (1) tap water, (2) autosoftener, (3) microfilter, (4) electrolytic cell, (5) anode, (6) DC power supply, (7) separation tank, (8) deodorising unit, (9) ozone concentration monitor, (10) ozone-containing water, (11) cathodic water.

water was studied (Naito et al. 2001b). A diagram of the ozone-containing water manufacturing machine is shown in Figure 9.4. The survival ratio of this strain was significantly decreased over 120 seconds, with increasing dissolved ozone concentration ranging from 0.5 to 2.0 mg/L (Figure 9.5) (Naito et al. 2001b). The treatment of 10^2–10^3/mL of these bacteria was usually more effective for ozone inactivation of this strain than was treatment of 10^4–10^5/mL at 0.5 mg/L ozone (Figure 9.6) (Naito et al. 2001b).

In general, smoked squid products are subject to a type of microbial spoilage similar to that for fish. For milder smoked and salted squid, a number of yeasts will survive; *Torulopsis lactis-condensi* is a common species of flora (Naito 1981). The main feature of salting smoked squid involves the removal of some of the water from the squid tissue, and its partial replacement by salt. Storage stability will be dependent primarily on the Aw, salt content, total heat input during smoking, and storage conditions. This product spoilage pattern will be similar to that of milder smoked and salted seafood. Salt-tolerant yeast such as *T. lactis-condens* will be the dominant organism for spoilage of smoked and salted squid (Aw 0.82, salt 6.0%, pH 5.60, yeast 2.5×10^6 g) (Naito 1981). For lightly salted and lightly smoked products, *Micrococcus* will be the dominant organism of squid spoilage (Naito 1983). This organism can be a part of the normal flora of smoked squid and marine sediments, growing at 0~24% salt concentration. *Micrococcus colpogenes* and *Paracoccus halodenitrificans* were detected from spoiled squid (Naito 1986). The predominant bacteria causing spoilage varies with the temperature at which the smoked squid products are held, but at the normal distribution temperature usually employed, species of

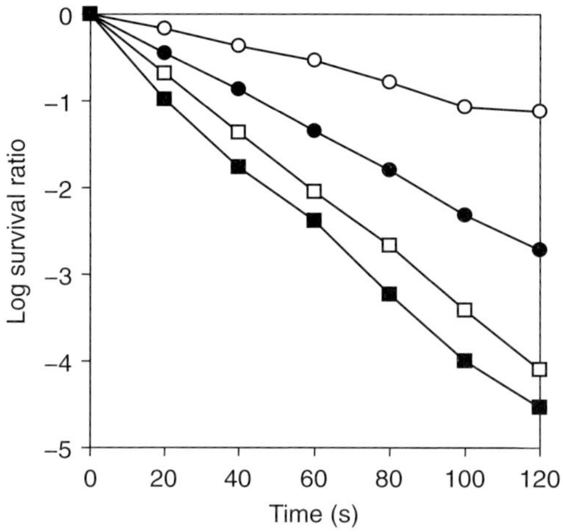

Figure 9.5 Survival curves for the inactivation of *L. fructivorans* isolated from packaged smoked squid by swelling spoilage. The bacterial cells (7.5×10^5 cells/mL) were treated with ozone-containing water containing 0.5 (○), 1.0 (●), 1.5 (□) and 2 (♦) mg/L dissolved ozone in water at an indicated period of up to 120 seconds at 20 °C.

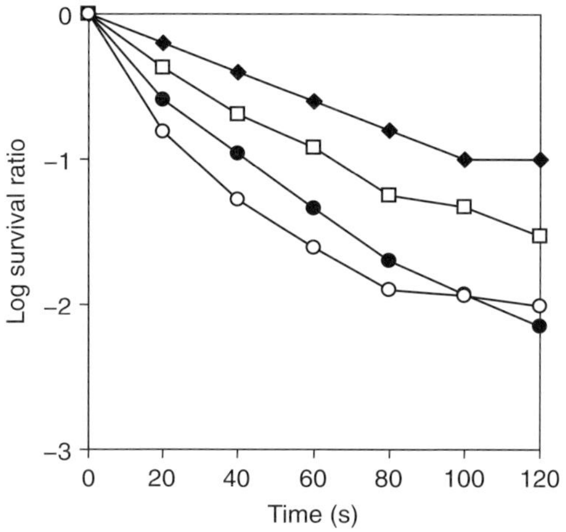

Figure 9.6 Effect of the microbial population on the survival curves for the inactivation by ozone-containing water of *L. fructivorans* isolated from packaged smoked squid affected by swelling spoilage (Naito et al. 2001). The bacterial cells (1.5×10^2 (○), 1.5×10^3 (●), 1.5 × 10^4 (□) and 1.5×10^5 (♦) cells/mL) were treated with water containing 0.5 mg/L dissolved ozone for an indicated period of up to 120 seconds at 20 °C and pH 5.0.

Micrococcus and *Candida* are most likely to be predominant, with *Bacillus* and *Pichia* species next in order of importance (Naito 1986).

Ozone gas treatment is the most commonly used method for preventing or delaying bacterial and yeast growth, and hence spoilage, until the squid is used or is otherwise processed (Naito 1986). The specifications of a sterilisation machine for dried and smoked products using ozone are shown in Table 9.1 (Naito 1986, 2001). *Micrococcus* and *Candida* levels are decreased by the ozone treatment for squid. Ozone gas treatment was effective for smoked squid with 0.2, 0.5 and 1.0 ppm (v/v) ozone concentration for 20 minutes at 10 days and 20 days.

Smoked squid after ozonation was stored for 1~10 days at 20 °C. The average initial number of bacteria after 1.0 ppm (v/v) ozone treatment was found to range from 10^2 to 10^3/g at 3 days storage at 20 °C, while the average initial number of yeasts after 0.2 ppm (v/v) of ozone treatment had counts from 0 to 10^2/g at 3 days storage at 20 °C (Table 9.2).

The flora of smoked squid were predominantly *Micrococcus, Bacillus, Candida* and *Picha. Mirococcus* and *Candida* were the most common spoilage

Table 9.1 Specifications of a sterilisation machine for seafood using ozone gas.

Components	Specification
Ozone gas treatment vessel	500 W × 500 D × 500 H mm
Retention contactor	2000 L × 800 W × 1800 H mm
Characteristics of machine capacity	Ventilation method: interior circulation, test vessel volume 3/4 ventilation per minute Ozone concentration range: 20~150 pphm (regulate range): 1~100 ppm Ozone control temperature: 5~35 °C
Disinfection method	Presterilisation of machine: dry ozone gas 10^2~10^4 ppm Sterilisation of sample: ozone lamp method: 20~150 pphm silent discharge method: 1~100 ppm
Ozone generator	Silent discharge, UV lamp (185 nm)
Monitor	UV absorption method ozone monitor

Table 9.2 Effect of ozone treatment on smoked squid storage (Naito and Sawairi 2000).

Ozone concentration(ppm)	Microorganisms	Storage time (days) 0	1	3	6	10
0	Bacteria	1.0×10^{4a}	1.2×10^5	3.0×10^5	5.2×10^5	7.1×10^5
	Yeast	2.0×10^{2a}	2.5×10^4	3.2×10^4	2.2×10^4	5.3×10^4
0.2	Bacteria	1.0×10^{4a}	8.7×10^3	5.0×10^3	5.2×10^3	5.1×10^3
	Yeast	2.0×10^{2a}	1.2×10^2	0	0	0
0.5	Bacteria	1.5×10^{3a}	1.2×10^2	3.0×10^2	2.2×10^3	5.1×10^3
	Yeast	1.0×10^{2a}	1.2×10^2	0	0	0
1.0	Bacteria	1.0×10^{3a}	3.2×10^2	3.0×10^2	2.2×10^2	3.1×10^2
	Yeast	1.0×10^{2a}	0	0	0	0

[a]Using ozone disinfection machine.
Ozone treatment: ozone concentration 0.2~1.0 ppm, time 20 minutes, temperature 20 °C; storage period: 1~10 days; storage temperature: 20 °C.

organisms isolated during the manufacturing process. The shelf life (the time of storage before microbial spoilage of smoked squid) was determined by the number of bacteria present and the storage temperature.

M. caseolyticus, *M. colpogenes*, *M. luteus*, *M. varians* and *M. halodenitrificans* on squid were remarkably decreased by ozone gas treatment with 0.2~0.5 ppm (v/v) ozone over 10~20 days. These microbial counts were decreased by longer storage periods. *Picha anomala*, *Candida edax*, *C. diddensiae*, *Rhodotorula rubra* and *C. lactis-condennsei* on squid were also remarkably decreased by the above ozone (gas) treatment (Naito 2001).

The treatment of dried cuttlefish with ozone gas results in improved sanitary conditions, and this has been a commercial process in Japan for decades. Salt fish are spoiled by salt-tolerant or halophilic bacteria of genera *Serratia*, *Micrococcus*, *Bacillus*, *Alcaligenes*, *Pseudomonas* and others, which often cause discolouration. *Micrococcus* sp. is the chief spoilage organism on dried cuttlefish. Naito and Sannomiya (1985) have investigated the ozone treatment of dried cuttlefish and showed dramatic reductions in the levels of *Micrococcus varians*, *M. caseoltticus* and *M. colpogenes*, which resulted in an increased storage period following ozone treatment at 0.2~0.5 ppm (v/v). They concluded that ozone gas added to dried cuttlefish in low concentrations reacts rapidly with *Micrococcus* and causes no discolourations of red, pink or yellow colours.

9.3 Application of ozone in seafood plant sanitation

Seafood processors are, by necessity, taking a second look at the new generation of ozone systems and their unequalled ability to maintain plant cleanliness and sanitation throughout the production day. The capacity of ozone as an effective degreaser and sanitiser on conveyor belts during production greatly reduces the risk of cross-contamination. Today, ozone sprays are used continuously and directly on food processing equipment surfaces, such as conveyors, knives and slicers, ensuring their cleanliness throughout production. Seafood processors have even begun interventional ozone cleaning of indirect surfaces at breaks and shift changes. Ozone's use during production yields a cleaner plant and decreases the labour time needed for full-plant sanitation. Ancillary benefits include reduced energy costs resulting from a large reduction in hot-water consumption and chemical cost reduction resulting from lessened chemical usage.

Recent seafood putrefaction studies with new strains of microorganisms, such as lactic acid bacteria and disinfectant-resistant strains of Gram-negative bacteria, moulds and yeasts, have increased interest in exploring different disinfection procedures for food sanitation. In Japan, food processing companies consider lactic acid bacteria and Gram-negative bacteria, moulds and yeasts to be of greatest concern because of the severity and extent of seafood putrefaction they cause.

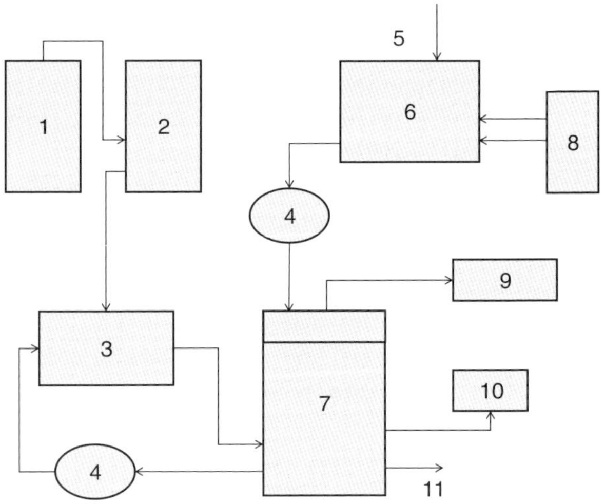

Figure 9.7 Diagram of an ozone-containing water manufacturing machine (Naito and Sawairi 2000). (1) oxygen cylinder, (2) ozone generator, (3) ozone dissolutor, (4) pump, (5) water, (6) water vessel, (7) dissolution vessel, (8) chiller, (9) ozone decomposition catalyst, (10) ozone monitor, (11) ozone-containing water.

There is also concern over the resistant strains and presence of chemical byproducts that are formed when chlorine is used as a disinfectant to control microorganisms. Because of these concerns, alternative disinfectants are being explored for food processing. Ozone should be useful in reducing the extent of microbial contamination caused by inadequate disinfection of new resistant strains. Use of ozone gas and ozone-containing water as alternatives to chlorine for treatment during food processing is increasing, mainly because they produce fewer chlorinated disinfection byproducts. There is a growing tendency to use ozone in the seafood industry in Japan as an effective means of disinfection without any additives. Many kinds of ozone generator and their ancillary application equipment have been developed, and establishment and improvement of their software is now the current impediment to their wider acceptance.

Based on the properties of ozone as a strong germicidal agent, the inactivation kinetics of lactic acid bacteria toward ozone-containing water were studied. A diagram of a sterilising machine is shown in Figure 9.7 and specifications are given in Table 9.3 (Naito and Sawairi 2000). Ozone was produced by passing high-purity oxygen through a silent-discharge ozone generator. The ozone was mixed with circulating water at 20 °C with MPG (micro porous glass). The ozone-containing water was put into a water bath at 20 °C and its ozone concentration was measured by means of a polarographic type of ozone monitor when it reached 20 °C.

The value of 90% inactivation (t90) obtained varied from 0.13 minutes (5 mg/L, *Enterococcus feacalis*) to 2.92 minutes (0.5 mg/L, *L. mesenteroides*). First-order inactivation kinetics with respect to both the concentration of

Table 9.3 Disinfection specifications for a seafood machine using ozone-containing water and ozone gas.

Components	Specification
Ozone-containing water tank	Two-stage, φ 300×600 H
Retention contactor	2380 L × 850 W × 1500 H
Machine capacity characteristics	Line speed: 0.5~2.0 m/min (variable), continuous transport of grip chain line width: 80~160 mm, line length: 2000 mm sample volume diameter: Macφ 300 mm Preformed product: 2~5 piece/min (variable), intermittent transport of grip chain?
Disinfection method	Presterilisation of machine: dry ozone gas 10^2~10^3 ppm Disinfection of product: dipped ozone-containing water (1.0~5.0 mg/L) Disinfection of product: ozone-containing water jet (1.0~5.0 mg/L)
Ozone generator	10 g/hr, silent discharge (oxygen)

dissolved ozone and microorganisms were found, resulting in overall second-order inactivation kinetics. Second-order inactivation rate constant (k2) values are a quantitative measurement of lactic acid bacteria resistance toward ozone. *L. mesenteroides* was the most resistant and *E. feacalis* the most sensitive to ozone. According to k2 values obtained, the ozone-containing water resistance of the strains studied may be ordered as *L. mesenteroides* > *L. fructivorans* > *L. plantarum* > *Weissella viridescence* > *E. faecium* > *E. feacalis*. However, the survival ratio of *L. mesenteroides* was found to decrease dramatically with increasing ozone-containing water concentrations, ranging from 0.5 to 5 mg/L, and temperature in the range 5–30 °C, and these two factors tended to act synergistically with each other. Furthermore, the survival ratio of *L. mesenteroides* as a function of ozone treatment in the presence of the organic radical scavenger *tert*-butyl alcohol was increased, suggesting that the death of lactic acid bacteria may be closely correlated with ozone-derived free radicals (such as the hydroxyl radical) by the rapid inactivation. The relative importance of oxidation by ozone or hydroxyl radical is also dependent on the concentration of bacterial cells. From these results, most of the lactic acid bacteria appear to have been killed by the combination of molecular ozone and ozone-derived free radicals such as hydroxyl radicals (Naito and Sawairi 2000).

Ozone treatment (gas phase) was applied to the insides of the Surimi Paste Product (chikuwa) manufacturing plant during the cooling process at 0.02~0.08 ppm (v/v) (5 °C, 1~1.5 years) and to chikuwa at 0.5, 5.0 and 50 ppm (5 °C, 30 minutes), respectively, to improve quality and extend shelf life (Naito and Yamazawa 1989). A diagram of the ozonation apparatus in the cooling room of the chikuwa manufactory is shown in Figure 9.8.

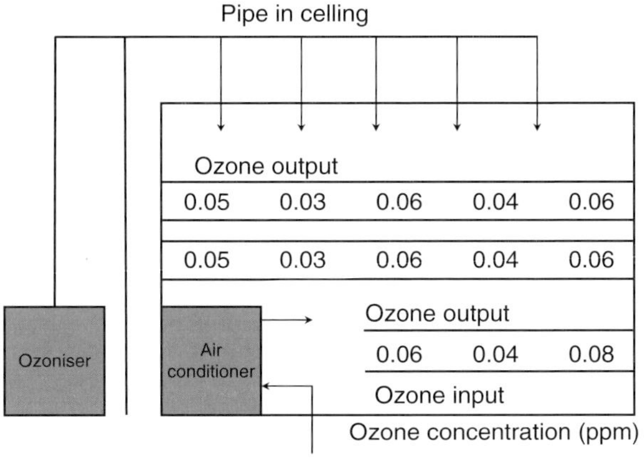

Figure 9.8 Diagram of ozonation apparatus in the cooling room of the Surimi Paste Product (chikuwa) manufactory (Naito and Yamazawa 1989).

Microbial examination of airborne microorganisms in the chikuwa manufactory showed they were remarkably reduced for every line by ozone treatment at 0.02~0.08 ppm over 1~1.5 years (Naito and Yamazawa 1989). The microbial flora of airborne microorganisms isolated from the chikuwa manufactory was changed by ozone treatment. The predominant strains of airborne microorganisms detected in the cooling process were *M. roseus*, *M. luteus*, *M. flavus*, and *M. colpogenes*. The microbial contents of these four *Micrococcus* spp. were 16, 69, 20 and 21 per 53 l of air, respectively, and after ozone treatment of the insides of the chikuwa manufactory the microbial contents were reduced to 3, 5, 1 and 1 per 53 l of air, respectively. The bacterial contents of chikuwa in the manufacturing processes of grinding, broiling, cooling and packaging were 1.6×10^6, 2.6×10^6, 7.6×10^2, 1.4×10^4 and 3.7×10^4 per gram, respectively, and after ozone treatment of the insides of the chikuwa manufactory, the microbial contents during the manufacturing process were reduced to 2.5×10^5, 1.7×10^5, 1.2×10^2, 1.9×10^2 and 2.5×10^2 per gram, respectively. Microbial growth in ozone-treated chikuwa during the cooling process was remarkably inhibited. Owing to this, the storage life at 5 and 10 °C of packaged chikuwa increased by 7 days for 5.0 and 50 ppm ozone treatments. In overall evaluation, organoleptic deterioration of odour and taste was detected in the 50 ppm ozone-treated sample. But there were only slight differences of odour in the 5 ppm ozone-treated samples and no significant differences were noted in the 0.5 ppm ozone-treated samples (Naito and Yamazawa 1989).

Ozone treatment (gas phase) was applied to the insides of the flakes of dried bonito (kezuri-bushi) manufactory during the manufacturing process at 0.5~1.0 ppm (v/v) for 1 year, except during working hours (about 10 hours per day), to try to establish a convenient method for decontamination and deodourisation. The levels of airborne microorganisms in the kezuri-bushi manufacturing process were reduced by ozone treatment. The predominant strains of airborne microorganisms detected in the overall process were *Micrococcus* and *Penicillium*. The microbial contents of the area around the steaming room were 263 per 531 air. *Micrococcus* growth was remarkably inhibited. After 30, 60, 90, 120, 150, 200 and 300 days of ozone treatment at 0.5~1.0 ppm (v/v) at 15~20°C and 55~80% RH, *Micrococcus* contents of the steaming room were remarkably decreased to 150, 102, 82, 63, 41, 31 and 14 per 531 of air, respectively (Naito 1994a,b). Levels of sulfur compounds were gradually decreased in all places by ozone treatment.

While ozone is a potent oxidant and can reduce bacterial levels in pure culture, its use in seafood processing operations where bacteria exist within organic material is more difficult. According to one study, the application of ozone also has many disadvantages in seafood processing (Crapo et al. 2004). Since ozone is the most powerful oxidising agent available, it is also potentially the most dangerous of oxidants. This danger was recognised in the early stages of ozone research, and techniques have been developed to ensure the absence of ozone accidents. The use of a gaseous ozone system under controlled conditions appears to be a viable option by which the seafood industry can improve catch quality and marketability.

9.4 Effects of ozone on microbial safety

Since its approval for use as a food additive by the Japanese Ministry of Health and Welfare on 10 August 1995 and 16 April 1996, ozone has been listed under 'Additives Already in Existence' due to Notification #160 (7) and #120 (8). In Japan, food additives have been regulated under Law #233, the Food Sanitation Hygiene and Nutrition Improvement Law (24 December 1947).

Because of the US Food and Drug Administration (FDA)'s approval of ozone as an antimicrobial agent for direct contact with all foods in 2001, ozone sanitation systems have been widely accepted as an effective means to improve food safety and quality. There is an enormous focus on food safety in connection with pathogens in fresh produce. Ozone-containing water manufacturing companies have installations in numerous agri-food facilities where sanitation is paramount. These facilities are involved in activities encompassing fish and seafood products. Some food processors started employing ozone after the FDA approval of ozone for use as an antimicrobial agent in June 2001. Many more processors are considering ozone as part of a multiple intervention approach to ensuring food safety based on recent highly publicised *E. coli* and other foodborne pathogen outbreaks in Japan during 2009 (pepper lunch trouble).

The recent rash of food safety warnings and the resultant recalls have brought financial disaster to affected companies. Ozone systems now operate in many food industries, including fish and seafood. Food processors, by necessity, are taking a second look at the new generation of ozone systems and their unequalled ability to maintain plant cleanliness and sanitation throughout a production day. The capacity of ozone as an effective degreaser and sanitiser on conveyor belts during production greatly reduces the risk of cross-contamination. Ozone sprays are applied continuously directly on to fish and seafood processing equipment surfaces such as conveyors, knives, slicers and portioners, ensuring their cleanliness throughout production.

The mutagenicities of products formed by the ozonation of amino acids and saccharomyces in aqueous solution were assayed with *Salmonella typhimurium* strains TA-98 and TA-100 in the presence or absence of S9-mix. An ozone–oxygen gas stream was fed into the aqueous solutions of amino acids, saccharides or mixtures of glucose and amino acids v/v at a rate of 26.6 L/min for 1~5 hours at 20 °C (average ozone concentration: 110~120 ppm (v/v). The most labile amino acids affected by ozonation were tryptophan, tyrosine, phenylalanine and methionine. Mutagenicity of these amino acids was not detected for freeze-dried amino acids after 1~5 hours of ozonation per 100~1000 µg with or without S9-mix. Also, mutagenicity by ozonation products of the other 14 amino acids was not observed clearly. The number of revertant colonies caused by the ozonation products ranged from 20 to 30 per 100~1000 µg with or without S9-mix. Mutagenicity of 10 freeze-dried saccharides after ozonation was investigated and was not detected after 1 hour of ozonation per 1000 µg with or without S9-mix. Mutagenicity of mixtures of freeze-dried amino acids and glucose after ozonation was investigated and was not detected after 1 hour of ozonation per 1000 µg with or without S9-mix (Naito 1992d).

The use of slurry ice, both alone and in combination with ozone, as compared with traditional flake ice, was investigated as a new refrigeration system for the storage of sardines (*Sardina pichardus*). Microbiological, chemical and sensory analyses were carried out throughout a storage period of 22 days. According to sensory analyses, sardine specimens stored in ozonised slurry ice had a shelf life of 19 days, while counterpart batches stored in slurry ice or flake ice had shelf lives of 15 and 8 days, respectively. Storage in slurry ice made from ozone-containing water led to significantly lower counts of aerobic mesophiles, psychrotrophic bacteria, anaerobes, coliforms and both lipolytic microorganisms in sardine muscle, and of surface counts of mesophiles and psychrotrophilic bacteria in sardine skin as compared with the slurry ice and the flake ice batches (Campos et al. 2005). Antimicrobial ice containing ozone was utilised to control foodborne pathogens in laboratory media and on fish skin. Aerobic methophiles, anaerobic coliforms, proteolytic bacteria and lypolytic bacteria strains were treated with ozonised slurry ice. The use of slurry ice resulted in significantly lower counts of all microbial populations in sardine muscle and skin as compared to flake ice. It should also be stressed that the use of slurry ice and ozone significantly decreased the average populations of mesophililes, anaerobes, coliforms, proteolytic and lipolytic bacteria in sardine muscle and skin (Kim et al. 1999). The initial load of foodborne pathogens was reduced by antimicrobial ice and the lowered microbial level was maintained during treatment. The application of antimicrobial ozone ice is a simple and effective method for the safe preservation of fish.

Seafood products have been widely consumed, especially in recent years, due to their low fat content, rapid preparation times and more economical cost compared to other foods. Seafood products are sold as whole or pieced, depending on the demands of consumers. There is an increase in demand for pieced seafood products, especially in large cities. Protection of quality and particularly safety during processing and storage of horse mackerel is important. Pathogenic and harmful bacteria that are present in horse mackerel's interior organs and skin surface can easily contaminate the meat during processing steps. Contamination is seen mostly at steps such as scalding, plucking and evisceration. In addition to this, cross-contamination in the carcasses and dirtiness of the process water and equipment increase the contamination level in the processing steps.

Emphasis has been placed on hazard analysis and critical control point (HACCP)-based programmes for the identification and prevention of possible microbiological risks that can originate from raw materials, processing stages, products and food plants. In order to prevent the microorganisms in the fish and seafood from proliferating, methods such as cooling, vapour-vacuum systems and vapour pasteurisation are being used. Along with these technologies, chemicals like chlorine and chlorine compounds, organic acids and trisodium phosphate are being widely used for decontamination purposes. Chlorine inhibits glucose oxidation in the bacteria and shows a bactericidal effect. It also decreases the activation of some enzymes that carry sulfide groups. However, excess usage of chlorine forms toxic and suspected carcinogenic compounds (called trihalomethanes) through reaction with the fish meat.

In 1982, ozone was declared to be Generally Recognised as Safe (GRAS) for bottled water disinfection by the FDA, and in 2001 ozone was approved for use by the FDA as an antimicrobial agent in direct contact with food products, including fish and seafood. Ozone, which is a strong oxidant, is effective against Gram-positive and Gram-negative bacteria, yeasts, moulds and viruses. Since ozone does not leave any residual material from its decomposition in food products, it does not change their taste and colour.

9.5 Effects of ozone on fish and seafood quality and shelf life

The use of ozone-containing slurry ice was investigated as a new refrigeration system for the storage of farmed turbot (*Psetta maxima*). With this purpose in mind, an ozone generator device was coupled to a slurry ice system working at subzero temperature (−1.5 °C). The ozone concentration was adjusted to provide a redox potential (reduction-oxidation potential) of 700 mV, and the slurry ice biphasic mixture was prepared at a 40% ice : 60% water ratio and 3.3% salinity. Certain biochemical parameters indicative of fish freshness, such as the rate of nucleotide degradation or TMA-N (trimethylamine-nitrogen) formation, were not significantly affected by the presence of ozone in the slurry ice mixture. However, storage in ozone-containing slurry ice significantly slowed down the mechanisms responsible for lipid hydrolysis and lipid oxidation in farmed turbot. Storage in ozone-containing slurry ice also led to significantly ($p < 0.05$) lower counts of both total aerobes and psychrotrophic bacteria in both turbot muscle and skin, as compared with the control batch stored without ozone. Sensory analyses confirmed an extended shelf life of turbot specimens stored in ozone-containing slurry ice. Samples stored in ozone-containing ice slurry maintained an 'A' sensory quality up to day 14 of storage, while the counterpart batch stored in slurry ice without ozone maintained this quality up to day 7 only of storage. The combination of ozone and slurry ice may be recommended for the chilling and storage of farmed turbot with a view to extending its shelf life (Campos et al. 2006).

The valid purpose of adding ozone to the water used to make ice is to ensure that the ice is chemically and microbiologically clean. However, when clean seawater is used to make ice with ozone, the disinfection effect is increased by hypobromous acid and hypobromite ion formed when treating seawater with ozone (or chlorine). Additionally, the formation of bromates and epoxides (ozone oxidises the fish directly) leads to the recommendation to use pure (fresh drinking) water, not seawater (which contains ~65 ppm of bromide ion), to make ice for the shelf life extension of fish stored in ozonated slurry ice.

The efficacy of ozone-containing water (0.6–1.5 ppm) was evaluated as a bactericidal agent for sanitising food contact surfaces and for treatment of raw seafood (Campos et al. 2006). The presence of ozone reduced the bacterial levels substantially on stainless steel surfaces and to a lesser extent on plastic cutting boards. Ozone was about as effective as chlorine in lowering levels of *Listeria innocua* on inoculated food contact surfaces.

Fish processing residuals present on the surface greatly reduced sanitiser effectiveness. Under high organic conditions, chlorinated water was slightly more effective than ozone-containing water. However, ozone-containing water applied to fish fillets and seafood products was effective for bacterial control. The presence of organic material, particularly with fillets, reduced the effectiveness of ozone. Ozone accelerated the development of rancidity in frozen fish, fillets and seafood, resulting in reduced shelf life. Ozone treatment of fish and seafood was recommended only as a sanitiser for cleaned fish and seafood contact surfaces (Crapo et al. 2004).

A beneficial decolourising effect resulted for horse mackerel mince washed with ozone-containing water within 10–20 minutes, but a longer washing time was required to improve the colour properties when cold water or alkaline solution was used (Chen et al. 1997). The CIELAB colour scale is an approximately uniform scale. Positive a* is red. Negative a* is green. The a* values of surimi prepared with washed mince also decreased with extended washing. Green hue (−a*) was observed and its value increased with elongation of washing time in all kamaboko samples. Ozone-containing water washing was most effective in diminishing greenness (Chen et al. 1997).

To demonstrate the potential of ozone in nori (seaweed), alga and *Porphyra* spp. processing, several research studies evaluated the efficacy of ozone as a disinfecting agent to reduce the microbial load brought into the plant during the nori manufacturing process. Wet nori preservation is the method of extending the shelf life of nori and other nori products by applying the principles of ozone chemistry, engineering and other branches of science in order to improve the quality of products. Bacteria were isolated from nori, alga and *Porphyra* spp., which had been grown healthily with chemically-defined media by the method of streaking on agar or homogeniser in an effort to identify accompanying bacteria. The bacteria isolated were identified at the genus level as Gram-positive – *Staphylococcus* and *Bacillus* – and Gram-negative – *Flavobacter, Pseudomonas, Enterobacter* and *Alcaligenes*.

The number of bacteria for dried green laver (seaweed), toasted laver (seaweed) and toasted and seasoned laver were as follows: 1.5×10^6, 2.5×10^3 and 5×10^5 per gram, respectively. The number of bacteria for dried green laver (seaweed) manufacturing processing of raw alga, minced alga, matured alga, cured alga, dehydrated alga and dried alga were as follows: 2.5×10^4, 6.2×10^5, 5×10^6, 7.0×10^6, 2.5×10^5, 6.2×10^5 and 1.5×10^6 per gram, respectively. Results from this analysis are conclusive because the variance of bacterial counts was not too high to make statistical measurements. The increase of bacterial counts was caused by contamination from the nori manufacturing processing. Installing ozonation at these points in the process line will take planning and innovation, since the nori product comes from the mince machine and a matured mince machine is slotted and allows the ozone-containing water to come in contact with all surfaces of the minced alga.

9.6 Current status and future trends for ozone and seafood

Among seaweeds, Shimane Izumo mozuku (*Nemacystus* sp.) and Okinawa mozuku (*Cladosiphon okamuranus*) are remarkable species with high contents of fucoidan, a kind of glycan. The term 'glycan' refers to a polysaccharide or oligosaccharide. It may also be used to refer to the carbohydrate portion of a glycoconjugate, such as a glycoprotein, glycolipid or proteoglycan. It has been reported that fucoidan is an effective substance for enhancing human immune responses. The production of Shimane Izumo mozuku is rapidly increasing in the Japanese prefecture of Shimane due to trends toward consumed healthy foods. The establishment of aquaculture species in Japan is influenced by consumers' taste preferences and market trending toward directly consumed products, such as mozuku, or those that are used as ingredients in health-supplement products, which are projected to increase in the future.

Quality and safety are the highest priorities of the mozuku industry. Ozone can help guarantee the microbiological quality of mozuku. However, there are many implied factors that can influence these results, such as method of use, temperature, pH, water activity and the quality of material used. Microbial cross-contamination in the workplace environment remains a leading cause of mozuku spoilage. To minimise the spread of spoilage microorganisms to mozuku products, better sanitation agents and enhanced sanitation regimens will be needed.

Self-sufficiency, a serious concern in Japan, has decreased considerably in recent years due to decreased domestic supply and increased spoiled products. Therefore, promotion of preventing the spoiling of mozuku is important to recovery of the self-sufficiency rate. Quality improvements of seasoned mozuku product have been studied (Naito 1987). Lately, there have also been some problems reported with seasoned mozuku product, of which the most serious is self-pollution by net cages. From the viewpoint of protection against microbial contamination, ozone-containing water rinse systems have been highlighted as potential rearing technologies. There are two types of seasoned mozuku product: mozuku in vinegar and mozuku in salt. The main objective of the method is to obtain practical guides for the simultaneous control of seasoned mozuku product spoilage bacteria by the ozone process. Mozuku in salt product usually develops a foul smell at 30 °C over 7 days and putrefies at 30 °C over 14 days. The mozuku in vinegar product usually has developed a foul smell at 30 °C over 14 days and has putrefied at 30 °C over 20 days. The storage lives of these products were doubled by ozone treatment (Naito 1987). We consider *Achromobacter aquamarine, M. colpogenes, Bacillus mycoides, B. cereus, Zygosaccharomyces bisporus* and *Z. bailii* to be of greatest concern because of the severity and number of seasoned mozuku product putrefactions they cause. Levels of these microorganisms decreased upon treatment with ozone-containing water (0.5~1.0 ppm) for 30~60 minutes at 5 °C. Further studies must be conducted to demonstrate the optimum concentrations and the best methods of ozone applications in diverse mozuku species, so that all governments can approve ozone application in the seaweed industry.

In Japan, seaweed production has been prosperously practiced since the early 17th century. The major mariculture species are seaweed boiled down in soy (tsukudani). Studies have shown that spoilage of seaweed (nori, tsukudani) by lactic acid bacteria is prevented by ozone treatment (Naito 1999a,b). Deterioration of laver (seaweed) tsukudani by lactic acid bacteria is prevented by ozone treatment. Lactic acid bacteria related to off-flavour generation of laver tsukudani survive in the manufacturing processing for long times. These lactic acid bacteria were identified as *Weissella viridesscence* and contaminate the manufacturing process. Tsukudani were cleaned during manufacturing processing with ozone gas and washed with ozone-containing water. These bacteria were disinfected easily by the combined use of ozone-containing water and ozone gas. It is possible to conclude that ozone use in the seaweed industry would be useful because there are ozone Gram-negative bacteria and lactic acid bacteria which spoil the seaweed that are not resistant to ozone. This presents an opportunity to guarantee the quality and shelf life of the seaweed and seafood products.

The effect of ozonation in aqueous solution (ozone concentration: 1 mg/L, time of ozonation: 60 and 90 minutes) on the shelf life of shucked, vacuum-packaged mussels stored under refrigeration was studied by monitoring the microbiological, chemical and sensory changes occurring in mussel samples over a period of 12 days. Non-ozonated vacuum-packaged mussels served as the control sample. Ozonation affected populations of bacteria, namely aerobic plate count (0.7–2.1 log cycle reduction), *Pseudomonas* spp. (0.5–1.1 log cycle reduction) and hydrogen sulfide producing bacteria (1.1–2.5 log cycle reduction), lactic acid bacteria (0.3–0.8 log cycle reduction) and *Enterobacteriaceae* (0.5–1.5 log cycle reduction). The effect of ozonation was more pronounced at the longer time of ozonation (Monousaridis et al. 2005). The microbial flora of mussels may be expected to vary considerably, depending on the quality of the seawater from which they are taken, the quality of washwater and the level of microbial cross-contamination in the workplace environment.

When exposed to adverse environmental stresses such as ozone, growth simulation for plants and animals will be repressed.

Goldfish and eels are very long-lived animals and consequently if they are maintained in the right environment they will grow to be very large. In Japan in 1988, monster-sized goldfish and eels were cultivated by ozone treatment. The ozone, however, does nothing to increase fish size other than to eliminate bacteria and parasites that might cause problems. Many people state that fish should be kept in sufficiently warm water, fed the best food possible, including plenty of protein and greens, and provided with large volumes of clean water that is changed regularly to remove the growth inhibitor secretion.

Ozone at concentrations of 0.5 ppm or less affects the respiration and growth of bean sprouts (Naito 1994b; Naito and Shiga 1989).

The elongation of hypocotyls of bean sprouts was ascribed mainly to the simultaneous treatment with 0.02~0.2 ppm ozone in air and 0.3~0.5 ppm of ozone-containing water. It was assumed that the most appropriate harvesting time was about 5 days after planting (length of hypocotyle: about 12~16 cm). Catalase and SOD (super oxide dismustase) activities increased rapidly 1~2 days after the beginning of germination, and then activities were maintained almost constant through 7 days. It is possible that some of these effects are produced by direct action of ozone on some of the enzyme within the cell. Although many enzymes are known to be susceptible to damage by oxidising agents such as hydrogen peroxide, the effect of ozone, which is an extremely active oxidising agent, on fish and seafood has not been investigated. New approaches to determining the principal ozone treatment for disinfection and odour removal from a seafood processing plant should be investigated (Naito and Takahara 2008). Disinfection or odour removal from air by ozone gas, and disinfection or odour removal on floors by ozone-containing water, are an important part of clean room technology in the seafood industry.

References

Ben Embarek, P. (1994) Presence, detection and growth of *Listeria monocytogenes* in seafoods: a review, *Int J Food Microbiol*, **23**: 17–34.

Campos, C.C., Losada, V., Rodriguez, O., Aubourg, S.A. and Velazquez, J.B. (2005) Effects of storage in ozonised slurry ice on the sensory and microbial quality of sardine (*Sardina pilchardus*), *Int J Food Microbiol*, **103**(2): 12–130.

Campos, C.C., Losada,V., Rodriguez, O., Aubourg, S.A. and Velazquez, J.B. (2006), Evaluation of an ozone–slurry ice combined refrigeration system for the storage of farmed turbot (*Psetta maxima*), *Food Chemistry*, **97**(2): 223–30.

Chawla, A., Bell, J.W. and Marlene, E.J. (2007) Optimization of ozonated water treatment of wild caught and mechanically peeled shrimp meat, *J Aquatic Food Product Technology*, **16**(2): 41–56.

Chen, H.C., Chang, S.O. and Ing. S.T. (1987) A study on the sterilization effect of ozone and its application for marine food processing, *J Fish Soc,Taiwan*, **14**: 79–89.

Chen, H.H., Chiu, E.M. and Huang, J.R. (1997) Color and gel formatting properties of horse mackerel (*Trachurus joponicus*) as related to washing conditions, *J Food Sci*, **62**: 985–91.

Crapo, C., Himelbloom, B., Vitt S. and Pedersen, L. (2004) Ozone efficacy as a bactericide in seafood processing, *J of Aquatic Food Product Technology*, **v1313**(3): 111–23. Retrieved from http://www.informaworld.com/smpp/title~db=all~content=t792303960~tab=issueslist~branches=13.

Dondo, A., Nachtman, C., Doglione, L., Rosso, A. and Genetti, A. (1992) Foods: their preservation by the combined use of refrigeration and ozone, *Ing Aliment. Conserve Anim*, **8**: 16–25.

Gelman, A., Sach, O., Khanin, Y., Drabkin, V. and Glatman, L. (2005) Effect of ozone pretreatment on fish storage life at low temperature, *J Food Prot*, **68**: 778–84.

Gram, L. and Huss, H.H. (1996) Microbiological spoilage of fish and fish products, *International Journal of Food Microbiology*, **33**: 121–37.

Haraguchi, T., Shimizu, U. and Aiso, K. (1969) Preservation of ozone on fish, *Bulletin of the Japanese Society of Scientific Fisheries*, **35**: 915–19.

Kim, J., Yousef, A. and Dave, S. (1999) Application of ozone for enhancing the microbiological safety and quality of foods: a review, *J Food Prot*, **62**: 1071–87.

Koetters, J., Prahst, A., Skura, B., Rosenthal, H., Black, E.A. and Rodigues-Lopez, J. (1997) Observation experiments on extending shelf-life of rockfish (*Sebastes* spp) products with ozone, *J Appl Ichthyol*, **13**: 1–8.

Mori, K., Sawada, H., Nabetani, O., Maruo, S. and Hirano, T. (1973) Softening spoilage of film packaged Kamaboko due to *Bacillus licheniformis*, *Bull of Japanese Society of Scientific Fisheries*, **39**(10): 1063–9.

Mori, K., Nabetani, O. and Hirono, T. (1974) Browning of Kamaboko by *Pseudomonas* sp, *Bull of Japanese Society of Scientific Fisheries*, **40**(9): 959–62.

Naito, S. (1981) Isolation and identification of yeast from smoked squid, *Annual Report of the Food Research Institute, Aichi Prefectural Government*, **22**: 86–93.

Naito, S. (1983) Isolation and identification of *Micrococcus* from smoked squid, *Annual Report of the Food Research Institute, Aichi Prefectural Government*, **24**, 65–75.

Naito, S. (1986) Ozone treatment of broiling squid for storage, *Annual Report of the Food Research Institute, Aichi Prefectural Government*, **27**: 39–50.

Naito, S. (1987) Microbial quality improvement of seasoned Mozuku product by ozone treatment, *Annual Report of the Food Research Institute, Aichi Prefectural Government*, **28**: 66–79.

Naito, S. (1992a) Synergistic sporicidal effects of gaseous ozone and UV on *Bacillus* spores and *Clostridium* spores, *J Antibacterial and Antifungal Agents, Japan*, **20**(6): 298–300.

Naito, S. (1992b) Synergistic sporicidal effects of gaseous ozone and ascorbic acid and isoascorbic acid on *Bacillus* spores, *J Antibacterial and Antifungal Agents, Japan*, **20**(11): 565–70.

Naito, S. (1992c) Effect of metallozeolite and ascorbic acid on the inactivation of *Bacillus subtilis* spores with gaseous ozone, *J Antibacterial and Antifungal Agents, Japan*, **20**(12): 629–32.

Naito, S. (1992d) Mutagenicity of ozone-treated amino acids and saccharides, *J Antibacterial and Antifungal Agents, Japan*, **20**(7): 365–72.

Naito, S. (1994a) Changes of microorganisms and volatile sulfur compounds derived from flakes of dried bonito (Kezuri-bushi) by ozone treatment, *Annual Report of the Food Research Institute, Aichi Prefectural Government*, **35**: 82–90.

Naito, S. (1994b) Utilization of ozone for physiological activities, *J Antibacterial and Antifungal Agents, Japan*, **22**(12): 23–30.

Naito, S. (1999a) On the spoilage of seaweed (Nori Tsukudani) by lactic acid bacteria and effects of ozone treatment, *Annual Report of the Food Research Institute, Aichi Prefectural Government*, **39**: 51–6.

Naito, S. (1999b) Sanitation of seaweed factory products by the utilization of ozone: utilization of ozone on seaweed factory product contamination by *Lactobacillus*, *Technical Journal on Food Chemistry and Chemicals*, **15**(6): 46–53.

Naito, S. (2001) Spoilage of smoked squid by microorganisms and ozone sterilization, *Technical Journal on Food Chemistry and Chemicals*, **17**(11): 53–8.

Naito, S. and Sannomiya, Y. (1985) Ozone treatment of dried cuttlefish for storage, *Annual Report of the Food Research Institute, Aichi Prefectural Government*, **26**: 104–12.

Naito, S. and Sawairi, A. (2000) Ozone inactivation of lactic acid bacteria, *Biocontrol Science*, **5**(2): 107–10.

Naito, S. and Shiga, I. (1982a) Microbiocidal properties of ozone on various microorganisms suspended in water, *J Japan Food Sci Technol*, **29**: 1–10.

Naito, S. and Shiga, I. (1982b) Effect of sulfur compounds on germicidal properties of ozone on *Bacillus* spores, *J Japan Food Sci Technol*, **29**: 63–9.

Naito, S. and Shiga, I. (1989) Effect of ozone treatment on elongation of hypocotyls and microbial counts of bean sprouts, *J Japan Food Sci Technol*, **35**: 181–8.

Naito, S. and Takahara, H. (2006) Ozone contribution in food industry in Japan, *Ozone: Science & Engineering*, **28**, 425–9.

Naito, S. and Takahara, H. (2008) Recent developments in food and agricultural uses of ozone as an antimicrobial agent in a food packaging film sterilization machine using ozone, *Ozone: Science & Engineering*, **30**: 81–7.

Naito, S. and Yamazawa, M. (1989) Effect of ozone treatment on storage of packaged chikuwa, *J Antibacterial and Antifungal Agents, Japan*, **17**(3): 111–18.

Naito, S., Seki, T. and Mizuno, Y. (2001a) Formation of yellow spots on Juten-Tofu by lactic acid bacteria and inactivation of these bacteria by ozonated water, *J Antibacterial and Antifungal Agents, Japan*, **29**(1): 23–30.

Naito, S., Okada, K. and Inoue, Y. (2001b) Swelling spoilage of smoked squid products caused by lactic acid bacteria and inactivation of three bacteria by ozonated water, *J Antibacterial and Antifungal Agents, Japan*, **29**(8): 497–505.

Nash, B. (2002) Ozone effective in preserving seafood freshness, *Marine Extension News, North Carolina Sea Grant*, Spring Issue.

Paranjpye, R.N., Peterson, M.E., Poysky, F.T. and Eklund, M.W. (2008) Incidence, growth, and inactivation of *Listeria monocytogenes* in cooked and peeled cold-water shrimp, *Journal of Aquatic Food Product Technology*, **17**(3): 266–84.

Powell, J., Anne, C., Moor, R. and Gow, J.A. (1979) Comparison of EC broth and medium A-1 for the recovery of *Escherichia coli* from frozen shucked Snow Crab, *Environ Microbiol*, **37**: 836–40.

Ravesi, E.M., Licciardello, J.J. and Racicot, L.D. (1988) Ozone treatment of fresh fish Atlantic Cod, *Gadus morhus*, *Marine Fisheries Review*, **49**(4): 37–42.

Rice, R.G., Farquhar, J.W. and Bollyky, L.J. (1982) Review of the applications of ozone for increasing storage times of perishable foods, *Ozone Science and Engineering*, **4**: 147–63.

Shewan, J.M. (1971) The microbiology of fish and fishery products: a progress report, *J Appl Bact*, **34**: 299–315.

Staehelin, J. and Hoigné, J. (1985) Decomposition of ozone in water in the presence of organic solutes acting as promoters and inhibitors of radical chain reactions, *Environ Sci Technol*, **19**(12): 1208–13.

Vaz-Velho, M., Silva, M., Pessoa, J. and Gibbs, P. (2006) Inactivation by ozone of *Listeria innocua* on salmon-trout during cold-smoke processing, *Food Control*, **17**: 609–16.

Yamayoshi, T. and Tatsumi, N. (1993) Microbiocidal effect of ozone solution on Methicillin-resistant *Staphylococcus aureus*, *Drugs Exptl Clin Res*, **XIX**(2): 59–64.

10 Ozone Sanitisation in the Food Industry

P.J. Cullen and Tomás Norton

10.1 Introduction

Effective cleaning and sanitation programmes are required to achieve a good level of hygiene in food handling or production facilities. Cleaning can be defined as the removal of soil (e.g. food residues) from food contact surfaces so that they are visually clean without signs of oxidation, don't have bad odours and are not greasy to touch. On the other hand, sanitisation is making sure a clean surface is substantially free from pathogenic microorganisms and an undesirable number of spoilage microorganisms (Marriott and Gravani 2006). Therefore, the cleaning and sanitation procedures used in the food industry must actively destroy vegetative cells of hazardous microorganisms and substantially reduce the numbers of other undesirable microorganisms without adversely affecting the product quality or its safety for the consumer (Pan et al. 2006). The interest in having highly effective cleaning and sanitation programmes is not solely a case of maintaining public responsibility but is also the law as food business operators must implement and maintain hygiene procedures based on hazard analysis and critical control point (HACCP) principles as part of EU Regulation 852/2004 (Article 5).

Food soil basically represents any unwanted material on food contact surfaces, with the primary source of this soil being the food product being handled. For example, residues of a perfectly edible product on a processing line, a film of yoghurt on the walls of an empty tank or a film on the internal surface of a heat exchanger can be considered food soil and must be removed, just as a person would remove mud from a tomato or sand from a strawberry (Berk 2009). However along with food soil, residue from water minerals and cleaning compounds, as well as microbiologically active residues (i.e. biofilms), which may or may not be visible, contribute to the soil build-up on surfaces. Such films are complex aggregations of materials that enhance the survival and growth of microorganisms, and when formed are very difficult to remove. Moreover, because such films are organic, their removal must be tackled by detergency rather than disinfection (Stanga 2010).

Ozone in Food Processing, First Edition. Edited by Colm O'Donnell, B.K. Tiwari, P.J. Cullen, and Rip G. Rice.

When food processers develop cleaning and sanitation programmes they must consider the influence of the contaminant, the food contact surface, temperature, fluid flow, the cleaning agent and the disinfectant on the effectiveness of the cleaning system. Therefore, developing suitable cleaning protocols can be very complicated, requiring in-depth knowledge of a wide range of associated phenomena. Regarding the effects of the contaminant, the types of soil and fouling of relevance during food processing may include a soil binding with different characteristics for different soil layers, resulting in different removal kinetics during the cleaning and sanitation programme. As regards the surface, the geometry of the contact surfaces and flow channels is of utmost importance with respect to cleaning, especially during the cleaning of closed systems, which depends on liquid flow. Finally, the characteristics of the cleaning agent must be considered for each type of application. In many cleaning applications in the food industry, physiochemical alteration mechanisms are utilised, whereby the cleaning agent physically interacts with soil through wetting, softening (plasticising), swelling, heating and so on. A chemical reaction such as saponification of lipids, hydrolysis of proteins and decomposition of salts alters the structure of the soil. The most commonly used detergents are dilute NaOH and dilute acids. In addition, fluid flow is often the only means of developing enough shear stress to dislodge the bound soil and transport it and microorganisms out of the system, in order to avoid contamination of already cleaned surfaces. All the above issues need to be taken into account during cleaning procedures in the food industry. On top of that, the effectiveness of surface sanitisers is extremely important as they must have a high microbial destruction capability against a wide range of microorganisms.

Once surface cleaning has been completed, sanitising agents are required to sanitise the surfaces. Sanitising agents include chlorine and its derivatives, iodine derivatives, ozone, hydrogen peroxide and quaternary amines (Marriott 1994). Some food industries use thermal sanitation methods and/or irradiation. Thermal sanitation can be effective in destroying microorganisms but the cost associated with developing steam and hot water is high and the excessive heat produced can be damaging to food processing equipment (Troller 1993). On the other hand, chemical sanitation methods are more economical to use and chemicals such as chlorine have found widespread use throughout the food industry, due to their ability to inactivate all types of vegetative cells. Chlorine compounds are broad spectrum germicides which act on microbial membranes, inhibit cellular enzymes involved in glucose metabolism, have a lethal effect on DNA and oxidise cellular protein. Chlorine has activity at low temperature, is relatively cheap and leaves minimal residue or film on surfaces. However, major disadvantages are also associated with chlorine byproducts, such as that they are corrosive to many metal surfaces (especially at higher temperatures), they can cause health and safety issues due to skin irritation and they can cause mucous membrane damage in confined areas. Moreover, under certain conditions chlorine can form chlorinated organic derivatives, some of which are the potentially carcinogenic trihalomethanes (THMs). Although there are hazards associated with the formation of carcinogenic compounds from chlorine that are of public concern, there are not many other suitable chemical sanitisers available with such a wide range of microbial action. Therefore, over the years, the search for new chemical sanitisers has become more widespread, with ozone a potential alternative to chlorine for use in the food industry given its special advantage of rapid inactivation and lower overall energy consumption.

The aim of the current chapter is to look at the issues specific to the use of ozone when applied as part of a cleaning regime in the food industry. First the chemical effectiveness and issues surrounding the use of ozone technology as part of a cleaning regime will be discussed. Then current cleaning procedures will be assessed and the industries showing the greatest possibilities for incorporating ozone technology will be examined.

10.2 Ozone as a sanitising agent

The bactericidal properties of ozone are mainly due to it being an unstable allotrope of oxygen, and for this reason it has been considered as a replacement for chlorine in the antimicrobial sanitisation of water, food and food processing surfaces and equipment. In practical use, ozone decomposes to a number of nanosecond half-lived free radicals, which can attack organic compounds indiscriminately, leaving no residual components when decomposition is complete, while at the same time liberating oxygen. This means that ozone is a very potent oxidising agent that readily inactivates microorganisms in aqueous solutions (Broadwater et al. 1973). Applications of ozone as a sanitisation technology could also include the sterilisation of surfaces of contaminated rooms, biosafety cabinets or entire buildings.

As mentioned above, ozone owes its excellent bactericidal, viricidal and sporicidal activities to its powerful oxidising properties. In sanitisation procedures within the food industry, ozone is generally used in an aqueous solution, in which it has a higher oxidation reduction potential (ORP) than the form of chlorine used in aqueous solutions: +2.07 V for ozone versus +1.49 V for chlorine. The presence of organic components will result in self-depletion of ozone and reduced biocidal effectiveness (Khadre et al. 2001). This means that, due to the strong oxidising power of ozone, the ORP values can be below those expected, and even negative (reducing) values can occur. Therefore, in order to ensure the effectiveness of ozone application, detection technology that works well at the limit of approved concentrations for cleaning and sanitisation operations must be used.

Of greatest significance to the efficacy of the ozone sanitisation procedure are variables such as temperature, pH and quantity of organic matter.

Table 10.1 **Summary of studies of surface disinfection using ozone. (Reprinted from *Trends in Food Science & Technology*, Volume 18, Issue Number 1, A. Pascual, I. Llorca, A. Canut, Use of ozone in food industries for reducing the environmental impact of cleaning and disinfection activities, S29–S35, 2007, with permission from Elsevier.)**

Application	Treatment	Microorganism	Results	Reference
Dairy biofilms on stainless steel surface	Ozonated water, 0.5 ppm for 10 minutes	*Pseudomonas fluorescens* and *Alcaligenes faecalis*	5.6 and 4.4 log reduction, respectively	Greene et al. (1993)
CIP system	Ozonated water	*Staphylococcus aureus*, *Pseudomonas aeruginosa* and *Candida albicans*	99% microbial count reduction	Lagrange et al. (2004)
Mixing kettle, table top and shroud (all stainless steel)	Ozonated water, 2 ppm at 10 gpm for 1 minute	Unspecified	Microbial plate count reduction ranging from 63.1 to 99.9% (depending on surface)	Hampson (2000)
'High-traffic' and 'low-traffic' floor areas	Ozonated water, 2 ppm at 10 gpm for 1 minute	Unspecified	Microbial plate count reductions 67.0–95.6% and 84.3–99.9%, respectively	Hampson (2000)
Plastic shipping container	Ozonated water, 2 ppm at 10 gpm for 1 minute	Unspecified	Microbial bioluminescence assay reduction 68.8–97.4%	Hampson (2000)
Stainless steel surfaces	2 ppm ozone gas at atmospheric pressure, 22 °C and 77% RH for 4 hours	*Escherichia coli*, *Serratia liquefaciens*, *Staphylococcus aureus*, *Listeria innocua* and *Rhodotorula rubra*	Reduction ranging from 7.56 to 2.41 log values	Moore et al. (2000)
Stainless steel surfaces in the presence of UHT milk	2 ppm ozone gas at atmospheric pressure, 22 °C and 77% RH for 4 hours	*Escherichia coli*, *Serratia liquefaciens*, *Staphylococcus aureus*, *Listeria innocua* and *Rhodotorula rubra*	Reduction ranging from 5.64 to 1.65 log values	Moore et al. (2000)
Stainless steel surfaces	2 ppm ozone gas in bioaerosol chamber at 20 °C and 50% RH for 1 hour	*Microccocus luteus*	2–3 log reduction	Bailey et al. (2001)
Surfaces	2 ppm ozone gas, 2-hour exposure	Unknown	2 log reduction	Taylor and Chana (2000)
Equipment, walls, floors, drains, tables and conveyors, previously well-cleaned	Ozonated water, 3.0–3.5 ppm	*Trichophyton mentagrophytes*, *Salmonella choleraesuis*, *Staphylococcus aureus*, *Pseudomonas aeruginosa*, *Campylobacter jejuni*, *Listeria monocytogenes*, *Aspergillus flavus*, *Brettanomyces bruxellensis*, *Escherichia coli*	Log reduction ranging from 6 to 4	Boisrobert (2002)

As regards temperature, it is widely accepted that the rate of inactivation of microorganisms with any disinfectant increases as temperature increases. However, when ozone is used as a sanitiser in water, as temperature increases ozone also becomes less soluble and less stable. Therefore, there is a threshold value of temperature at which ozonation is most efficacious, which is significantly less than typical sanitation temperatures. On the other hand, in some cases the increase in ozone reactivity can compensate for the decrease in its stability when the medium temperature is increased. As regards pH, ozone is more stable at low than at high pH values, but at higher pH ozone decomposes, with the radicals produced apparently enhancing its efficacy. Therefore, the actual pH that is required should be conditioned by the target microorganism (Khadre et al. 2001). Finally, some of the organic substances that are present in the aqueous ozone medium may compete with microorganisms for ozone. Therefore, unless additional ozone is added to compensate for this potential loss, the efficacy of the ozone sanitising process may be reduced. Clean water absent of organic matter is desirable for use in the sanitisation of food processing equipment with ozone.

Ozone disinfection kinetics in air at low relative humidity (RH) can be orders of magnitude slower than in aqueous media. While ozone does not react significantly with water or air in the absence of ultraviolet (UV) radiation, under UV irradiation ozone reacts with water and decomposes into various short-lived radicals, such as the highly reactive hydroxyl radical. Theoretical and empirical evidence suggests that most of the sterilisation effect results from the radicals produced, and not the ozone itself (Beltran 1995). The decomposition reaction can be enhanced in air by the use of UV irradiation and through controlled humidity. Theoretically, therefore, the effects of ozone in air, under controlled conditions, should parallel the effects of ozone in water, and the effectiveness of ozone for eliminating airborne pathogens in either medium may be comparable at high RH.

The threshold concentrations at which ozone inactivates viruses and bacteria in water are remarkably low. For example, the threshold for *Escherichia coli* lies between 0.1 and 0.2 ppm (Broadwater et al. 1973). Viruses are also sensitive to low levels of ozone, an advantage for an air-based system, since these small microbes are especially difficult to remove by filtration.

While ozone has a wide antimicrobial spectrum, its effectiveness is a function of the target microorganism. For example, in the case of bacteria, while results are varied in the literature, it can be assumed that Gram-negative bacteria are more sensitive to ozonation than Gram-positive bacteria and that ozone is more effective against vegetative bacterial cells than against bacterial and fungal spores (Khadre et al. 2001). In another study, Lagrange et al. (2004) found that ozone-containing water was not effective in disinfecting food contact surfaces in the presence of protein and consequently recommended efficient cleaning before using ozone-containing water for disinfection purposes. To illustrate the wide spectrum of ozone, Table 10.1 lists some results of the inactivation of bacteria by ozone in water.

10.3 Health and safety issues

At low concentrations, ozone (~0.1 mg/L) can cause irritation to the nose, throat and eyes, and an hour's exposure to ozone concentrations of 2, 4, 15 and 95 mg/′ induces symptomatic, irritant, toxic and irreversible lethal effects, respectively, in humans (Khadre et al. 2001). Given that health and safety are of prime importance for the practical application of ozone in food surface sanitisation procedures, it is necessary for processors to have some means of detecting and destructing ozone in the plant. Ozone can be used as a sanitiser in both gaseous and aqueous forms. Naturally, greater care is required in the gaseous phase. For example, in gas applications of ozone, analyser equipment can be installed in ozonation rooms and can be designed to trigger both a displayed and an acoustic warning signal as soon as the ozone content in the ambient air exceeds 0.1 ppm (Khadre et al. 2001). Moreover, there should be plans for remedial action in case of accidents, response procedures for accidental ozone inhalation and training of personnel covering the nature and dangers of ozone, precautions and first aid for ozone inhalation. These issues are discussed in detail in Chapter 15.

10.4 Using ozone in industrial cleaning procedures

Ozone is an unstable gas and readily reacts with many organic substances. It sanitises by interacting with microbial membranes and denaturing metabolic enzymes. It does not leave a chemical residue. Ozone must be electrically generated on demand and cannot be stored for later use. An advantage of ozone is its ability to readily oxidise microbes in solution. Because ozone requires no storage or special handling or mixing considerations, it may be viewed as advantageous over other chemical sanitisers.

As for the usage of ozone in gaseous form, studies on its efficiency and effectiveness in disinfecting stainless steel surfaces have been reported in the literature. Moore et al. (2000) found that ozone applications were less successful when dried biofilms were present on the food contact surface. The authors concluded that ozone may prove more effective, therefore, in disinfecting product contact surfaces, which should typically be cleaned several times a day, rather than environmental surfaces, such as walls, which need be cleaned only once a week. Kowalski et al. (1998) found that a high level of bacterial sterilisation was possible with airborne ozone. They found inactivation levels had strong parallels with the ozonation in an aqueous medium; however, a high RH is required.

Another way of cleaning food processing equipment is by spraying ozonated water directly on floors, drains, walls, wet equipment, tanks (externally or internally) and clean rooms via a mobile or centralised system with handheld, drop-down or low-pressure sprayers. Such systems are especially suitable for equipment cleaning-out-of-place and water-containing ozone cleaning systems have been employed successfully in this way for both the dairy (Greene et al. 1999) and the fresh produce industries.

However, the main benefits of ozonation emerge when used as part of a clean-in-place (CIP) system. A CIP system is an automatically operated cleaning system that delivers a number of wash and rinse cycles to the internal surfaces of processing equipment such as cleaning tanks, piping, filling machines, pasteurisers and homogenisers. CIP systems largely eliminate human contact with cleaning agents and can save significant labour costs. One of the main advantages of CIP systems is that they can recirculate the cleaning solution through a series of holding tanks and associated pumps and piping to allow the reuse of chemicals and water, thereby reducing water and chemical consumption. Ozone-containing water can be used in CIP systems by directly injecting it into a facility's fluid distribution network and circulating it for a set amount of time.

The advantages of using ozone in CIP systems, compared to traditional disinfectants, are that it leaves no residues and it is applied in cold water. As discussed above, there is an upper limit to the temperature that the cleaning water can attain due to ozone, in order to maintain a balance in stability, solubility and disinfection power. Moreover, unlike in chlorine sanitisation, no residues are left, and consequently multiple rinses, which are standard practice in industry, are not needed to remove product residues. Therefore, the quantity of water consumed is less when ozone is employed and there is no risk due to residue formation on the food contact surfaces. Moreover, by using a low-temperature water medium, the operating costs of such a system are much reduced with minimum energy consumption. Also, the expansion and contraction of welds on the processing line will not occur, thereby preventing untimely deterioration.

Another major attribute of ozone sanitisation is the reduced quantity of chemicals discharged into sewer systems, together with the large amounts of water necessary to rinse out residual chemicals from the machines. Pascual et al (2007) reported that in typical cleaning and disinfection practices the wastewaters contain soluble organic material, fat, oil and grease (FOG), suspended solids (SS), nitrate, nitrite, ammonia and phosphate from product remnants and removed deposit soil, combined with residues of cleaning agents; for example, acid or alkali solutions in the wastewaters which may have high or low pH and high conductivity. Moreover, the use of phosphoric and nitric acids will increase the phosphate and nitrate content of the wastewaters. However, according to Pascual et al. (2007), replacing chemical products with ozone lowers the concentration of salts and, therefore, the electrical conductivity of discharges. The use of ozone can save water in comparison to other biocides, as it does not leave residues; therefore it does not require a final rinse to remove any residual disinfectant that might remain. Also, ozonated water, which has been used for disinfection, can potentially be reused for the initial cleaning stages.

10.5 Ozone applications in food processing

10.5.1 Dairy industry

While the multifunctionality of ozone application makes it a promising sanitising agent, ozone has not been commonly used in the dairy industry, and research is still needed to prove its effectiveness. One of the first reported studies of ozone in the dairy industry was by Greene et al. (1993), who compared the effectiveness of ozonated water and chlorinated sanitiser for the disinfection of stainless steel surfaces which had been incubated with UHT milk inoculated with either *Pseudomonas fluorescens* (ATCC 949) or *Alcaligenes faecalis* (ATCC 337) at 32 °C for 4–24 hours. They found that ozone was as effective as chlorination against dairy surface-attached bacteria, as both treatments reduced bacterial populations by 99%. However, while ozone is an effective sanitiser, it is also known for being corrosive. Therefore, in a follow-up study Greene et al. (1994) investigated the effects of chlorine and ozone sanitisers on gasket appearance, tensile strength, elongation at failure and elastic modulus. They found that, with the exception of some slight discolouration, ozone treatment does not appear to be detrimental to Buna, silicone and polyethylene gasket materials. However, long-term exposure to ozone does affect PTFE (Teflon) by increasing the tensile strength and elongation.

In a later study Greene et al. (1999) found no corrosive effect of ozone on stainless steel when compared to a warm water control. Güzel-Seydim et al. (2000) studied the use of ozonated water as a pre-rinse technique in dairy equipment by soiling stainless steel coupons and then treating with ozone-containing water as a pre-rinse. The authors found that the ozone-containing water (10 °C) removed 84% of dairy soil when compared to warm water (40 °C), which removed 51%. Scanning electron microscopy images further validated their results and it was therefore concluded that when ozone is used in the pre-rinse stage, decreased detergent use in the cleaning solution recirculation step may be achievable.

Recent studies have looked into the removal of biofilms from food processing equipment, as these can lead to a variety of problems including reduced flow rates, reduced heat transfer and contamination of the finished product. A major concern with the contamination of a finished product is that human pathogens such as *E.coli* and *Listeria monocytogenes* may prevail. This problem is particularly associated with the dairy industry, as shown by Jacquet et al. (1993), who were able to isolate the same strains on ripening cheeses as were found on the plant equipment in a 2-year study of a dairy plant. Baumann et al. (2009) investigated removal of *L. monocytogenes* biofilms from stainless steel chips by using individual or combined treatments of ozone and ultrasound. They found that when both ozone and ultrasound were used synchronously, the sanitisation was more efficacious than with their individual application. The authors recommended that due to the synergistic attributes that combined use of ozonation and sonication demonstrated, there is a potential for further exploration of these effects for the removal of biofilms. It was found that higher concentrations of ozone resulted in a synergistic effect when used with sonication for a 60-second exposure time. Nonetheless, the simultaneous use of low ozone concentrations with sonication appeared to have similar additive effects for all exposure times. Therefore, their results indicated that the combination of power ultrasound and ozonation may be an effective treatment for biofilm removal from stainless steel food contact surfaces.

Bott and Tianqing (2004) also investigated the synergistc effects of synchronous application of ultrasound and ozone in order to reduce biofilm accumulation, due to its adverse effect on efficiency of the associated heat exchangers, such as those required in the dairy industry. They also found that there is an enhancement of ozone biocidal efficiency through the application of ultrasound. While the results were positive, they concluded that there is a need to compare the combined cost of ozone and ultrasound treatment in relation to the necessary treatment of the water before final discharge.

When moulds occur in a food processing plant, they can result in spoilage of food products. As cheese ripening rooms possess an environment that encourages mould growth, it is likely that cheese will become mouldy if the room is contaminated with mould spores. Ozone is effective in the inactivation of airborne moulds in this environment. Serra et al. (2003) indicated that it was necessary to wipe the surfaces with a commercial sanitiser in order to decrease the viable mould load on these surfaces. This finding was in correspondence with the report of Ingram and Barnes (1954), who concluded that ozone was able to destroy most aerial spores but failed to prevent their germination when they were lodged in apple surface wounds.

Güzel-Seydim et al. (2000) compared the cleaning efficiency of ozone-containing water at 10 °C and warm water at 40 °C for dairy soils on stainless steel. Results indicated that the ozone treatment removed 84% of soil from metal plates versus 51% soil removal by the warm water treatment, but the effectiveness of the two treatments did not differ ($p > 0.05$). The authors concluded from this study that ozonated water is effective in removing heated milk soil materials.

Dosti et al. (2005) studied the efficacy of ozone (0.6 ppm for 10 minutes) and chlorine (100 ppm for 2 minutes) for the removal of bacterial biofilm. They observed that both ozone and chlorine significantly reduced biofilms formed by *Pseudomonas* spp. adhered to stainless steel compared to the control, as shown in Figure 10.1. They reported no significant difference between ozone and chlorine inactivation of the bacteria, with the exception of *P. putida*, where ozone was found to be more effective compared to chlorine. This study shows that ozone can be a potential alternative to chlorine as a sanitising agent for dairy processing equipment.

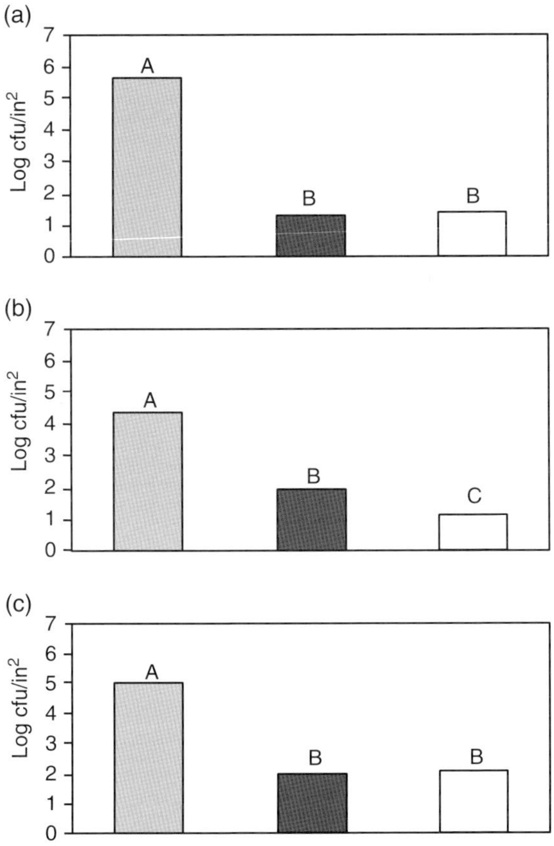

Figure 10.1 Efficacy of sanitising agents on the removal efficiency of (a) *Pseudomonas fragi*, (b) *P. putida*, (c) *P. fluorescens* from stainless steel (Dosti et al. 2005). Bars labelled with same letter A,B,C are not significantly different at $p < 0.05$.

10.5.2 Wine industry

Maintaining a sterile and clean environment in the wine industry is extremely important. Cross-contamination between batches of wine is a major concern, as is the management of the active yeast. The yeast is a major ingredient in the fermentation process and without it the fermentation process would not occur. However, *Brettanomyces* (a non-spore forming genus of yeast in the family *Saccharomycetaceae*) can contaminate the finished wine product and give it undesired off-flavours. It is a well known fact that the best recognised wine brands often obtain their distinctive trademark flavour and smell by being stored in oak barrels as part of the aging process. Cantacuzene et al. (2003) found ozone to be very effective at removing *Brettanomyces* from wine barrels. Moreover, due to the ability of ozone-containing water to disinfect oak barrels, thereby removing the threat of *Brettanomyces*, this technology is now being used on an industrial scale in Australia as an alternative to chlorine. Unlike chlorine, ozone will not leach into the oak barrels, and it will revert quickly back to its more stable oxygen form shortly after treatment. Another important advantage of using ozone disinfection is that it prevents the occurrence of substances such as trichloroanisole (TCA), which is responsible for cork taint problems in many wines. Ozone is also considered to provide cost savings as it reduces the need to buy and store chlorine.

In order to enhance wine production, a recent study conducted by Guillen et al. (2010) tested the efficacy of ozone in a CIP system in the wine industry. In the study, cleaning and sanitising treatments were performed in a hose used to transport wine and the effects of ozone were compared to other solutions that are currently used in wineries. The authors found that the use of a CIP system with ozone-containing water was more efficient for microbial load reduction than the use of peracetic acid alone or the combined use of soda and peracetic acid. Sanitation with ozone-containing water also tended to be much faster because contact time was reduced and washing was not necessary. The authors also noted that, for health and safety reasons, a CIP system with ozone must be carefully implemented, with particular attention being paid to recommended dosages and limits of human exposure.

10.5.3 Brewing industry

Beer spoilage has been a long-standing problem for the brewing industry and its retailers. Issues may be caused by wild yeasts, such as *Candida*, *Brettanomyces* and *Zygosaccharomyces*, which cause the beer to become turbid or ropey, or to develop a yeasty aroma. Moreover, breweries use high volumes of fresh water. For every 1 barrel of beer produced, 10 barrels of water are used. Consequently, water treatment is critical. Many of the most important problems brewery plant managers face are system inefficiencies that can cause process problems, downtime and poor product quality. Treatment of the boilers, cooling towers, water preparation, wastewater and the most important phase of the brewing operation, the pasteurisation process, is necessary. Chemicals prevent microbial growth, scale, corrosion and 'dome staining' or rust rings from forming on beer cans as they pass through the pasteurising tunnel. Ozone-containing water can obviously reduce the amount of water that is required to sanitise processing lines, given that no residual byproducts will occur during its application.

Porter (2002) reported the effectiveness of applying UV light followed by ozonation to rinsing water to be used in a brewery; the ozonated water is used for bottle rinsing and at various CIP locations throughout the plant, such as final rinsing of stainless steel fermentation tanks. A more recent investigation has looked into beer line sanitisation at retail outlets such as pubs, bars and hotels, as contamination of beer lines occurs as beer

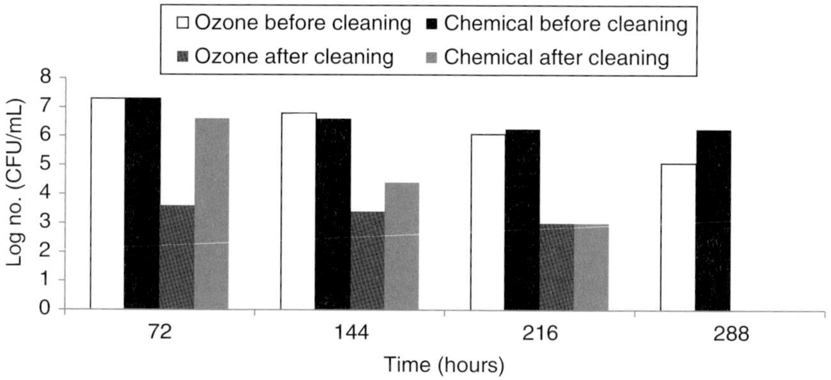

Figure 10.2 Effect of repeated cleaning cycles with ozone-containing water or a chemical beer line cleaner on the aerobic and anaerobic colony counts of tubing.

contaminated with low levels of bacteria and yeast passes over the tubing surfaces as a drink is dispensed (Fielding et al. 2007). The authors evaluated the use of ozone-containing water as an alternative to a traditional, chemical beer line cleaning agent against the consortium of microorganisms in a model beer line. From their results, they noted that even though a reduction in the number of both aerobic and anaerobic organisms was found with both the ozone-containing water and the chemical beer line cleaner, the log reductions observed were greater for the system that had been cleaned with the ozone-containing water than for that which had been cleaned with the chemical beer line cleaner during the initial cleans only; that is, only when the system was at its most contaminated (Figure 10.2).

References

Bailey, R.A., Young, A., Fielding, L. and Griffiths, C. (2001) Use of a bioaerosol test chamber to assess the antibacterial properties of ozone on *micrococcus luteus*. Poster at EUROCAFT 2001.

Baumann, A.R., Martin, S.E. and Feng, H. (2009) Removal of Listeria *monocytogenes* biofilms from stainless steel by use of ultrasound and ozone, *Journal of Food Protection*, **72**: 1306–9.

Beltran, F.J. (1995) Theoretical aspects of the kinetics of competitive reactions of ozone in water, *Ozone Science and Engineering*, **17**: 163–81.

Berk, Z. (2009) *Food Process Engineering and Technology*, USA: Academic Press, Elsevier.

Boisrobert C. (2002) *Ozone III: Agricultural & Food Processing Applications of Ozone as an Antimicrobial Agent*, US regulatory review of ozone use in the food industry.

Bott, T.R. and Tianqing, L. (2004) Ultrasound enhancement of biocide efficiency, *Ultrasonic Sonochemistry*, **11**: 323–6.

Broadwater, W.T., Hoehn, R.C. and King, P.H. (1973) Sensitivity of three selected bacterial species to ozone, *Applied Microbiology*, **26**: 391–3.

Cantacuzene, N.O., Dormedy, E.S., Smilanick, J.L., Fugelsang, K.C., Wample, R.L., Bacon, D.J. et al. (2003) Treating *Brettanomyces* in oak cubes with gaseous and aqueous ozone, *ASEV 54th annual meeting*, Reno, NV.

Dosti, B., Güzel-Seydim, Z.B., Wyffels, J.T. and Greene, A.K. (2005) Effectiveness of ozone, heat and chlorine for destroying common food spoilage bacteria in synthetic media and biofilms, *International Journal of Dairy Technology*, **58**(1), 19–24.

Fielding, L.M., Hall, A. and Peters, A.C. (2007) An evaluation of ozonated water as an alternative to chemical cleaning and sanitisation of beer lines, *Journal of Foodservice*, **18**: 59–68.

Greene, A.K., Few, B.K., Joao, C. and Serafini, J.C. (1993) A comparison of ozonation and chlorination for the disinfection of stainless steel surfaces, *Journal of Dairy Science*, **76**: 3617–20.

Greene, A.K., Vergano, P.J., Few, B.K. and Serafini, J.C. (1994) Effects of ozonated water sanitization on gasket materials used in fluid food processing, *Journal of Food Engineering*, **21**: 439–46.

Greene, A.K., Smith, G.W. and Knight, C.S. (1999) Ozone in dairy chilling water systems: effect on metal materials, *International Journal of Dairy Technology*, **52**(4): 126–8.

Guillen, A., Kechinski, C. and Manfroi, V. (2010) The use of ozone in a CIP system in the wine industry, *Ozone: Science & Engineering: The Journal of the International Ozone Association*, **32**(5): 355–60.

Güzel-Seydim, Z.B., Wyffels, J.T., Greene, A.K. and Bodine, A.B. (2000) Removal of dairy soil from heated stainless steel surfaces: use of ozonated water as a prerinse, *Journal of Dairy Science*, **83**(8): 1887–91.

Hampson, B.C. (2000) Use of ozone for winery and environmental sanitation, *Practical Winery & Vineyard*, Jan.–Feb.: 27–30.

Ingram, M. and Barnes, E.M. (1954) Sterilization by means of ozone, *Journal of Applied Microbiology*, **17**: 246–71.

Jacquet, C., Rocourt, J. and Renaud, A. (1993) Study of *Listeria monocytogenes* contamination in a dairy plant and characterization of the strains isolated, *International Journal of Food Microbiology*, **20**: 13–22.

Khadre, M.A., Yousef, A.E. and Kim, J.G. (2001) Microbiological aspects of ozone applications in food: a review, *Journal of Food Science*, **66**: 1242–52.

Kowalski, W.J., Bahnfleth, W.P. and Whittam, T.S. (1998) Bactericidal effects of high airborne ozone concentrations on *Escherichia coli* and *Staphylococcus aureus*, *Ozone-Science & Engineering*, **20**(3): 205–21.

Lagrange, F., Reiprich, W. and Hoffmann, M. (2004) CIP-cleaning and disinfection with ozone water, *Fleischwirtschaft*, **84**(2): 112–14.

Marriott, M.J. (1994) Self-cleansing sewer gradient, *Journal of the Institution of Water and Environmental Management*, **8**(4): 360–1.

Marriott, N.G. and Gravani, R.B. (2006) *Principles of Food Sanitation*, UK: Springer.

Moore, G., Griffith, C. and Peters, A. (2000) Bactericidal properties of ozone and its potential application as a terminal disinfectant, *Journal of Food Protection*, **63**(8): 1100–6.

Pan, Y., Breidt, F. and Kathariou, S. (2006) Resistance of *Listeria monocytogenes* biofilms to sanitizing agents in a simulated food processing environment, *Applied and Environmental Microbiology*, **72**(12): 7711–17.

Pascual, A., Llorca, I. and Canut, A. (2007) Use of ozone in food industries for reducing the environmental impact of cleaning and disinfection activities, *Trends in Food Science and Technology*, **18**: S29–S35.

Porter, S. (2002) Case study: ozone system to provide sterile rinsewater, *Ozone III Conference*, Fresno Proceedings, CA, October 28–30.

Serra, R., Abrunhosa, L., Kozakiewicz, Z., Venancio, A. and Lima, N. (2003) Use of ozone to reduce molds in a cheese ripening room, *Journal of Food Protection*, **66**: 2355–8.

Stanga, M. (2010) Laboratory tests, in Stanga, M. (ed) *Sanitation: Cleaning and Disinfection in the Food Industry*, Weinheim, Germany: Wiley-VCH Verlag GmbH & Co. KGaA.

Taylor, J. and Chana, D. (2000) *The Evaluation of Ozone for Airborne and Surface Disinfection*, Chipping Campden, UK: Campden and Chorleywood Food Research Association, R&D Report No. 109.

Troller, A.J. (1993) *Sanitation in Food Processing*, USA: Academy Press, Elsevier.

11 Ozone for Water Treatment and its Potential for Process Water Reuse in the Food Industry

Tomás Norton and Paula Misiewicz

Nomenclature

- *Ha*: Hatta number;
- k_1: chemical reaction rate (M^{-1}/s);
- k_L: mass transfer coefficient for the liquid phase (m/s);
- KL_a: liquid phase volumetric mass transfer coefficient, per second;
- v: velocity;
- α: volumetric fraction;
- ρ: density (m^3/s).

11.1 Introduction

For many years ozone, which has the highest thermodynamic oxidation potential of all the common oxidants, has been known for its powerful oxidising and disinfection properties (Glaze et al. 1987). Because its oxidative properties can be used as reaction partners in gas, liquid or solid phases, the applications of ozone are manifold and stretch across a wide range of different industries. From the decontamination of oil-polluted soil to the remediation of odorous chemicals in liquid animal manure, ozone can help tackle many industrial problems that were often deemed intractable in the past. In the food industry, ozone has already been used for microbial reduction during food preservation and packaging (Tiwari et al. 2009), as a sanitiser of fresh produce before packaging (Beltrán et al. 2005), as a fumigant for insect and fungal control in grain storage (Mendez et al. 2003; Tiwari et al. 2010), and as a sterilisation agent for food plant equipment (Güzel-Seydim et al. 2004a). Another powerful use of ozone, which has direct application in the food industry, is in the treatment of process water so that it can be reused or recycled during plant operations.

Water is a key ingredient in the food industry, playing a fundamental role in many of the common food processing methods and unit operations, such

Ozone in Food Processing, First Edition. Edited by Colm O'Donnell, B.K. Tiwari, P.J. Cullen, and Rip G. Rice.

as soaking, washing, rinsing, blanching, heating, pasteurising, chilling, cooling, steam production, as an ingredient and for general cleaning, sanitation and disinfection purposes (Casani et al. 2005; Poretti 1990). Consequently, the food and beverage industry is well known for its high consumption of water, with nearly 157 million m^3 of water being used annually in the UK (Envirowise Web site, www.envirowise.gov.uk). However, the industry is not so well known for its use of water-saving devices and practices (Casani et al. 2005). For example, even though the effluent water from some food processing operations may not be heavily contaminated, for example condensates from product drying or evaporation processes, or even cooling or rinsing water, this water is generally released without treatment into the sewage system (Fähnrich et al. 1998). This practice is fuelling a growing concern over the continued cost and availability of high-quality fresh water into the future, coupled with concerns over the need to minimise environmental pollution with such wastewaters (Casani et al. 2006). Therefore, many food processors are now seeking to use different technologies to raise the quality of process water so that process water can be effectively reused within the plant, thereby improving their overall process efficiencies and ensuring greater profitability from their end products (Envirowise 2010a).

For the last several years researchers have investigated suitable treatment processes for reducing TOC (total organic carbon), BOD (biochemical oxygen demand), COD (chemical oxygen demand) and microbial levels in process water down to accepted drinking standards. Treatment processes based on nanofiltration (NF), membrane bioreactors, reverse osmosis (RO), adsorption and ultraviolet (UV) oxidation have been analysed under both laboratory and pilot plant conditions (Fähnrich et al. 1998; Noronha et al. 2002), with no technology achieving a suitable drinking water quality without being part of, at the very minimum, a double-staged process. In recent years, it has been shown that an efficient degradation of contaminants can be achieved through direct ozonation and via secondary oxidants such as OH molecules, for both wastewater (Huber et al. 2003, Ternes et al. 2003) and drinking water treatment (Ternes et al. 2003; Vieno et al. 2007). As a result, ozone has been a popular treatment process with the UK water utilities. A recent survey of six of the UK utility companies showed that 31 water treatment works use ozone, ranging from a few MLD (million litres per day) through to one 360 MLD plant, and primarily for the destruction of refractory organic compounds in the water (Parsons and Jerrersen 2005). While ozone, which can be used in combination with other technologies and reagents, has been successfully employed globally for water treatment, the literature shows that it has not been largely adopted as a water treatment technology by the food industry, even though it was approved for application in the reconditioning of recycled poultry chilling water by the US Department of Agriculture in 1997 (Güzel-Seydim et al. 2004b).

This chapter begins by looking at the importance of water treatment and reuse in various sectors of the food industry. It continues by discussing the fundamental aspects of ozone in water treatment by concentrating on its chemistry, kinetics and hydrodynamics in typical water treatment scenarios. The chapter serves as an introduction to how ozone can be used for water treatment in order to inspire increased application of ozone in the food industry into the future.

11.2 Water in the food industry

Water forms a significant natural component of foods, can be added as an ingredient for the production of cooked or baked foods, and can be also used as a heat and mass transfer medium during processing and cleaning. Therefore, water plays an important role in food processing, with the quality of the water used being dependent on many factors, going all the way back to the source of the water, how it has been treated and how it has been distributed up until the point at which it is consumed. It is therefore a prerequisite that the contaminants in water that are harmful or unpleasant to humans are removed by a series of treatment steps in order to produce safe potable water which meets food industry requirements.

Most of the water used by the UK food industry comes directly from the mains supply (70% dependence on mains water in UK (Envirowise Web site, www.envirowise.gov.uk)) and is therefore treated offsite by large water companies. While this means that the water should be of high quality, the mains supply is becoming an increasingly expensive resource, with charges rising year on year. The average cost of mains water for industrial uses increased by 18% between 1997 and 1999 in the UK (Envirowise Web site, www.envirowise.gov.uk). Moreover, trade effluent charges have increased, on average, by 16%, with the trend likely to continue as water companies seek to recoup the large investment required to upgrade sewage treatment works to meet higher environmental standards. Reducing water consumption can be an inexpensive way of achieving cost savings, with many companies saving up to 30% of their water costs by implementing simple and inexpensive water minimisation measures (Envirowise 2010b), which are often based on simple mass balances, considering where water enters and leaves the factory. However, given that some process operations in the food industry require large amounts of water to operate successfully, water minimisation measures may have little overall effect on water consumption, unless they are used in conjunction with water treatment technologies to allow water to be reused in the plant operations.

Reuse of process water has been recommended for many years by the European Union, but the older regulations, which only permitted potable water for use in food industries (EC 1980) were often too restrictive to allow investment in reuse/recycling techniques. Flexibility was introduced when a new drinking water directive on the water quality intended for human consumption was introduced (EC 1998), which provided justification for

Table 11.1 Summary of treatment processes. (Reprinted from *Encyclopedia of Food Sciences and Nutrition*, J. Manning, 6105–6111, 2003, with permission from Elsevier.)

Example	Uses
Disinfection	
Chlorination, ozonation, chlorine dioxide, UV irradiation	Oxidants used to kill bacteria and viruses
Coagulation	
Aluminum sulfate, ferric chloride or polyaluminium chloride, ozonation	Combines small, suspended and colloidal particles into larger, flocculated particles by chemical attraction. Algae, iron, manganese, colour and turbidity can be entrained in floc particles, which are then removed by solid/liquid separation processes
Solid/liquid separation	
Microstraining	Large (20–100 mm) particles such as algae and leaves removed by straining
Dissolved air flotation	Removal of coagulant sludges by floating on a curtain of small air bubbles
Sedimentation	Removal of coagulant sludges by settlement. Suited to sludge with high inorganic content
Slow sand filtration	Removal of colloidal, organic and suspended material by straining and biological activity on the surface of sand filter beds
Rapid gravity filtration	Removal of suspended material by straining and entrapment within the depth of sand filter beds
Membrane micro- and ultrafiltration (size exclusion)	Removal of submicron-sized particles using membranes manufactured from polymeric material
Membrane nanofiltration and reverse osmosis (diffusion control)	Removal of submicron particles and dissolved inorganic species of high molecular weight
Membrane ion exchange or electrodialysis (charge-controlled)	Removal of salts; that is, desalination of sea water or nitrate removal
Adsorption	
Granular-activated carbon filtration (fixed bed) or powdered activated carbon (requires filtration process)	Adsorption of pesticides and volatile organic compounds, removal of taste and odours
Aeration	
Packed towers or cascades	Removal of H_2S, volatile organic compounds or carbon dioxide for corrosivity control. Oxidation of iron and manganese
Chemical addition	
Acid and alkali for pH adjustment	Mineral acid and/or sodium hydroxide, sodium carbonate or lime for optimum coagulation
Phosphoric acid	Provides phosphate coating on lead pipes to control lead concentrations in water
Softening	
Lime addition	Removal of calcium carbonate by precipitation at high pH. Reduces scale formation in downstream structures

dairy wastewater is distinguished by a high BOD and high levels of dissolved or suspended solids including fats and oils, nutrients such as ammonia or minerals, and phosphate (Guillen-Jimenez et al. 2000). Dairy wastewater also gives off bad odours as the components quickly decompose and rancidify in storage. Among the food industries, the dairy industry is one of the most polluting. However, in contrast to fresh produce processing, mains water in the dairy industry only comes into direct contact with waste milk and milk-derived products before disposal. Therefore, the need to disinfect wastewater before reuse is not generally a high priority.

Over recent years, investigators have examined different technologies for the recycling or reuse of at least a reasonable quantity of the wastewater produced in the plant (Balannec et al. 2002; Güzel-Seydim et al. 2000). The main focus of the current research is to use various types of filtration technologies, such as RO, NF and ultrafiltration (UF), all of which have the ability to remove enough organic matter for the process water to be reused, to help in solving the problem of attaining a quality of water that can be recycled back to the processing plant (Vourch et al. 2008). For example, Chmiel et al. (2000) found that NF and RO of low-contaminated vapour condensate from milk processing can produce reusable water. For process wastewater it was shown that RO operation gave better water quality than NF, with an RO + RO treatment recycling water to drinking water standards in the dairy plant (Vourch et al. 2008). However, the proteinous materials of the dairy process water were found to be severe foulants for the existing membrane materials (Sarkar et al. 2006). In fact, fouling is one of the main problems in any membrane separation, but for RO and NF it might be somewhat more complex because the interactions leading to fouling take place at extremely small scales, and therefore are difficult to understand (van der Bruggen et al. 2008).

Biofouling occurs due to biologically active organisms, such as bacteria and fungi, fouling membranes during filtration (Flemming et al. 1997). Biofouling then results in the growth of biological species on the membrane surface, and in the case of spiral wound membranes, on the feed spacer as well. During filtration of dairy wastewater, this fouling reduces productivity of the membrane and increases the feed concentrate pressure drop, while the surfactants may change the filterability by concentration polarisation or micelle formation. A solution to biofouling can be attained through pretreatment of the wastewater via such techniques as UF and microfiltration, ozonation or UV/H_2O_2 oxidation, adsorption (PAC) and flocculation (Tanninen et al. 2005). One of the most promising applications of ozonation is in reducing organic pollutant levels in the water stream via its coagulation–flocculation effects, and it is widely accepted that a preozonation process can improve TOC, COD or turbidity removal, with improvements including extending filter run length due to lower pressure head-loss build-up (Kowalska et al. 2006; Camel and Bermond 1998).

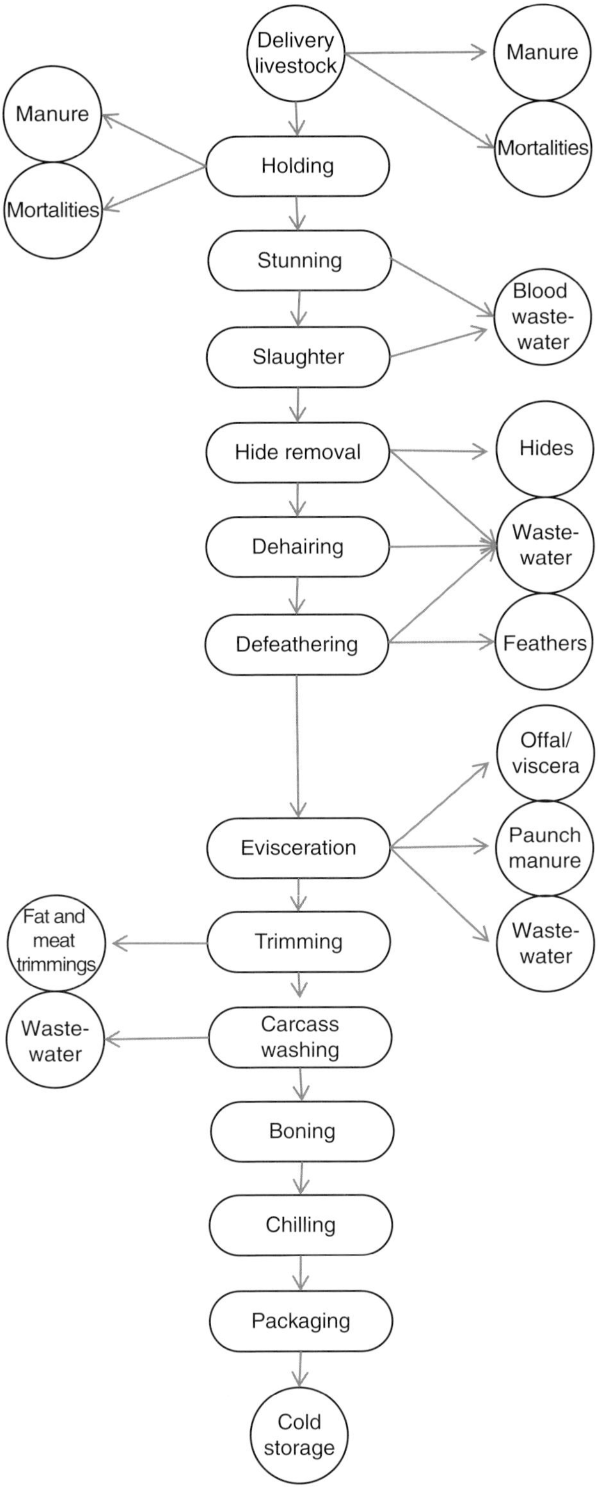

Figure 11.1 The main processes involved in meat processing operations (Arvanitoyannis and Ladas 2008).

11.2.3 Meat and poultry processing

Both meat and poultry abattoirs generate significant amounts of wastewater, not only because of the many processing steps involved (Figure 11.1) but also due to the washing of equipment and facilities. As can be seen in Figure 11.1, such wastewaters are characterised by high organic rates and suspended solids concentrations, with the quantities of each being very much dependent on the technicalities of the process and the amount of water consumed at each stage (Del Nery et al. 2001). In response to worldwide recommendations on the management of water and effluents from the food industry, some studies have highlighted the importance of reducing water consumption and hydraulic and organic loads in abattoirs by employing various techniques, such as training of personnel, optimisation of the blood collection system, dry collection of slaughter byproducts and implementation of efficient methodologies for slaughterhouse cleaning and sanitation (Johns 1995; Sheldon et al. 2010). Pipe-to-pipe potable reuse of water in the meat industry is generally not favoured due to health and safety concerns, arising from the fact that such wastewater is rich in oils and greases, sanitisers and blood – substances that need to be removed by effective physicochemical and biological treatment (Bohdziewicz et al. 2002; Johns 1995; Sroka et al. 2004). The opportunity for water reuse only comes about when the water has a low-contamination load, and even when this occurs many stages are often required to remove the suspended solids and reduce the BOD, COD and microbial loads during the water treatment process (Fähnrich et al. 1997).

Water can also be used in combination with a sanitising agent in order to wash parts of carcasses which are traditionally known for high-density microbial loads. In this regard, a sanitiser, such as ozone-containing water, has to be extremely effective at killing environmental and faecal contaminants such as *Enterococcus faecalis* and *Escherichia coli*, and foodborne pathogens such as *Listeria monocytogenes*, *Bacillus cereus*, *Salmonella typhimurium*, *Yersinia enterocolitica* and *Staphylococcus aureus*, which are effectively killed in low-ozone-demand substrates (Khadre et al. 2001; Kim et al. 1999; Restaino et al. 1995).

11.3 Ozonation as a water treatment process

Ozone can be used in three functional ways: as a biocide, as a classic oxidant and as a pretreatment for improving subsequent treatment processes. As discussed above, ozonation can be used in water treatment to achieve a variety of treatment goals, including (Langlais et al. 1991):

(1) Disinfection.
(2) Oxidation of inorganic and organic pollutants:
  (a) iron and manganese;
  (b) taste and odour compounds;
  (c) phenolic pollutants;
  (d) pesticides.

Table 11.2 Summary of the advantages and limitations of different chemical treatment processes. (Reprinted from *Food Control*, Volume 17, Issue Number 7, Sandra Casani, Tommas Leth, Susanne Knøchel, Water reuse in a shrimp processing line: safety considerations using a HACCP approach, 540–550, 2006, with permission from Elsevier.)

Water treatment method	Limitations	Advantages
Chemical		
Chlorine	Reacts with organic matter Safety measures required due to the formation of chlorine gas, which may harm workers and equipment Potentially hazardous byproducts Sensory changes	Cheap and available Monitoring of free residuals is simple
Chlorine dioxide	Requires onsite generation Potentially hazardous byproducts	Easy to use and generate Low concentrations required,[a] resulting in less corrosion of the equipment Biocidal properties not influenced by pH
Chloramines	Poor biocidal effect Requires long contact times	Stable and lasting effect
Ozone	Short half-life Lack of stable residual, which limits online testing efficacy Toxic when inhaled	Great biocidal activity at short contact times Byproducts are similar to normal oxidation products GRAS for treatment of bottled water and approved as an antimicrobial agent for direct applications in food[b]
Hydrogen peroxide	Requires high concentrations and long contact times[c]	Simplicity of application and low costs Decomposes into water and oxygen (nontoxic)
Peracetic acid	Corrosive (equipment)	Effective at low concentrations and short contact times Low sensitivity to organic matter[d]
Physical		
Membrane processes	High investment costs Infectious potential of surviving organisms found in filtration concentrates	May be used for both purifying water and recovering byproducts[e] Removes particles, preventing support of microbial regrowth Absence of residual toxicity
UV	Pretreatment normally necessary, increasing costs Difficulties in accurately determining the UV dose	Absence of residual toxicity

[a]When compared to chlorine for treating poultry processing water, for example (Lillard 1978).
[b]US Food and Drug Administration (Khadre et al. 2001).
[c]For example, for poultry chiller water (Lillard and Thomson 1983).
[d]When compared to chlorine and hydrogen peroxide (Stampi et al. 2001).
[e]For example, proteins and fat from poultry retentate waters (Hart et al. 1988).

(3) Oxidation of organic macropollutants (i.e. nonspecific target organics):
 (a) bleaching of colour;
 (b) increasing the biodegradability of organics;
 (c) destruction of trihalomethane formation potential (THMFP), total organic halide formation potential (TOXFP) and chlorine demand.
(4) Improvement of coagulation.

Other advantages, along with the limitations, of using ozonation and other technologies during water treatment are listed in Table 11.2, where it is clearly outlined why ozone is one of the best water treatment techniques currently available. The wealth of applications of ozonation led to an explosion of US water treatment plants employing this technology throughout the 1990s. A further reason comes from the fact that ozonation can also provide cost benefits, as it can be used for multiple applications and therefore is introduced at various stages of the process. For example, when used to oxidise manganese/iron or for its coagulation properties, ozone will be applied at the head of the treatment plant. The use of ozone to improve granular activated carbon (GAC) filtration or taste and odour remediation requires ozonation to occur midway through the process. On the other hand, ozonation for disinfection can be used both at the head and the end of the treatment plant, where a concentration of 0.4 mg/L should be maintained for 5 minutes; for sporicidal activity, 2 mg/L is needed (EHEDG 2007). The reasons why ozone is a powerful water treatment technology are due to its physicochemical properties, which are outlined in Chapter 3.

11.4 The kinetics of ozonation

The kinetics of ozone reactions in water are dependent on both the physical and the chemical (physiochemical) attributes of the water treatment system. When ozone is introduced into water, it diffuses into the liquid phase, undergoing numerous reactions with substances during this diffusion time, which is also called heterogeneous parallel–series gas–liquid reaction processing. While it is necessary to know the typical reactions that occur, it is also desirable to understand kinetic parameters, such as reaction order, n, and reaction rate constant, k, as these help us to assess the feasibility of using ozonation to treat water and to design an appropriate reactor system (Gottschalk et al. 2010). As noted by Beltran (2004), kinetic laws are empirical and the associated rate constants must be determined from experiments. The complete kinetic equations can then be formed from mass balances of reacting species, which depend on the type of flow (both gas and water) through the reactor.

An important index which is used to quantify the rate at which reactions progress during ozonation is called the Hatta number, which can be represented as follows:

$$Ha = \frac{\sqrt{k_1 D_A}}{k_L} \quad (11.1)$$

Fast kinetic regimes occur when the concentration of substances reacting with ozone in the liquid phase is high enough to maintain the Hatta number greater than 0.3; that is, when there is a fast reaction rate with respect to the mass transfer of ozone from gaseous to liquid phase. This generally limits the occurrence of fast kinetic regimes to the ozonation of wastewater, which has high concentrations of fast ozone-reacting substances (phenols, dyes, etc.) alongside the presence of catalysts (e.g. salts). This means that the ozonation rate is controlled by mass transfer until a stage at which the presence of ozone-inhibiting compounds becomes restrictive, mainly due to pollutant formation as a result of indirect reactions. On the other hand, when lower concentrations of water pollutants are found (e.g. in ppb), a low Hatta number is achieved and therefore reactions are slower. If the Hatta number is much less than 0.03, the liquid can become saturated with ozone and the overall rate of ozone consumption is controlled by the rate of reaction of ozone in the liquid bulk. As the Hatta number approaches 0.3, the liquid will contain dissolved ozone, but at a lower concentration than the saturation level (Yan Lan et al. 2008).

11.4.1 The kinetics of mass transfer

Before ozone can react with water or any of the substances in the liquid phase, whether the liquid is water or an organic solvent, the ozone must transfer through the interface between the two phases and reach equilibrium. Therefore, the rate at which this can occur has a large overall effect on the system performance. The overall mechanism of the gas–liquid reaction system can be considered as consisting of several steps (Kuo and Yocum 1982):

- diffusion of ozone through the gas phase into an interface between the gas and liquid phases;
- transport across the interface between the gas and liquid phases;
- transport across the interface to the liquid phase boundary, and then into the bulk liquid.

Of course, the amount of dissolved ozone may be depleted during any of these steps because the ozone is decomposing and because it is reacting with compounds in the liquid. Therefore, in order to understand the phenomena of ozonation, it is first necessary to quantify the fundamental mass transfer principles. The diffusion of two immiscible fluids has been a subject of study for many years, and for this reason a number of theories have emerged, each with their own specific limitations (Beltran 2004).

Film theory

This was the first theory used to predict intertransport of components between gaseous and liquid phases. Lewis and Whitman (1924) postulated that when two immiscible fluids come into contact with each other, a thin layer of stagnant fluid is formed at each side of the phase boundary. Because the boundary layer is very thin, no accumulation of mass occurs and therefore a steady-state equilibrium in concentration is achieved on both sides of the phase boundary.

It is then possible to simply quantify the absorption rate of ozone by assuming that one component of ozone (gas phase) is transferred to the water (liquid phase) as follows:

$$N_A = k_g\left(P_{AB} - P_i\right) = k_L\left(C_A^* - C_{Ab}\right) \tag{11.2}$$

where N_A is in mol/m²/s; k_G and k_L are the individual mass transfer coefficients for the gas and liquid phases, respectively; P_{Ab} and P_i are the partial pressures of A in the bulk gas and at the interface, respectively; and C_A and C_{Ab} are the concentrations of A at the interface and in the bulk of the liquid, respectively. While this rate law is simple, it is only accurate if a mathematical expression is available to predict the mass transfer coefficients and if the interfacial concentrations can be determined.

Using the film theory, the expression for k_L can be derived as:

$$k_L = \frac{D_A}{L} \tag{11.3}$$

where D_A is the molecular diffusivity of component A (mol/m²/s) and L is the thickness of the film (m).

Surface renewal theory

In the surface renewal theory of Danckwerts (1968), the liquid phase is considered to be completely disturbed, containing numerous infinitesimally small eddies (elements) which are exposed to the interface for a certain period of time, and then are recycled by other elements from the bulk liquid. While the distribution of eddies at the interphase boundary may vary, the fresh phase (in microscopic amounts) is brought to the boundary. By taking the boundary eddy replacement rate to be s, Danckwerts (1968) showed that the distribution of surface age can by calculated by:

$$\Phi(t) = s\exp(-st) \tag{11.4}$$

By using this theory in unsteady molecular diffusion for transport via turbulent eddies, the following expression for the liquid-side mass coefficient can be developed:

$$k_L = \sqrt{D_A s} \tag{11.5}$$

11.4.2 Determining the chemical reaction kinetics

By feeding an ozone-containing gas into the substrate aqueous solution, the transfer process becomes heterogeneous, as it is dependent on the hydrodynamics as well as the influence of local concentrations of ozone/

substrate on mass transfer and the associated ozone decomposition and substrate decay rates. Therefore, such systems can be used to develop kinetic equations that express the amount of ozone absorbed per unit time and volume in order to quantify the relative importance of chemical and physical processes on the efficiency of the overall treatment system. The reactor design is an important feature in determining the kinetics of ozone and substrate decomposition. For the generalised reactor, where the concentration of all species is assumed to be uniformly distributed in space, the following equation holds (Beltran 2004):

$$F_{i0} - F_i + G_i\,\beta\Delta V = \frac{\Delta n_i}{\Delta t} \tag{11.6}$$

where F_{i0} and F_i represent the molar rates of species i, at the entrance and exit of the reaction volume, ΔV, respectively; G_i is the generation rate, which represents the i mole rate per unit of volume that is formed or removed; $\Delta n_i/\Delta t$ is the accumulation rate of i in that volume; and β is the liquid holdup or liquid fraction of the reaction volume. Since water ozonation systems do not involve variations in temperature, ozonation can be considered an isothermic system, and an energy balance is not required.

A batch reactor is often operated by feeding in two different solutions, one containing ozone at a known concentration and the other containing the substrate, and bringing them together under controlled mixing and temperature conditions. In this way it is different from conventional ozonation, in which the ozone is bubbled continuously through an aqueous substrate solution. This is known as the homogenous approach to determining rate constants, and its main advantage is that it does not have the problem of mass transfer, with rate constants being obtained straightforwardly from experimental data of concentration with time. In these systems, the rate of agitation is set such that the time for complete mixing of the solutions is shorter than the characteristic time of the ozone reactions, which is sometimes difficult to achieve in practice. Similar to the batch reactors, semibatch systems (Figure 11.2) contain a fixed volume of water solution. However, the ozone is introduced continuously via a feed of ozone-containing air or oxygen and it is therefore a heterogeneous approach to determining the rate constants of the kinetic equations. As well as providing estimations for reaction rate constants, batch and semibatch reactors are generally used to assist in limiting the range of parameters to be evaluated at pilot scale (Langlais et al. 1991).

Once the rate constants have been experimentally determined, the ozonation of a heterogeneous batch or semibatch reactor can usually be represented by the following model, which represents the decomposition of ozone in the gas and liquid phases as well as the decay of nonvolatile components:

$$\frac{1}{\tau_i}(C_{io} - C_i) + G_i = \frac{dC_i}{dt} \tag{11.7}$$

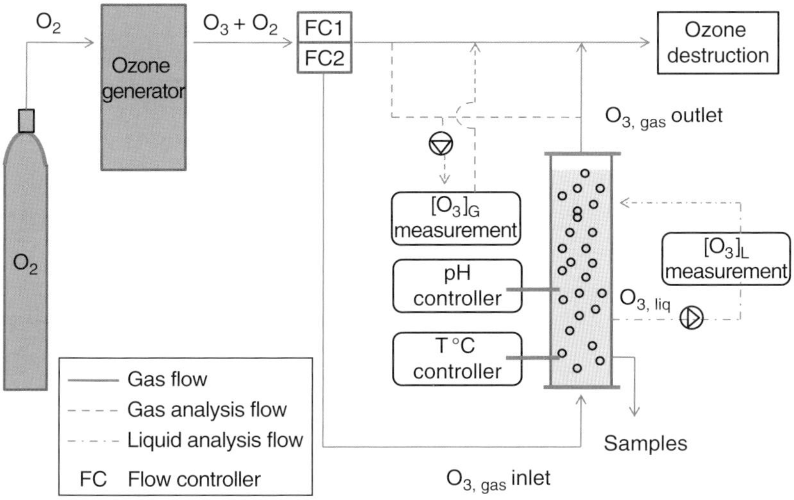

Figure 11.2 A semibatch reactor system. (Reprinted from *Chemosphere*, Volume 66, Issue Number 11, A. López-López, J.S. Pic, H. Debellefontaine, Ozonation of azo dye in a semi-batch reactor: a determination of the molecular and radical contributions, 2120–2126, 2007, with permission from Elsevier.)

where the subscript i can represent liquid (L), gas (g) or substrate (s). When ozone is in the gas phase, $G_g = -h_{mg}\, a(C_{O3} - C^{*}_{O3})\dfrac{\beta}{1-\beta}$; when ozone is in liquid phase, $G_L = KL_a\, a(C_{O3} - C^{*}_{O3}) + k_{O3}$; and for the nonvolatile components, $G_g = k_s$. Therefore, a system of three equations is obtained and this system can be solved using standard numerical solvers, such as Runge–Kutta or Euler solvers.

11.4.3 Hydrodynamics

One of the main issues in industrial ozone contactors, which are basically two-phase flow reactors, is to accurately determine hydrodynamics and ozone mass transfer. The design of commercial systems can often lead to back-mixing and short-circuiting in the flow, which inevitably reduces the efficiency of the ozonation process. Therefore, in order to understand the various phenomena and better design commercial ozone contactors, two-phase flow models can be developed based on the continuity and momentum balance of two phases in the flow regime.

The multiphase Eulerian–Eulerian approach is regularly used to model hydrodynamic systems containing gas and fluid phases, from fluidised beds to spraying. With the Eulerian– Eulerian model a more complete representation of a two-phase system can be realised because the transport of ozone from gaseous to liquid phase, and the consequential effects this has on the flow field, can be directly modelled. The most important thing to note about the Eulerian–Eulerian approach is that a single pressure is shared by all phases, but momentum and continuity equations are solved for each phase. Both phases are also considered to be interpenetrating continua, incorporating phasic volume fractions (α_A). The phases considered in the following will be denoted by phase *A* and phase *B*.

The continuity equation of phase *A*:

$$\frac{\partial}{\partial t}(\alpha_A \rho_A) + \nabla \cdot (\alpha_A \rho_A \vec{v}_A) = \sum_{B=1}^{n} (\dot{m}_{BA} - \dot{m}_{AB}) + S_B \quad (11.8)$$

where $\vec{v}_A$ is the velocity of phase *A*; $\dot{m}_{AB}$ denotes the mass transfer from phase *A* to phase *B*; similar holds for m_{BA}. The source term S_A on the right-hand side of Equation 11.8 is zero, but, if necessary, it is possible to specify a mass source for each phase. This source term appears in the momentum and enthalpy equations. The momentum balance for phase *A* is given by:

$$\frac{\partial}{\partial t}(\alpha_A \rho_A \vec{v}_A) + \nabla \cdot (\alpha_A \rho_A \vec{v}_A \vec{v}_A) = -\alpha_A \nabla p + \nabla \cdot \bar{\bar{\tau}}_B$$
$$+ \alpha_A \rho_A \vec{g} + \sum_{A=1}^{n} (\vec{R}_{AB} + \dot{m}_{AB}\, \vec{v}_{BA} - \dot{m}_{AB} \vec{v}_{BA}) + (\vec{F}_A + \vec{F}_{lift,A} + \vec{F}_{vm,A}) \quad (11.9)$$

where $\bar{\bar{\tau}}$ is the stress–strain tensor of phase *B*:

$$\bar{\bar{\tau}}_B = \alpha_B \mu_B (\nabla \vec{v}_B + \nabla \vec{v}_B^{\,T}) + \alpha_B \left(\lambda_B - \frac{2}{3} \mu_B \right) \nabla \cdot \vec{v}_B \bar{\bar{I}} \quad (11.10)$$

where μ_A and λ_A are the shear and bulk viscosity of phase *A*, $\vec{F}_A$ is the external body force, $\vec{F}_{lift,A}$ is the lift force, $\vec{F}_{vm,A}$ is the virtual mass force, $\vec{R}_{BA}$ is the interaction force between phases and *P* is the pressure shared by all phases.

The interphase velocity, $\vec{v}_{AB}$, is defined as follows: if $\dot{m}_{BA}>0$ (i.e. mass is being transferred from phase *A* to phase *B*), $\vec{v}_{BA}=\vec{v}_B$; if $\dot{m}_{BA}<0$ (i.e. phase *B* mass is being transferred to phase *A*), $\vec{v}_{BA}=\vec{v}_A$; and if $\dot{m}_{BA}>0$ then $\vec{v}_{BA}=\vec{v}_A$.

Equation 11.10 needs to be closed using an appropriate expression for the interphase force $\vec{R}_{AB}$. This force depends on the friction, pressure and cohesion. It is subject to the conditions that $\vec{R}_{AB}=\vec{R}_{BA}$ and $\vec{R}_{AB}=0$. Often, a simple interaction terms of the following form is used to describe the interphase forces:

$$\sum_{A=1}^{n} \vec{R}_{AB} = \sum_{A=1}^{n} K_{AB} (\vec{v}_A - \vec{v}_B) \quad (11.11)$$

where K_{AB} (= K_{BA}) is the interphase momentum exchange coefficient. This equation is in effect similar to Equation 11.1 except that it considers momentum rather than concentration and shows the importance of considering the multiphase effects on hydrodynamics. For bubbly gas–liquid mixtures, the Schiller Neumann model is often used. The interfacial momentum transfer model considers the spherical shape and uniform size of the dispersed gas phase. It accounts for drag, lift and added mass forces, and the local drift velocity takes into account the dispersion of the gas phase transported by the liquid phase.

The ozone must become dissolved into the liquid phase before it can react with substances dissolved in the water. Therefore, there are three issues that must be considered when modelling the ozonation process: the interfacial mass transfer of ozone from gas to liquid phase; the decay of the dissolved ozone; and the microbial inactivation. These considerations can be represented by the following species transport equation:

$$\frac{\partial}{\partial t}(\alpha_A \rho_A \phi_A) + \nabla \cdot (\alpha_A \rho_A \vec{v}_A \, \phi_A) - \nabla\left(\alpha_A\left(\rho_A D_A^{\phi} + \frac{\mu_{t\alpha}}{Sc_{t\alpha}}\right)\nabla \phi_A\right) = S_{\alpha}^{\Phi} \quad (11.12)$$

where $\vec{v}_A$ is the flow velocity, ϕ_A is the conserved quantity per unit mass of phase A, D_A^{ϕ} is the kinematic diffusivity for the scalar in phase α_A, $Sc_{t\alpha}$ is the turbulent Schmidt number and S_{α}^{Φ} is the external volumetric source term, with units of conserved quantity per unit volume per unit time.

Transport equations for all the important chemical or biological components of an ozone disinfection system must be included in a full hydrodynamic model, which can then be solved using commercial or in-house CFD (computational fluid dynamics) programs. A full model will include the fast-reacting natural organic matter, ozone concentrations in gas/liquid phases (*Cg* and *Cl*), the ozone exposure (*Ct*) and pathogen concentrations (*N*). For more details on the physical process and sources terms associated with each species, the reader is referred to Yang et al. (2007).

11.4.4 Applications of hydrodynamic modelling to investigate ozone water treatment

The most commonly used technique for estimating the removal or the inactivation of waterborne pathogens during ozone disinfection involves the well-known *Ct* concept, where *C* is defined as average ozone residual concentration at the outlet of a contactor and can be determined by measurements at the contactor sampling ports or by using US EPA mathematical functions, and *T* represents the residence time, which is the other important metric for characterising the effectiveness and efficacy of an ozone contactor. In common usage, the effective contact time is taken to be T_{10} rather than the mean hydraulic retention time, as this represents the earliest 10% of microorganisms to travel from the contactor inlet to the outlet. While the residence time is a function of contactor geometry and operating conditions, due to its oversimplicity, T_{10} is overconservative as it cannot take into consideration the governing hydrodynamic behaviour of the ozone system (Greene 2002). Therefore, more detailed modelling and design procedures, taking into account the flow regime as well as the distribution of chemical reactions, are necessary. Hydrodynamic modelling of ozonation processes via CFD is starting to become more prevalent in the literature, as this technique can consider phenomena such as interphase mass transfer, multiphase and reactions.

When CFD was first recognised as a tool, many researchers (Huang and Brouckaert 2002) tried to simplify the problem by developing monophase CFD models to simulate the ozone contactors and applying them in order to optimise the ozone contactor hydraulics. However, because these models did not consider the effects of the gas phase on contactor hydraulics, they were not capable of adequately describing the influence of gas flow rate, concentration and the injection locations on the contactor flow patterns. The most advanced single-phase model of ozonation was developed by Kamimura et al. (2002), who developed a CFD model of an O_3/UV reactor in order to determine the distribution of the concentration of TOC, ozone and other substances in the reactor. Using the model, the reactor configuration and operational conditions were optimised in order to reduce the ozone demand required for decomposing TOC. However, because it was only a single-phase model, it could not accurately account for the distribution of other phases.

One of the first multiphase modelling studies of an ozone contactor, utilising the hydrodynamics theory for multiphase flows, was carried out by Henry and Freeman (1995), who used a CFD model to track ozone bubble trajectories and their influence on the velocity and pressure distribution of the mean flow field in the contactor. Following validation, the effects of the gas : liquid ratio and the addition of vanes and corner fillets were modified in order to optimise the contactor design. A more recent study of contactor hydrodynamics was carried out by Ta and Hague (2004). However, the focus of their study was limited to contactor hydraulics and therefore they did not simulate ozone profile and pathogen inactivation performance. Zhang (2006) overcame the previous limitations by developing a 'CFD-based integrated disinfection design approach' for full-scale ozone disinfection processes. This modelling was done via a three-dimensional multiphase CFD model in which the authors addressed all the major factors associated with the ozone disinfection process, such as ozone mass transfer, ozone decay, microbial inactivation and disinfection byproduct formation kinetics. The CFD model was then utilised to investigate the contactor hydraulics under various operational conditions.

In order to discuss the power of cutting-edge multiphase modelling, like that developed by Zhang (2006), the typical flow field inside the contactor that drives the overall performance of the ozone water treatment process will be presented. As seen in Plate 11.1, the velocity field that was predicted found large recirculation zones in the two cells of the ozone contactor, which in combination caused short-circuiting and zones of stagnated fluid. The existence of these zones reduces the contacting opportunities between ozone and pathogens and therefore renders the disinfection significantly less efficient than is generally acknowledged.

Using a CFD-based technique to determine *CT*, Plate 11.1c shows how the flow field could significantly affect the ozone residual and CT distributions. For example, a portion of the water might experience short-circuiting phenomena and have a much smaller CT value than other portions exiting the contactor. By being able to predict and optimise such phenomena, it is possible to reduce the risks of poorly disinfected drinking water.

11.5 Conclusion

Water plays a fundamental role in many of the common processing methods and unit operations employed in the food industry. However, water recycling and reuse practices in the food industry generally take place on a small scale, being dependent on the type of water treatment process employed. Nevertheless, opportunities for increased recycling and reuse of water in a number of industries do currently exist and have been discussed in this chapter. The importance of water treatment and reuse in various sectors of the food industry have been highlighted. The suitability of ozone for water treatment was demonstrated by reviewing its powerful oxidising and disinfection properties. The fundamental aspects (chemistry, kinetics and hydrodynamics) of ozonation were discussed with regard to typical water treatment scenarios, in order to serve as an introduction to how ozone can be used for water treatment and to encourage increased process water reuse in the food industry into the future.

References

Arvanitoyannis, I.S. and Ladas, D. (2008) Meat waste treatment methods and potential uses, *International Journal of Food Science & Technology*, **43**(3): 543–59.

Balannec, B., Gesan-Guiziou, G., Chaufer, B., Rabiller-Baudry, M. and Daufin, G. (2002) Treatment of dairy process waters by membrane operations for water reuse and milk constituents concentration, *Desalination*, **147**: 89–94.

Beltrán, F.J. (2004) *Ozone Reaction Kinetics for Water and Wastewater System*, New York, NY: CRC Press CLL.

Beltran, D., Selma, M.V., Marin, A. and Gil, M.I. (2005) Ozonated water extends the shelf life of fresh-cut lettuce, *Journal of Agricultural and Food Chemistry*, **53**: 5654–63.

Bohdziewicz, J., Sroka, E. and Lobos, E. (2002) Application of the system which combines coagulation, activated sludge and reverse osmosis to the treatment of the wastewater produced by the meat industry, *Desalination*, **144**: 393–8.

Camel, V. and Bermond, A. (1998) The use of ozone and associated oxidation processes in drinking water treatment, *Water Research*, **32**: 3208–22.

Casani, S., Rouhany, M. and Knøchel, S. (2005) A discussion paper on challenges and limitations to water reuse and hygiene in the food industry, *Water Research*, **39**(6): 1134–46.

Casani, S., Leth, T. and Knøchel, S. (2006) Water reuse in a shrimp processing line: safety considerations using a HACCP approach, *Food Control*, **17**(7): 540–50.

CDC (2009) Centers for Disease Control and Prevention Investigation update: outbreak of *Salmonella typhimurium* infections, 2008–2009. Retrieved from http://www.cdc.gov/print.do?url=http%3A//www.cdc.gov/salmonella/typhimurium/. Last accessed 24 May 2009.

Chmiel, H., Mavrov, V. and Belieres., E. (2000) Reuse of vapour condensate from milk processing using nanofiltration, *Filtration & Separation*, **37**(33): 24–7.

Danckwerts, P.V. (1968). Gas absorption with instantaneous reaction, *Chemical Engineering Science*, **23**(9): 1045–51.

Delaquis, P.J. (2005) Fresh-cut vegetables, in Sapers, G.M., Gorny, J.R. and Yousef, A.E. (eds) *Microbiology of Fresh Fruits and Vegetables*, CRC Press.

Del Nery, V., Damianovic, M.H.Z. and Barros, F.G. (2001) The use of upflow anaerobic sludge blanket reactors in the treatment of poultry slaughterhouse wastewater, *Water Science and Technology*, **44**: 83–8.

EC (1980) Council Directive of 15 July 1980 relating to the quality of water intended for human consumption (80/778/EEC), *Official Journal of the European Communities*, No. L 229/11-29.

EC (1989) Council Directive of 21 December 1988 on the approximation of the laws of the Member States concerning food additives authorized for use in foodstuffs intended for human consumption, *Official Journal of the European Communities*, OJ L 40.

EHEDG (2007) Safe and hygienic water treatment in food factories, *Trends in Food Science & Technology*, **18**: 93–98.

Envirowise (2010a) Reducing water savings at Cadbury brings environmental benefits and cost savings, Case Study CS950. Retrieved from www.envirowise.gov.uk. Last accessed 24 May 2010.

Envirowise (2010b) Water minimisation in the food and drinks industry, Case Study GG349R. Retrieved from www.envirowise.gov.uk. Last accessed 24 May 2010.

EPRI (1999) Membrane filtration and ozonation of poultry chiller overflow water: study of membrane treatment to reduce water use and ozonation for sanitation at a poultry processing plant, Technical Report 114435, Palo Alto, CA: EPRI.

Fähnrich, A., Mavrov, V. and Chmiel, H. (1998) Membrane processes for water reuse in the food industry, *Desalination*, **119**(1–3): 213–16.

Fawell, J. (2000) Risk assessment case study: chloroform and related substances, *Food Chemistry Toxicology*, **38**: S91–S95.

Flemming, H.C., Schaule, G., Griebe, T. and Schmitt, J. (1997) Biofouling: the Achilles heel of membrane processes, *Desalination*, **113**: 215–25.

Gil, M., Selma, M.V., López-Gálvez, F. and Allende, A. (2009) Fresh-cut product sanitation and wash water disinfection: problems and solutions, *International Journal of Food Microbiology*, **134**: 37–45.

Glaze, W.H., Kang, J.-W. and Chapin, D.H. (1987) Chemistry of water treatment and processes involving ozone, hydrogen peroxide, and ultraviolet radiation, *Ozone: Science & Engineering*, **9**(4): 335–2.

Gorny, J.R. (2005) Microbial contamination of fresh fruits and vegetables, in G.M. Sapers, J.R. Gorny and A.E. Yousef (eds) *Microbiology of Fruits and Vegetables*, Boca Raton, FL: CRC Press/Taylor and Francis, pp. 3–32.

Gottschalk, C., Libra, J.A. and Saupe, A. (2010) *Ozonation of Water and Waste Water*, 2 edn, Weinheim, Germany: Wiley-VCH Verlag GmbH & Co. KGaA.

Greene, D.J. (2002) Numerical simulation of chlorine disinfection process in non-ideal reactors, PhD thesis, Drexel University, PA.

Guillen-Jimenez, E., Alvarez-Mateos, P., Romero-Guzman, F. and Pereda-Marin, J. (2000). Biomineralization of organic matter in dairy wastewater, as affected by pH: the evolution of ammonia and phosphates, *Water Research*, **37**: 1454–63.

Güzel-Seydim, Z.B., Wyffels, J.T., Greene, A.K. and Bodine, A.B. (2000) Removal of dairy soil from heated stainless steel surfaces: use of ozonated water as a prerinse, *Journal of Dairy Science*, **83**: 1887–91.

Güzel-Seydim, Z.-B., Greene, A.K. and Seydim, A.C. (2004a) Use of ozone in the food industry, *Lebensmittel-Wissenschaft und Technologie*, **37**(4): 453–60.

Güzel-Seydim, Z., Bever P.L. and Greene, A.K. (2004b) Efficacy of ozone to reduce bacterial populations in the presence of food components, *Food Microbiology*, **21**(4): 475–9.

Hampson, B.C. and Fiori, S.R. (1997) Applications of ozone in food processing operations, *Proceedings of 1997 IOA PAG Conf.*, Lake Tahoe, NV, pp. 261–7.

Hansen, C.L. and Cheong, D.-Y. (1997) Agricultural waste management in food processing, in *Handbook of Farm, Dairy, and Food Machinery*, UK: Elsevier.

Harris, L.J., Farber, J.N., Beuchat, L.R., Parish, M.E., Suslow, T.V., Garrett, E.H. and Busta, F.F. (2003) Outbreaks associated with fresh produce, incidence, growth and survival of pathogens in fresh and fresh-cut produce, *Comprehensive Reviews in Food Science and Food Safety*, **2**: 78–141.

Hart, M.S., Huxsoll, C.C., Tsai, L.-S. and Ng, K.C. (1988) Preliminary studies of microfiltration for food processing water reuse, *J Food Prot*, **51**(4): 269–76.

Henry, D.J. and Freeman, E.M. (1995) Finite element analysis and T_{10} optimization of ozone contactor, *Ozone: Science & Engineering*, **17**: 587–606.

Huang, T. and Brouckaert, C.J. (2002) A CFD and experimental study of an ozone contactor, *Water Science and Technology*, **46**(9): 87–93.

Huber, M.M., Canonica, S., Park, G.Y. and von Gunten, U. (2003) Oxidation of pharmaceuticals during ozonation and advanced oxidation processes, *Environmental Science & Technology*, **37**(5): 1016–24.

Huxley, R.R. Lean, M., Crozier, A., John, J.H. and Neil, H.A.W. (2004) Effect of dietary advice to increase fruit and vegetable consumption on plasma flavonol concentrations: results from a randomised controlled intervention trial, *J Epidemiol Community Health*, **58**: 288–9.

Johns, M.R. (1995) Developments in wastewater treatment in the meat processing industry: a review, *Bioresource Technology*, **54**: 203–16.

Kamimura, M., Furukawa, S. and Hirotsuji, J. (2002) Development of a simulator for ozone/UV reactor based on CFD analysis, *Water Science and Technology*, **46**: 13–19.

Khadre, M.A., Yousef, A.E. and Kim, J.-G. (2001) Microbiological aspects of ozone applications in food: a review, *Journal of Food Science*, **66**(9): 1242–52.

Kim, J.-G., Yousef, A.E. and Dave, S. (1999) Application of ozone for enhancing the microbiological safety and quality of foods: a review, *Journal of Food Protection*, **62**(9): 1071–87.

Kirby, R.M., Bartram, J. and Carr, R. (2003) Water in food production and processing: quantity and quality concerns, *Food Control*, **14**(5): 283–99.

Kowalska, I., Majewska-Nowak, K. and Kabsch-Korbutowicz, M. (2006) Influence of temperature on anionic surface active agent removal from a water solution by ultrafiltration, *Desalination*, **198**: 124–31.

Kuo, C.H. and Yocum, F.H. (1982) Mass transfer of ozone into aqueous systems, in Rice, R.G. and Netzer, A. (eds) *Handbook of Ozone Technology and Applications*, Vol. 1, Ann Arbor, MI: Ann Arbor Science, pp. 105–41.

Langlais, B., Reckhow, D.A. and Brink, D.R. (1991) *Ozone in Water Treatment: Application and Engineering*, Denver, CO: American Water Works Association Research Foundation.

Lewis, W.K. and Whitman, W.G. (1924) Principles of gas absorption, *Industrial Engineering Chemistry*, **16**: 1215–20.

Lillard, H.S. (1978) Improving quality of bird chiller water for recycling by diatomaceous earth filtration and chlorination, *J Food Sci*, **43**(5): 1528–31.

Lillard, H.S. and Thomson, J.E. (1983) Efficacy of hydrogen peroxide as a bactericide in poultry chiller water, *J Food Sci*, **48**(1): 125–6.

Mendez, F., Maier, D.E., Mason, L.J. and Woloshuk, C.P. (2003) Penetration of ozone into columns of stored grains and effects on chemical composition and processing performance, *Journal of Stored Products Research*, **39**: 33–44.

Noronha, M., Britz, T., Mavrov, V., Janke, H.D. and Chmiel, H. (2002) Treatment of spent process water from a fruit juice company for purposes of reuse: hybrid process concept and on-site test operation of a pilot plant, *Desalination*, **143**(2): 183–96.

Norton, T. and Sun, D.-W. (2008) Recent advances in the use of high pressure as an effective processing technique in the food industry, *Food and Bioprocess Technology*, **1**(1): 2–34.

Ölmez, H. and Kretzschmar, U. (2009) Potential alternative disinfection methods for organic fresh-cut industry for minimizing water consumption and environmental impact, *Lebensmittel-Wissenschaft und Technologie*, **42**: 686–93.

Parsons, S.A. and Jefferson, B. (2005) *Introduction to Potable Water Treatment Processes*, Oxford, UK: Blackwell Publishing Ltd.

Poretti, M. (1990) Quality control of water as raw material in the food industry, *Food Control*, **1**(2): 79–83.

Restaino, L., Frampton, E.W., Hemphill, J.B. and Palnikar, P. (1995) Efficacy of ozonated water against various food-related microorganisms, *Applied and Environmental Microbiology*, **61**(9): 3471–5.

Sarkar, B., Chakrabarti, P.P., Vijaykumar, A. and Kale, V. (2006) Wastewater treatment in dairy industries: possibility of reuse, *Desalination*, **195**: 141–52.

Sheldon, B.W. (1986) Effects of ozonation of poultry processing water on the characteristics of process water and potential for recycling, Industry Summary, Southwestern Poultry and Egg Association.

Sheldon, B.W., Carawan, R.E. and Merka, W.C. (2010) Water use and wastewater discharge patterns in a turkey processing plant. Retrieved from http://www.p2pays.org/ref/13/12926.pdf. Last accessed 26 May 2010.

Sroka, A., Kaminski, W. and Bohdziewicz, J. (2004) Biological treatment of meat industry wastewater, *Desalination*, **162**: 85–91.

Stampi, S., De Luca, G. and Zanetti, F. (2001) Evaluation of the efficiency of peracetic acid in the disinfection of sewage effluents, *J Appl Microbiol*, **91**: 833–8.

Suslow, T.V. (2001) Water disinfection: a practical approach to calculating dose values for preharvest and postharvest applications, Publication 7256, University of California, Agriculture and Natural Resources. Retrieved from http://vric.ucdavis.edu.

Ta, C.T. and Hague, J. (2004) A two-phase computational fluid dynamics model for ozone tank design and troubleshooting in water treatment, *Ozone: Science & Engineering*, **26**(4): 403–11.

Tanninen, J., Kamppinen, L. and Nyström, M. (2005) Pretreatment and hybrid processes, in Schäfer, A.I., Fane, A.G. and Waite, T.D. (eds), *Nanofiltration: Principles and Applications*, Oxford, UK: Elsevier.

Ternes, T.A., Stüber, J., Herrmann, N., McDowell, D., Ried, A., Kampmann, M. and Teiser, B. (2003) Ozonation: a tool for removal of pharmaceuticals, contrast media and musk fragrances from wastewater? *Water Research*, **37**(8): 1976–82.

Tiwari, B.K., O'Donnell, C.P., Patras, A., Brunton, N. and Cullen, P.J. (2009) Effect of ozone processing on anthocyanins and ascorbic acid degradation of strawberry juice, *Food Chemistry*, **113**(4): 1119–26.

Tiwari, B.K., Brennan, C.S., Curran, T., Gallagher, E., Cullen, P.J. and O'Donnell, C.P. (2010) Application of ozone in grain processing, *Journal of Cereal Science*, **51**(3): 248–55.

van der Bruggen, B., Mänttäri, M. and Nyström, M. (2008) Drawbacks of applying nanofiltration and how to avoid them: a review, *Separation and Purification Technology*, **63**, 2(22): 251–63.

Vieno, N.M., Harkki, H., Tuhkanen, T. and Kronberg, L. (2007) Occurrence of pharmaceuticals in river water and their elimination in a pilot-scale drinking water treatment plant, *Environmental Science & Technology*, **41**(14): 5077–84.

Vourch, M., Balannec, B., Chaufer, B. and Dorange, G. (2008) Treatment of dairy industry wastewater by reverse osmosis for water reuse, *Desalination*, **219**: 190–202.

Yan Lan, B., Nigmatullin, R. and Puma, G. (2008) Ozonation kinetics of cork-processing water in a bubble column reactor, *Water Research*, **42**(10–11): 2473–82.

Yang, Z., Peng, X.F., Lee, D.J. and Su, A. (2007) Reynolds number-dependent permeability of wastewater sludge flocs, *J Chin Inst Chem Eng*, **38**: 135.

Zhang, J. (2006) An integrated design approach for improving drinking water ozone disinfection treatment based on computational fluid dynamics, PhD thesis, University of Waterloo, ON, Canada.

12 Ozone for Food Waste and Odour Treatment

Ioannis S. Arvanitoyannis

12.1 Introduction

The food industry is currently in need of innovative processing technologies to meet consumer demands for fresh and safe ready-to-eat products. Ozone appears to be a powerful sanitiser which meets the expectations of industry and has the approval of regulatory agencies (Khadre et al. 2001). It is a highly unstable triatomic oxygen molecule that can be generated commercially when oxygen molecules are subjected to a high-voltage electric discharge. Molecular oxygen splitting into two atoms of oxygen results in highly reactive moieties, some of which combine to produce ozone (Cullen et al. 2009). Due to its high instability, ozone rapidly degrades to molecular oxygen, with the released free oxygen atom reacting with another free oxygen atom to form molecular oxygen or combining with other chemical moieties for oxidation purposes. Upon release of the third oxygen atom, ozone acts as a strong oxidising agent. It is due to this strong oxidative capacity that ozone is highly effective in destroying microorganisms (Rice et al. 1981).

Ozone is a gas with a pungent odour at ordinary temperature. At concentrations at which it is normally produced, the colour of ozone is not noticeable. Ozone was been shown to kill a wide range of viruses such as Venezuelan equine encephalomyelitis virus, hepatitis A, influenza A, vesicular stomatitis virus and infectious bovine rhinotracheitis virus, as well as several bacteriophage strains. Ozone is also effective in destroying oocysts of the protozoan *Cryptosporidium parvum*. The bactericidal effects of ozone have been repeatedly reported and are well documented on a wide variety of organisms, including both Gram-positive and Gram-negative bacteria as well as spores and vegetative cells (Güzel-Seydim et al. 2004b).

Despite these advantages, ozonated water technology has not been widely adopted in the food industry and in particular for the treatment of food wastes or odours. While much information is available to food processors on the capital and operational costs of ozone technology, there is a lack

Ozone in Food Processing, First Edition. Edited by Colm O'Donnell, B.K. Tiwari, P.J. Cullen, and Rip G. Rice.

of information on O_3 solubility, including the effects of temperature and water quality (Chawla 2006).

Reported ozone applications in the food industry are mostly related to decontamination of product surfaces and water treatment. Ozone has been used with reasonable success to inactivate or destroy contaminant microflora on foods of both plant and animal origin. Other applications of ozone are in the elimination of mycotoxins and pesticide residues from agricultural products. However, excessive use of ozone may induce oxidation of ingredients residing on food surfaces, thereby causing discolouration and deterioration of food flavour (Jin-Gab et al. 1999).

The antimicrobial effects of ozone-containing water have been confirmed for Gram-positive and Gram-negative bacteria (Pascual et al. 2007):

- *L. monocytogenes;*
- *S. aureus;*
- *B. cereus;*
- *E. faecalis;*
- *P. aeruginosa;*
- *Y. enterocolitica.*

Ozone also destroyed the nests and spores of *Candida albicans, Zygosaccharomyces bacilli* and *A. niger*.

According to Güzel-Seydim et al. (2004a), microorganism destruction by ozone is attributed to gradual oxidation of cellular components, which occurs by means of the following three oxidation mechanisms:

(1) sulfhydryl groups, aminoacids, peptides and proteins;
(2) polyunsaturated fatty acids and peroxides;
(3) destruction of viral RNA and conversion of polypeptide chains into viral protein coats.

Ozone is also very reactive towards compounds incorporating conjugated double bonds associated with colour and functional groups with high electron densities (Coca et al. 2005).

There are many methods for the production of ozone, such as electrical discharge in oxygen, electrolysis of water, and thermal, photochemical and radiochemical methods. For industrial use, ozone is generated mainly from high-purity oxygen or atmospheric oxygen in a corona discharge process. In corona discharge, air or high-purity oxygen is fed into a unit that converts the oxygen to ozone using high voltage. The major advantage of ozone for food waste treatment is that it decomposes rapidly to molecular oxygen without any residues (Inan et al. 2007).

The two main limitations of ozone as a disinfection agent are: (1) its instability in water (decomposition to oxygen depends on water pH); and

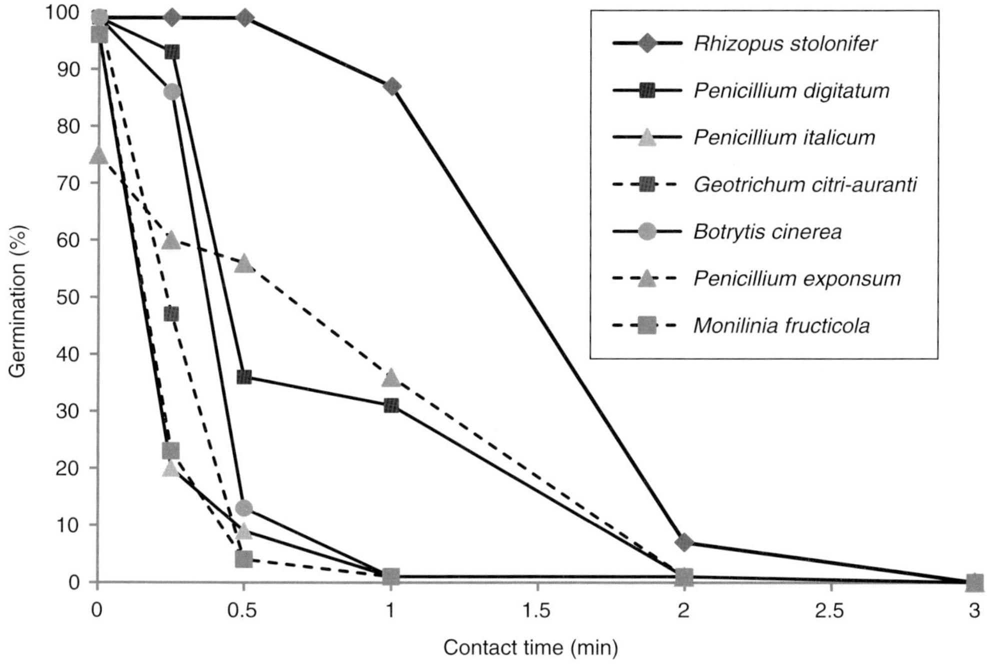

Figure 12.1 Germination of spores of various postharvest pathogenic fungi after exposure to 1.5 ppm (μg/mL) ozone in water at pH 6.4 and temperature 16.5 °C (Khadre et al. 2001).

(2) that organic substances are converted to low-molecular-weight biodegradable compounds, which have to be filtered (Cullen et al. 2009).

Ozone, apart from oxidising many nuisance compounds or potential toxicants in water supplies, can also coagulate natural water constituents. Comparison tests revealed that ozone is more effective than chlorine for these purposes. Although the coagulation mechanism induced by ozone is not clear, several suggestions have been made, among which the most important are: (1) metal ions oxidation (becoming insoluble); (2) humic materials oxidation (polar and/or chelating group formation); and (3) colloid destabilisation by means of oxidative dissociation of organic material contained in colloidal clay particles (Glaze et al. 1987).

Ozone is considered one of the most effective sanitisers known. Ozone treatment has no heat requirements, thereby saves energy. Although the initial cost of ozone generators may be of concern to small processors in particular, ozone is widely used in many industries in both the EU and the USA (Khadre et al. 2001).

Figure 12.1 shows the germination rates of spores of various postharvest pathogenic fungi after exposure to 1.5 ppm (μg/mL) ozone in water at pH 6.4 and temperature 16.5 °C.

Distillery wastewater has been effectively treated with ozone as a chemical oxidant in combination with conventional aerobic oxidation.

Table 12.1 Literature data on the efficiency of sludge solubilisation (Chu et al. 2009).

Initial TSS conc. (mg/L)	Volume (L)	Ozone conc. (mg/L)	Ozone dose ($gO_3/gTSS$)	Operation mode	Solubilisation efficiency (%)	References
3790–4570	630	–	0.025–0.035	Continuous	20–30%	Manterola et al. 2008
1200–4000	2	20–90	0.03–0.04	Semibatch	30	Saktaywin et al. 2005
–	6000	–	0.03–0.06	Semibatch	9.2–13.3[a]	Sievers et al. 2004
5000–5300	350	50	0.05	Semibatch	30	Lee et al. 2005
18900	1	30	0.1–0.16	Semibatch	20–25	Bougrier et al. 2006
8000–12000	1000	150	0.1	Semibatch	45	Park et al. 2003
			0.2		55	
			0.5		66	
–	2		0.57–1.09	Semibatch	50–56[b]	Mines et al. 2008
2850	2.1	50	1.2	Semibatch	58.9[b]	Cui and Jahng 2004

[a]Recalculated assuming the conversion biomass to COD as 1.2 g COD/g SS.
[b]TSS elimination.
TSS = total soluble solids.

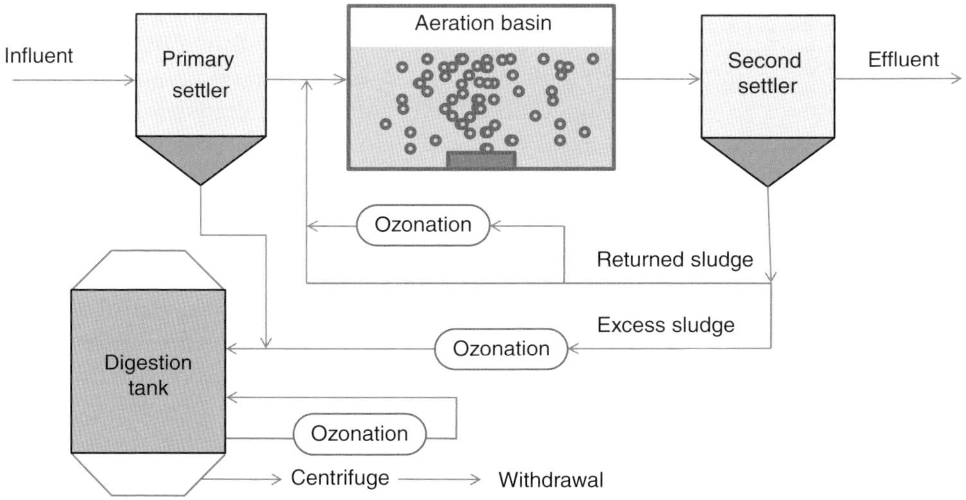

Figure 12.2 Application of ozonation for sludge reduction. (Reprinted from *Water Research*, Volume 43, Issue Number 7, Libing Chu, Sangtian Yan, Xin-Hui Xing, Xulin Sun, Benjamin Jurcik, Progress and perspectives of sludge ozonation as a powerful pretreatment method for minimization of excess sludge production, 1811–1822, 2009, with permission from Elsevier.)

Ozone has been mainly used in the pretreatment and post-treatment stages. Application of both chemical and biological processes resulted in increased destruction of organic contaminants present in the effluent (Sangave et al. 2007). Ozone in wastewater treatment can also be useful in sludge solubilisation (Chu et al. 2009). Ozone efficiency of sludge solubilisation, as shown in Table 12.1, depends on various factors, including sludge properties and extrinsic and intrinsic parameters governing the ozone process. Studies show that sludge can be reduced by 40–100% by employing ozone at several stages, as shown in Figure 12.2 (Chu et al. 2009).

12.2 Application of ozonation to waste treatment

12.2.1 Wastewater of plant origin

Fruit

In the fruit and vegetable processing industries wastewater and solid wastes are the main effluents. The wastewater is high in suspended solids, sugars and starches and may contain varying amounts of residual pesticides. The solid waste comprises organic materials including rinds, seeds and skins from raw materials removed to mechanical processes. The greatest part of solid waste not resold as animal feed is treated with conventional biological or composting processes.

Several oxidising agents such as ozone, H_2O_2 and ultraviolet (UV) have been successfully employed to (1) decrease the chemical oxygen demand (COD) and polyphenol contents of table olive debittering wastewaters (Garcia and Beltrán-Heredia 2008) and (2) degrade and treat distillery and tomato processing wastewaters (Beltrán et al. 1997). It was reported that O_3/H_2O_2 oxidation leads to important increases in COD degradation rates (86% at pH 6 in tomato wastewaters). The differences between the oxidation types (O_3 and O_3/H_2O_2) diminish with increase in pH. With distillery wastewater, the presence of hydrogen peroxide hardly increased the oxidation rate. However, the combination of O_3/UV radiation was the best oxidation method applied because of the improvements achieved in both COD and total organic carbon (TOC) disappearance rates compared to those of ozonation alone, regardless of wastewater type treated.

The use of ozone, H_2O_2 and granular activated carbon (GAC) in drinking water treatment is widespread throughout Western Europe, Canada, the USA and Japan. Ozone, H_2O_2 and activated carbon were applied either solely or in combination to assess the effectiveness as post-treatment options for UASB (upflow anaerobic sludge blanket) reactor-treated alkaline fruit cannery effluent. Colour reduction in the effluents ranged from 15 to 92% and the corresponding COD reductions were in the range 26–91%. Combinations of ozone and H_2O_2 gave better results than employing either oxidant solely. When a triple combination of reactants (ozone,

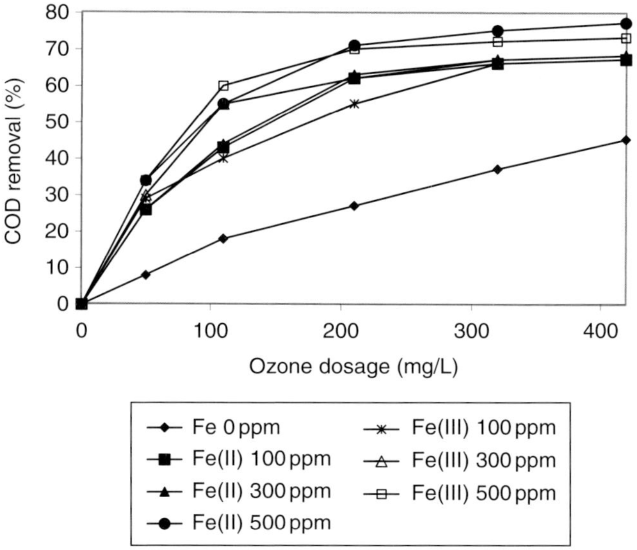

Figure 12.3 COD removal versus ozone dosage (for thin stillage from Fe(III)-catalysed ozonation experiments using an ozone application rate of 7 mg/min and 40× dilution). (Reprinted from *Bioresource Technology*, Volume 99, Issue 6, Shilpi Singh, Maohong Fan, Robert C. Brown, Ozone treatment of process water from a dry-mill ethanol plant, 1801–1805, 2008, with permission from Elsevier.)

H_2O_2 and GAC) was applied, the lowest COD levels were obtained (COD < 75 mg/L limit) (Sigge et al. 2001).

Corn mill

Treatment of wet corn-milling wastewater with filamentous fungi was investigated as a means of obtaining fungal biomass as an additional byproduct. Competitive bacterial growth is a common problem occurring with this nonaseptic treatment process. Selective disinfection with ozone was evaluated for elimination of bacterial populations during fungal cultivation (Sankaran et al. 2008).

Figure 12.3 shows the COD removal results for thin stillage from Fe(III)-catalysed ozonation experiments using an ozone application rate of 7 mg/min and 40× dilution (Singh et al. 2008). Figure 12.4 displays a comparison of COD removal using ozone application rates of 7 or 21 mg/min on 20× and 40× dilutions (Singh et al. 2008).

The possibility of applying ozone (O_3) instead of sulfur dioxide (SO_2) in corn steeping has been investigated recently. Steep water is often enriched with 0.1–0.2% sulfur dioxide to increase starch–protein separation, in order to achieve high starch yields and to control microbial growth. However, residual SO_2 in starch products has a strong impact on product quality and can eventually endanger 'organic products' claims. Furthermore, SO_2

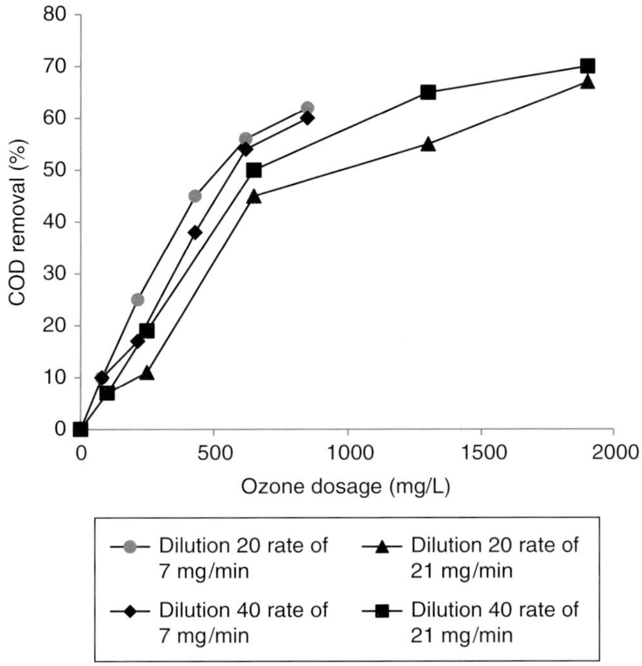

Figure 12.4 COD removal versus ozone dosage (7 or 21 mg/min on dilutions 20× and 40×). (Reprinted from *Bioresource Technology*, Volume 99, Issue 6, Shilpi Singh, Maohong Fan, Robert C. Brown, Ozone treatment of process water from a dry-mill ethanol plant, 1801–1805, 2008, with permission from Elsevier.)

released to the environment can heavily pollute both water and air. Therefore, application of O_3 to corn steeping appears to be a promising alternative.

Olive oil mill

The manufacture of olive oil production is conducted, in most cases, by a plethora of small and medium-sized plants which operate seasonally. Water is used to wash and process the original olives and produces what is known as 'olive mill wastewater' (OMW) (Benitez 1997). Up to 8.4 million m^3 of OMW are produced annually (Sassi et al. 2006). These wastes are usually acidic and vary in colour from dark red to black, depending on their state of degradation and the climatic conditions. They contain a high concentration of phenolic compounds, which have antimicrobial properties. OMW treatments to date have involved physical, chemical and physicochemical processes (Belaid et al. 2006).

OMW is often concentrated in evaporation ponds and left to dry throughout the summer season. This procedure negatively affects the regional environment through strong and unpleasant odours after anaerobic digestion, and the potential hazards to surface and groundwater sources. Anaerobic treatment is a very suitable and promising approach for this type of aqueous waste, considering the seasonal production aspects and the high organic load of OMW (Basheer et al. 2004).

Ozonation prior to anaerobic treatment is an effective pretreatment step due to the oxidising power of ozone. Ozonation experiments on undiluted OMW revealed that after ozonation for 5 hours, phenols and lipids were significantly reduced, leading to lower-molecular weight (MW) acids, whereas the total COD of the sample was hardly affected (Paraskeva and Diamadopoulos 2006).

The wastewater coming from olive oil mills (OMW) has a COD of around 3000 mg/dm. CDEO (conductive-diamond electrochemical oxidation) allowed the complete mineralisation of the waste with high efficiency. Similarly, both ozonation and Fenton oxidation (Fenton's reagent is a solution of hydrogen peroxide and an iron catalyst) were used to treat the wastes, but the obtained results varied considerably with regard to efficiency and mineralisation. The increase of oxidation-refractory compounds as final products excludes the use of ozonation and Fenton oxidation as exclusive treatment technologies (Canizares et al. 2007).

Wine distilleries

Nowadays, there is an increasing awareness of the negative impacts of seasonal discharges of waters containing high nutrient and organic loadings into water sources. In the USA, disposal of such waters by irrigation has been adequately documented and regulated, mainly by the National Water Act (Du Plessis et al. 2007).

Wine distilleries produce large volumes of wastewaters (wine distillery wastewater, WDW), called 'vinasses', the composition of which varies significantly based on the raw materials distilled: wine, lies and pressed grapes (Basu 1975). Effluent discharging into public sewages is a large-scale acute environmental issue because of its polluting effect on both underground and surface waters. Benitez et al. (1999) claimed that the ozonation treatment of WDW results in a reduction in the organic substrate concentration of approximately 15%, which increases up to 26 and 21% with the action of hydroxyl radicals at pH 9 or in the presence of UV radiation and hydrogen peroxide in addition to ozone, respectively. Ozonation of the aerobically pretreated wastes largely augments the organic matter removal in comparison to ozonation of unpretreated wastes, and substrate conversions in the range 40–67% are obtained (Benitez et al. 1999; Gimeno et al. 2007).

According to Lucas et al. (2010), the effectiveness of different ozone-based advanced oxidation processes (AOPs) (O_3, O_3/UV and O_3/UV/H_2O_2) on the treatment of WDW was investigated in a pilot-scale bubble column reactor. At the natural pH of the wastewater (pH 4), the effectiveness of each AOP followed the sequence: $O_3/UV/H_2O_2 > O_3/UV > O_3 >$ UVC. The rate of COD and TOC removal was enhanced by operation at neutral (pH 7) and at alkaline pH (pH 10). The rate of ozone consumption in the reactor with the O_3/UV and O_3/UV/H_2O_2 processes was in the range 70–95% during the experiments, suggesting an effective use of the ozone supplied to the system. In all the experiments the disappearance of the WDW organic load was described by pseudo-first-order apparent reaction kinetics.

The ozonation of WDW in the bubble column was analysed in terms of a mole balance coupled with ozonation kinetics modelled by the two-film theory of mass transfer and chemical reaction. It was determined that the ozonation reaction can develop both in and across different kinetic regimes – fast, moderate and slow – depending on the experimental conditions. The dynamic change of the rate coefficient estimated by the model was correlated with changes in the water composition and oxidant species (Lucas et al. 2009).

12.2.2 Wastewater of animal origin

Dairy products

The dairy industry is one of the largest sources of industrial effluents in Europe. The dairy wastewater effluents have a high biochemical oxygen demand (BOD) and COD due to their high organic content (milk, residues, proteins, fats and cleaning agents, among others).

The effectiveness of membrane separation techniques in conjunction with ozone for removal of surfactant from dairy wastewater was investigated by Laszlo et al. (2009). It was found that the most suitable conditions for surfactant solutions were 40 bar and 20 °C. Introduction of a preozonation step considerably enhanced the retention of both COD and BOD and surfactants from the wastewater. Since nanofiltration on its own could not effectively eliminate the waste materials, it was concluded that application of ozonation was a promising alternative.

Ozone was applied to model dairy wastewater (prepared from milk powder by dilution), and the effects of ozonation time and surfactant concentration on the flux, membrane resistances, membrane fouling and gel formation were investigated. Laszlo et al. (2009) found that the microflocculation effect of ozone may play a significant role at a higher gas flow rate, causing a decreased level of fouling and increased gel formation. At a lower flow rate the effect of the degradation of large molecules was more marked, causing a higher flux but decreasing the retention. Although ozonation diminished the fouling effect, it enhanced the gel formation resistance, probably due to the ozonation-microflocculation effect. Microflocs do not foul the membrane pores because of layer formation on the membrane surface. Moreover, it was demonstrated that the presence of detergent considerably enhanced gel formation but had no effect whatsoever on flux.

Laszlo et al. (2007) investigated the applicability of membrane filtration in conjunction with preozonation in dairy wastewater treatment technology.

The best performance of surfactant removal from model anionic surfactant solution with nanofiltration occurred at 20 °C and 40 bar. Further studies on the effects of ozone treatment of the wastewater indicated that preozonation decreased the flux and increased the COD and surfactant removal efficiencies. Ozone treatment greatly enhanced the biodegradability of the retentate from 68.8 to 96.4%.

The logical combination of preozonation and membrane filtration performed by Laszlo et al. (2007) was to reduce the surfactant content of waste water to below the legally regulated limit. They further observed that the surfactant content from waste water the required level of retention at a surfactant concentration of 0.01% was 50%, while at a surfactant concentration of 0.1%, it was observed to be 95%. Significant increase in the retention of waste materials was observed as a result of preozonation treatment. The possible mechanism responsible for the microflocculation effect of preozonation of organic matter may be because of a reaction between the components present in dairy wastes, the ozonation by-products and metal ions e.g., calcium (present in considerable amount in dairy wastewaters) may preclude the formation of aggregates, the decline of the average flux during nanofiltration. Studies of Laszlo et al. (2007) and Laszlo et al. (2009) indicate that preozonation may enhance the treatability of dairy waste waters with nanofiltration with clear industrial application however, further large scale experiments are required to optimalize the ozone dosage and the ozonation time.

Meat

Ozonation of red meat processing wastewater was conducted in a semibatch reactor to investigate the possibility of water reuse. The experimental results revealed that ozone was very effective in the disinfection of the red meat processing wastewater. After 8 minutes of ozonation with an applied ozone dose of 23.09 mg/min/L of wastewater, 99% of aerobic bacteria, total coliforms and *E. coli* were inactivated. Empirical models were developed to predict the microbial inactivation efficacy of ozone from the *Ct* values for the red meat processing wastewater (Wu and Doan 2005).

Initial experiments using 250 mL samples of unscreened poultry overflow chiller water were designed to evaluate the beneficial bactericidal and oxidative effects of four different treatments (namely, O_2/O_3, O_2/UV, O_2/O_3/UV and O_2 as the control). Following foam removal, synergistic reductions higher than 1.5-log colony-forming units (CFU)/mL for aerobic plate counts (APCs) were additionally achieved after 4 minutes for all O_3/UV treatment combinations, as compared to serially applied treatments of O_3 and UV acting separately (Diaz et al. 2002).

Seafood

In a study by Campos et al. (2009) the wastewater treatment plant consisted of two coagulation–flocculation units and a biological unit and generated around 6550 kg/day of sludge. Ozone was applied to sludge coming from flotation units (110 g total soluble solids (TSS)/L) at doses up to 0.03 g O_3/g TSS during batch tests, and no solids solubilisation were observed. Ozone doses within the range 0.007–0.02 g O_3/g TSS were applied to the raw wastewater in a bubble column, achieving a rate of 6.8% TSS removal for the highest ozone dose. Moreover, the impact of the preozonation (0.05 g O_3/g TSS) of wastewater from the first flotation unit was investigated in two activated sludge systems over 70 days. The authors reported that ozonation caused a marked decrease in the yield coefficient of biomass from 0.14 down to 0.07 g TSS/g COD removed and a rather minor amelioration of COD removal efficiencies. Taking into account the industrial ozone production capacity, the maximum anticipated reduction of sludge by the wastewater treatment plant was 7.5%.

Oakes et al. (1979) investigated ozone disinfection of fish hatchery waters. Plate counts for total bacteria were taken at various points in the pilot plant. Similar plate counts were taken from existing UV disinfection equipment. Ozone residual levels entering and within the recycle system, ammonia, nitrate, nitrite, total organic nitrogen, suspended solids, turbidity and BOD were mentioned as well. Analysis of the plate counts showed that ozone consistently provided more effective disinfection of the makeup water than the existing UV system. Nitrite levels after ozonation were significantly decreased, while nitrate levels increased. Ozone disinfection of recycled water was also satisfactorily demonstrated. Moreover, batch studies indicated that ozone effectively destroyed algae present.

Some of the most representative ozone treatments of wastewater effluents of various industries are summarised in Table 12.2.

12.3 Application of ozonation to odour removal

In the last decades, several countries have reported an increase in complaints due to agriculture and food processing industries causing (or emitting) odours (Both 2001; Mahin 2001). Recent environmental legislation has obliged industries to search for effective technologies to reduce the emissions of pollutants from their plants. However, currently there are no defined acceptable limits for odours that if surpassed, would be considered violations. Therefore, most attempts to reduce levels of volatile organic compounds consider little or not at all the emission of malodorous volatile organic compounds (Domeno et al. 2010).

Odour problems are limited to the areas nearby the emitting source. Malodours can have negative effects on property values and the quality of rural life in the areas in which they occur (Tyndall and Colletti 2007). There are two basic ways to control odours: (1) by preventing or reducing the

Table 12.2 Ozone applications in various sectors.

Sector	Examples
Wastewater treatment	Domestic and/or municipal water treatment Pulp and paper industry wastewater treatment Pharmaceutical (e.g. phenol degradation) Textile and leather wastewater treatment Landfill leacheates
Animal husbandry	Animal waste treatment Animal drinking water Swine house (odour removal)
Agriculture	Cooling towers treatment Boiler water treatment Chilled water treatment Cutting fluids recycling Water dripping treatment Irrigation water disinfecting
Food industry	Drinking and water bottling Grain silo disinfecting Fruit and vegetable storage Meat storage Slaughterhouse disinfecting Fruits and vegetable wash Food containers sterilisation Chicken egg wash
Other industries	Smoke and odour treatment Semiconductor wafers clean Laundry water recycling Medical instrument sterilisation Hospital air sterilisation Aquaculture Paper pulp bleach Sour gas desulfurisation Zebra mussels treatment Rubber recycling

generation of odours at the sources; and (2) by treating odours from collected gaseous streams before they are released into the environment.

The control of generated odours is always the first choice for odour control (Kreis 1978; Rappert and Muller 2005). Odour control generally starts with identifying the sources that produce the odorous compounds, measuring the odour fluxes, ranking the sources to define treatment priorities and lastly identifying the type of odorous compound in order to choose the most appropriate method of control (Ramel and Nomine 2000). Odours can be controlled by changing the chemical nature of the compound, inhibiting anaerobic decomposition of wastes or confining the generated odour. Site selection and management are also integral parts of odour control endeavours. Odour control with chemicals is expensive because of the large amount of chemical required at frequent intervals in order for it to be

Table 12.3 Volatile compounds responsible for odours emitted from food processing industries.

Processing facilities	Emitted volatile compounds	References
Facilities for canning fruit and vegetables	Aldehydes, ketones, terpenes, acids, alcohols, esters, sulfur and aromatic compounds	Safriet 1995c
Facilities for fishmeal production	Carbonyls, sulfur compounds, amines and ammonia	Olafsdottir et al. 2000
Facilities for meat rendering	Ammonia, amines, amide, quinoline, organic acids, alcohols, ketones, pyrazines, aliphatic hydrocarbons, sulfur and aromatic compounds	Safriet 1995b

effective (Ritter 1989). Although ozone is used primarily as a disinfectant, it can also be used for the removal of odours. It has been used to treat odours since the early 1900s and is widely applied in developed countries (Balakrishnana et al. 2002; Gottschalk et al. 2000).

12.3.1 *Odours originating from food industry processes*

Odours produced by the food industry are usually caused by the biological or chemical reactions occurring due to the physical processing of foods. Most of the compounds produced in food processing are reduced carbon, nitrogen and sulfur compounds, such as aldehydes, ketones, alcohols, acids, ammonia, amines, sulfides, mercaptans and hydrogen sulfide. In some cases, the odours are the result of the production of volatile organic compounds. The previously mentioned reactions are usually associated with common processes in the food industry such as cooking, condensation, drying, smoking and so on. Processes conducted in closed tanks do not produce significant emissions (Safriet 1995a,b). Another source of odour is the stored materials. Storage conditions which are not suitable for specific materials (e.g. raw materials, food and waste products) can result in their decomposition and therefore the production of malodours (Olafsdottir et al. 2000; Miljøstyrelsen 2002). Table 12.3 summarises the odourous compounds that can be emitted by food industries.

12.3.2 *Odours originating from agricultural operations*

Significant sources of odour are manure and biosolids, which are used as fertilisers on agricultural land. Odours at land application sites often result in complaints and even public opposition. Odour production from land applications is considered a priority for the management of biosolids and of manure (O'Neill and Phillips 1991; Persaud et al. 1996). Another significant source of odour is from animal production facilities (Persaud et al.

1996). Such odours may be released from livestock buildings and feedlots, waste storage facilities and slurry processing equipment (Ritter 1989). The intensity of the odour in waste air from livestock buildings increases from cattle through hens to pigs and is affected even more by the type of building, the age of the grown animals and the purpose of the facility (Hartung 1992). Odour produced from manure is primarily a result of the incomplete anaerobic degradation of organic matter (Nahm 2002; Mackie et al. 1998), which results in the production of malodorous compounds that can be categorised into nitrogen compounds, sulfur compounds, aromatic compounds and volatile fatty acids (Varel and Miller 2001; Persaud et al. 1996; Zhu 2000; Nahm 2002; Mackie et al. 1998).

Kim-Yang et al. (2005) studied the effects of ozonation on odours emitted from a swine housing facility. The results revealed that ozone treatment reduced the levels of indolic compounds to a great extent. However, the reduction in the levels of volatile fatty acids and phenolic compounds in the air was small. Results from sensory evaluation showed that there was a small increase in the odour detection thresholds for air in the ozonated rooms.

In another study by Schrader et al. (2010) the effect of ozonation on the off-flavour metabolites geosmin and 2-methylisoborneol in rainbow trout (*Oncorhynchus mykiss*) cultured in a recirculating aquaculture system was studied. Results revealed that ozonation reduced the levels of the previously mentioned off-flavour compounds both in the water and in fish flesh, but not significantly. Higher dosages of ozone might be beneficial in removing the off-flavour compounds but would increase the risk of ozone toxicity.

Ozonation of swine manure slurry resulted in the elimination of malodorous bacterial metabolites (phenol, p-cresol, p-ethylphenol and skatole). The odour intensity of the manure slurry was also significantly reduced after ozonation at a dosage of 0.5 g/L. The use of hydrogen peroxide in conjunction with ozone did not result in a more pronounced effect than that obtained with ozone treatment alone. Furthermore, treated manure did not regain its initial malodour after one month of storage. Temperature appeared to have no significant effect on the efficiency of ozonation (Wu et al. 1999). Priem (1977) found that ozone in doses up to 0.2 ppm was able to reduce ammonia levels in a swine barn by 50 and 15% under winter and summer ventilation conditions, respectively.

Arsovic and Burchard (1973) studied the treatment of gaseous and liquid odour emissions from animal rendering plants. Treatment with chlorine or ozone resulted in negligible effects. However, when chlorine dioxide or chlorine was used in conjunction with ozone, significant reductions of malodours were achieved. The same results were obtained when wastewater from these facilities was tested by using the same oxidising agents. Mohana et al. (2009) suggested that ozonation has potential for removing the emitted malodours from distillery spent wash, caused by the presence of skatole,

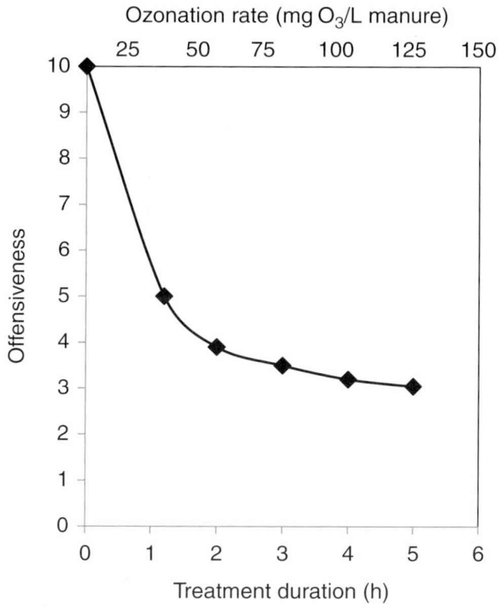

Figure 12.5 Effect of treatment time on odour offensiveness (Alkoaik 2009).

indole and sulfur compounds, which are not effectively decomposed by yeast during distillation.

Watkins et al. (1997) investigated the effects of ozonation on stored swine manure slurry. Treatment resulted in a significant reduction in odours in ozonated samples. However, volatile fatty acids, nitrate, phosphate and ammonia concentrations were not affected by ozonation. The concentrations of odorous phenolic and indolic microbial metabolites were reduced to nondetectable levels. Moreover, ozone treatment was able to reduce *E. coli* counts and total coliforms.

Elenbaas-Thomas (2005) found that ozonation of a swine facility at the maximum safety concentration of 0.1 ppm did not have any statistically significant effects on dust mass concentration, odour concentration and emission rate, sulfur compound concentrations (dimethyl sulfide, dimethyldisulfide and dimethyltrisulfide) or bacteria counts. Moreover, ozone treatment increased ammonia concentration. Li et al. (2008) investigated the effects of ozonation on NH_3 emissions from manure. The results showed that ozone has no effect on NH_3 concentration, and with ozone treatment, high concentrations of particles were generated, which could lead to health problems. According to Alkoaik (2009), the use of ozone for the treatment of animal manure proved to be effective in reducing odour offensiveness (Figure 12.5). This author also reported that when increasing the level of ozone, odour offensiveness increased due to the presence of intermediate products caused by the interaction between ozone and hydrogen sulfide and methylamine.

12.4 Conclusions

The application of ozone to food industry wastewater effluents is considered promising in view of its greater effectiveness compared to other chemical agents (i.e. chlorine). The best results are obtained in the presence of a synergistic action of ozone and other oxidising agents (H_2O_2, UV, etc.). Ozone treatment can potentially reduce odours produced from livestock facilities (1) by killing the odour-producing microorganisms and (2) by oxidising the odorous metabolites. However, only a limited number of published studies have evaluated the use of ozone for odour reduction in animal production facilities. Further research is required to investigate the use of ozonation in conjunction with other methodologies in order to produce more pronounced effects. All research carried out must take into account the effects of ozone on the safety of employees.

References

Alkoaik, F.N. (2009) Ozone treatment of animal manure for odour control, *American Journal of Environmental Sciences*, **5**(6): 765–71.

Arsovic, H.M. and Burchard, H. (1973) The purification of gaseous and liquid emissions from animal body disposal plants (Arsox process), *Gesundheitstechnik*, **7**(4): 67–70.

Balakrishnana, P.A., Arunagiria, A. and Rao, P.G. (2002) Ozone generation by silent electric discharge and its application in tertiary treatment of tannery effluent, *Journal of Electrostatics*, **56**: 77–86.

Basheer, S., Sabbah, I. and Marzook, T. (2004) The effect of pretreatment on anaerobic activity of olive mill wastewater using batch and continuous systems, *Process Biochem*, **39**: 1947–51.

Basu, A.K. (1975) Characteristics of distillery wastewater, *Journal of Water Pollution Control Federation*, **47**: 2184–90.

Belaid, C., Kallel, M., Khadhraou, M., Lalleve, G., Boubaker Elleuch, B. and Fauvarque, J.-F. (2006) Electrochemical treatment of olive mill wastewaters: removal of phenolic compounds and decolourization, *Journal of Applied Electrochemistry*, **36**(10): 28–36.

Beltrán, F.J., García-Araya, J.F. and Álvarez, P. (1997) Impact of chemical oxidation on biological treatment of a primary municipal wastewater. 2. Effects of ozonation on kinetics of biological oxidation, *Ozone: Science and Engineering*, **19**(6): 513–26.

Benitez, F.J., Beltrán-Heredia, J., Torregrosa, J. and Acero, J.L. (1997) Improvement of the anaerobic biodegradation of olive mill wastewaters by prior ozonation pretreatment, *Bioprocess Engineering*, **17**: 169–75.

Benitez, F.G., Beltrán-Heredia, J., Torregrosa, J. and Acero, J.L. (1999) Treatment of olive mill wastewaters by ozonation, aerobic degradation and the combination of both treatments, *Chemical Technology and Biotechnology*, **74**: 639–46.

Both, R. (2001) Directive on odour in ambient air: an established system of odour measurement and odour regulation in Germany, *Wat Sci Technol*, **44**(9): 119–26.

Bougrier, C., Albasi, C., Delgenès, J.P. and Carrère, H. (2006) Effect of ultrasonic, thermal and ozone pre-treatments on waste activated sludge solubilisation and anaerobic biodegradability, *Chemical Engineering and Processing*, **45**(8): 711–18.

Campos, J.L., Otero, L., Franco, A., Mosquera-Corral, A. and Roca, E. (2009) Ozonation strategies to reduce sludge production of a seafood industry WWTP, *Bioresource Technology*, **100**: 1069–73.

Canizares, P., Lobato, J., Paz, R., Rodrigo, M.A. and Saez, C. (2007) Advanced oxidation processes for the treatment of olive-oil mill wastewater, *Chemosphere*, **67**: 832–8.

Chawla, A.C. (2006) Application of ozonated water technology for improving quality and safety of peeled shrimp meat. Submitted to the Graduate Faculty of the Louisiana State University and Agricultural and Mechanical College in partial fulfilment of the requirements for the degree of Master of Science in The Department of Food Science.

Chu, L.B., Wang, J.L., Wang, B., Xing, X.H., Yan, S.T., Sun, X.L. and Jurcik, B. (2009) Changes in biomass activity and characteristics of activated sludge exposed to low ozone dose, *Chemosphere*, **77**(2): 269–72.

Coca, M., Pena, M. and Gonzalez, G. (2005) Variables affecting efficiency of molasses fermentation wastewater ozonation, *Chemosphere*, **60**: 1408–15.

Cui, R. and Jahng, D. (2004) Nitrogen control in AO process with recirculation of solubilized excess sludge, *Water Research*, **38**(5): 1159–72.

Cullen, P.J., Tiwari, B.K., O'Donnell, C.P. and Muthukumarappan, K. (2009) Modelling approaches to ozone processing of liquid foods, *Trends Food Sci Technol*, **20**(3–4): 125–36.

Diaz, M.E., Law, L.E. and Birt, D.M. (2002) Microbiological benefits of removing foam formed after UV-enhanced ozonation of poultry-processing chiller water for recycling, *Food Engineering and Physical Properties*, **67**(3): 1036–42.

Domeno, C., Rodriguez-Lafuente, A., Martos, J.M., Bilbao, R. and Nerin, N. (2010) VOC removal and deodorization of effluent gases from an industrial plant by photo-oxidation, chemical oxidation and ozonization, *Environmental Science and Technology*, **44**: 2585–91.

Dunnick, J.K. and Melnick, R.L. (1993) Assessment of the carcinogenic potential of chlorinated water: experimental studies of chlorine, chloramine, and trihalomethanes, *Journal of the National Cancer Institute*, **85**(10): 817–22.

Du Plessis, D., Green, K., Sigge, J. and Britz, G. (2007) Effect of ozone on distillery wastewater treatment in wetlands: investigating the effect of ozone on the biodegradability of distillery wastewater, *South African Journal of Enology and Viticulture*, **28**(2): 155–62.

Elenbaas-Thomas, A.M., Zhao, L.Y., Hyun, Y., Wang, X., Anderson, B., Riskowski, G.L., Ellis, M. and Heber, A.J. (2005) Effects of room ozonation on air quality and pig performance, *Transactions of the ASAE*, **48**(3): 1–7.

Garcia, J. and Beltrán-Heredia, J. (2008) Chemical treatment by Fenton's reagent oxidation and ozonation of wastewaters from olive oil mill, *Journal of Advanced Oxidation Technologies*, **11**(3): 568–72.

Gimeno, O., Rivas, F.J., Beltrán, F.J. and Carbajo, M. (2007) Photocatalytic ozonation of winery wastewaters, *Journal of Agricultural and Food Chemistry*, **55**(24): 9944–50.

Glaze, W.H., Kang, J.W. and Chapin, D.H. (1987) The chemistry of water treatment processes involving ozone, hydrogen peroxide, and ultraviolet radiation, *Ozone: Science & Engineering*, **9**(4): 335–52.

Gottschalk, C., Libra, J.A. and Saupe, A. (2000) *Ozonation of Water and Wastewater: A Practical Guide to Understanding Ozone and its Application*, Weinheim, Germany: Wiley-VCH Verlag GmbH & Co. KGaA, pp. 25–48.

Güzel-Seydim, Z.B., Greene, A.K. and Seydim, A.C. (2004a) Use of ozone in the food industry, *Lebensmittel-Wissenschaft und Technologie*, **37**: 453–60.

Güzel-Seydim, Z.B., Bever, P.I. Jr and Greene, A.K. (2004b) Efficacy of ozone to reduce bacterial populations in the presence of food components, *Food Microbiology*, **2**: 475–9.

Hartung, J. (1992) Emission and control of gases and odorous substances from animal housing and manure stores, *Zbl Hyg*, **192**(5): 389–418.

Inan, F., Pala, M. and Doymaz, I. (2007) Use of ozone in detoxification of aflatoxin B1 in red pepper, *Journal of Stored Products Research*, **43**: 425–9.

Jin-Gab, K., Ahmed, E.Y. and Sandhya, D. (1999) Application of ozone for enhancing the microbiological safety and quality of foods: a review, *Journal of Food Protection*, **62**: 1071–87.

Khadre, M.A., Yousef, A.E. and Kim, J.G. (2001) Microbiological aspects of ozone applications in food: a review, *Journal of Food Science*, **66**(9): 1242–52.

Kim-Yang, H., Von Bernuth, R.D., Hill, J.D. and Davies, S.H. (2005) Effect of ozonation on odor and the concentration of odorous organic compounds in air in a swine housing facility, *Transactions of the American Society of Agricultural Engineers*, **48**(6): 2297–302.

Kreis, R.D. (1978) Limiting the environmental impact of animal production odors, *Environment International*, **1**: 247–75.

Laszlo, Z., Kertesz, S., Mlinkovics, E. and Hodur, S. (2007) Dairy waste water treatment by combining ozonation and nanofiltration, *Separation Science and Technology*, **42**(7): 1627–37.

Laszlo, Z., Kertesz, S., Beszedes, S., Hovorka-Horvath, Z., Szabo, G. and Hodur, S. (2009) Effect of preozonation on the filterability of model dairy waste water in nanofiltration, *Desalination*, **240**: 170–7.

Lee, J.W., Cha, H.Y., Park, K.Y., Song, K.G. and Ahn, K.H. (2005) Operational strategies for an activated sludge process in conjunction with ozone oxidation for zero excess sludge production during winter season, *Water Research*, **39**(7): 1199–204.

Li, Q., Wang, L., Kamens, R.M. and Liu, Z. (2008) Could ozonation technology really work for mitigating air emissions from animal feeding operations? A short review and a new study, *American Society of Agricultural and Biological Engineers Annual International Meeting*, **5**: 2963–76.

Lucas, M.S., Peres, J.A., Lan, Y.B. and Li Puma, G. (2009) Ozonation kinetics of winery wastewater in a pilot-scale bubble column reactor, *Water Research*, **43**: 1523–32.

Lucas, M.S., Peres, J.A., Lan, Y.B. and Li Puma, G. (2010) Treatment of winery wastewater by ozone-based advanced oxidation processes (O3, O3/UV and

O3/UV/H2O2) in a pilot-scale bubble column reactor and process economics, *Separation Purification Technology*, doi:10.1016/j.seppur.2010.01.016.

Mackie, R.I., Stroot, P.G. and Varel, V.H. (1998) Biochemical identification and biological origin of key odour components in livestock waste, *J Anim Sci*, **76**: 1331–42.

Mahin, T.D. (2001) Comparison of different approaches used to regulate odours around the world, *Wat Sci Tech*, **44**(9): 87–102.

Manterola, G., Uriarte, I. and Sancho, L. (2008) The effect of operational parameters of the process of sludge ozonation on the solubilisation of organic and nitrogenous compounds, *Water Research*, **42**(12): 3191–7.

Miljøstyrelsen (2002) Industrial odour control, Environmental Guideline, No. 9, Danish Environmental Protection Agency, Danish Ministry of the Environment, Denmark, Vejledning nr. 9/2002.

Mines, R.O., Northenor, C.B. and Murchison, M. (2008) Oxidation and ozonation of waste activated sludge, *Journal of Environmental Science and Health Part A: Toxic/Hazardous Substances & Environmental Engineering*, **43**(6): 610–18.

Mohana, M., Acharya, B.K. and Madamwar, D. (2009) Distillery spent wash: treatment technologies and potential applications, *Journal of Hazardous Materials*, **163**: 12–25.

Nahm, K.H. (2002) Efficient feed nutrient utilization to reduce pollutants in poultry and swine manure, *Crit Rev Environ Sci Technol*, **32**(1): 1–16.

Oakes, D., Cooley, P., Edwards, L.L., Hirsch, R.W. and Miller, V.G. (1979) Ozone disinfection of fish hatchery waters: pilot plant results, prototype design and control considerations, *Proceedings of the World Mariculture Society*, **10**: 854–70.

Olafsdottir, G., Hognadottir, A., Martinsdottir, E. and Jonsdottir, H. (2000) Application of an electronic nose to predict total volatile bases in Capelin (*Mallotus villosus*) for fishmeal production, *J Agric Food Chem*, **48**: 2353–9.

O'Neill, D.H. and Phillips, V.R. (1991) A review of the control of odour nuisance from livestock buildings. Part 1. Influence of the techniques for managing waste within the building, *J Agric Eng Res*, **50**: 1–10.

Paraskeva, P. and Diamadopoulos, E. (2006) Technologies for olive mill wastewater (OMW) treatment: a review, *Journal of Chemical Technology and Biotechnology*, **81**: 1475–85.

Park, K.Y., Ahn, K.H., Maeng, S.K., Hwang, J.H. and Kwon, J.H. (2003) Feasibility of sludge ozonation for stabilization and conditioning, *Ozone: Science & Engineering*, **25**(1): 73–80.

Pascual, A., Llorca, L. and Canut, A. (2007) Use of ozone in food industries for reducing the environmental impact of cleaning and disinfection activities, *Trends in Food Science & Technology*, **18**(Suppl. 1): S29–S35.

Persaud, K.C., Khaffaf, S.M., Hobbs, P.J. and Sneath, R.W. (1996) Assessment of conducting polymer odour sensors for agricultural malodour measurements, *Chem Senses*, **21**(5): 495–505.

Priem, R. (1977) Deodorization by means of ozone, *Agric and Environ*, **3**(2): 227–37.

Ramel, M. and Nomine, M. (2000) Physicochemical characterisation of odours, *Analysis*, **28**(3): 171–9.

Rappert, S. and Muller, R. (2005) Odor compounds in waste gas emissions from agricultural operations and food industries, *Waste Management*, **25**: 887–907.

Rice, R.G., Robson, C.M., Miller, G.W. and Hill, A.G. (1981) Uses of ozone in drinking water treatment, *Journal of the American Water Works Association*, **73**(1): 44–57.

Ritter, W.F.J. (1989) Odour control of livestock wastes: state-of-the-art in North America, *Journal of Agricultural Engineering Research*, **42**(1): 51–62.

Safriet, D. (1995a) Emission factor documentation for AP-42, Section 9.8.3: pickles, sauces, and salad dressings, EPA-68-D2-0159, US Environmental Protection Agency, Office of Air Quality Planning and Standards, Research Triangle Park, NC.

Safriet, D. (1995b) Emission factor documentation for AP-42, Section 9.5.3: meat rendering plants, EPA-68-D2-0159, US Environmental Protection Agency, Office of Air Quality Planning and Standards, Research Triangle Park, NC.

Safriet, D. (1995c) Emission factor documentation for AP-42, Section 9.8.1: canned fruits and vegetables, EPA-68-D2-0159, US Environmental Protection Agency, Office of Air Quality Planning and Standards, Research Triangle Park, NC.

Saktaywin, W., Tsuno, H., Nagare, H., Soyama, T. and Weerapakkaroon, J. (2005) Advanced sewage treatment process with excess sludge reduction and phosphorus recovery, *Water Research*, **39**(5): 902–10.

Sangave, P.C., Gogate, P.R. and Pandit, A.B. (2007) Ultrasound and ozone assisted biological degradation of thermally pretreated and anaerobically pretreated distillery wastewater, *Chemosphere*, **68**: 42–50.

Sankaran, S., Khanal, S.K., Pometto, A.L. and Leeuwen, J.H. (2008) Ozone as a selective disinfectant for nonaseptic fungal cultivation on corn-processing wastewater, *Bioresource Technology*, **99**: 8265–72.

Sassi, A.B., Boularbah, A., Jaouad, A., Walker, G. and Boussaid, A. (2006) A comparison of Olive oil Mill Wastewaters (OMW) from three different processes in Morocco, *Process Biochemistry*, **41**: 74–8.

Schrader, K.K., Davidson, J.W. and Agnes, M. (2010) Evaluation of ozonation on levels of the off-flavor compounds geosmin and 2-methylisoborneol in water and rainbow trout *oncorhynchus mykiss* from recirculating aquaculture systems, *Aquacultural Engineering*, **43**(2): 46–50.

Sievers, M., Ried, A. and Koll, R. (2004) Sludge treatment by ozonation: evaluation of full-scale results, *Water Science and Technology*, **49**(4): 247–53.

Sigge, G.O., Britz, T.J, Fourie, P.C., Barnardt, C.A. and Strydom, R. (2001) Use of ozone and hydrogen peroxide in the post-treatment of UASB-treated alkaline fruit cannery effluent, *Water Science and Technology*, **44**(5): 69–74.

Singh, S., Fan, M. and Brown, R.C. (2008) Ozone treatment of process water from a dry-mill ethanol plant, *Bioresource Technology*, **99**: 1801–5.

Tyndall, J. and Colletti, J. (2007) Mitigating swine odor with strategically designed shelterbelt systems: a review, *Agroforest Systems*, **69**: 45–65.

Varel, V.H. and Miller, D.N. (2001) Effect of carvacrol and thymol on odour emissions from livestock wastes, *Wat Sci Tech*, **44**(9): 143–8.

Watkins, B.D., Hengemuehle, S.M., Person, H.L., Yokoyama, M.T. and Masten, S.J. (1997) Ozonation of swine manure wastes to control odors and reduce the concentrations of pathogens and toxic fermentation metabolites, *Ozone: Science and Engineering*, **19**(5): 425–37.

Wu, J. and Doan, H. (2005) Disinfection of recycled red meat processing wastewater by ozone, *Journal of Chemical Technology and Biotechnology*, **80**: 828–33.

Wu, J.J., Park, S.H., Hengemuehle, S.M., Yokoyama, M.T., Person, H.L., Gerrish, J.B. and Masten, S.J. (1999) The use of ozone to reduce the concentration of malodorous metabolites in swine manure slurry, *Journal of Agricultural Engineering Research*, **72**(4): 317–27.

Zhu, J. (2000) A review of microbiology in swine manure odour control, *Agricutlural Ecosystems and Environment*, **78**: 93–106.

13 Efficacy of Ozone on Pesticide Residues

Gilbert Y.S. Chan and J.G. Wu

13.1 Introduction

Modern agricultural production relies on the use of pesticides to improve the quality, quantity and diversity of crop yields. Figure 13.1 shows world and US amounts of pesticide active ingredients used by pesticide type in 2006 and 2007 (Grube et al. 2011). About 172 million pounds of pesticide were used in California in 2007 (Schwarzenegger et al. 2008), while 1.2 million tons of pesticide were consumed annually in China (Li 2006). Agricultural pesticides include herbicides, insecticides, fungicides, nematocides and rodenticides (Table 13.1). Pesticides applied to agriculture may affect nontarget organisms and contaminate soil and water bodies. An extra 318 million pounds of herbicides were used due to the switch to gene-modified crops, since herbicide-tolerant crops allow broad-spectrum herbicides to be used. Furthermore, the increase in seed price and decrease in herbicide cost causes farmers to consume more herbicides with less economic pinch (Benbrook 2009).

The type and quantity of pesticides in agriculture and aquaculture varies according to country. Variation in use also determines the pesticide maximum residual limits (MRLs) in international trade. Excessive application of pesticides in agriculture is absorbed either by the plant roots or foliar parts and incorporated into the plant tissues. These pesticides have a carryover effect in the resulting food and food product (Furlani et al. 2011). Toxic residues from pesticides can be found everywhere, including water systems, animal tissue and even human breast milk (Polder et al. 2009; Colles et al. 2008).

Chronic effects of pesticides are neurotoxicity, carcinogenesis, abnormal reproduction and cell development, mostly in the early stages of life (Burrows et al. 2002). The health implications of pesticide residues in food and consumer demands for chemical-free and safe food have resulted in the development of strategies to reduce pesticide residues from agricultural commodities through effective and sustainable post-harvest technologies. However, increasing food production in response to population growth and food storage without the use of pesticides and chemical

Ozone in Food Processing, First Edition. Edited by Colm O'Donnell, B.K. Tiwari, P.J. Cullen, and Rip G. Rice.

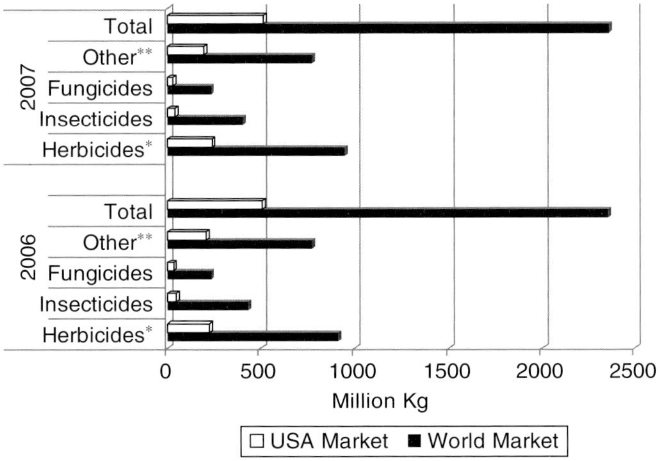

Figure 13.1 World and US pesticide expenditure at end user level in 2006–2007 (Grube et al. 2011). *'Herbicides' includes herbicides and plant growth regulators. **'Other' includes nematicides, fumigants and other miscellaneous conventional pesticides, and other chemicals used as pesticides such as sulfur, petroleum oil and sulfuric acid.

Table 13.1 Different types of pesticides.

Insecticides	Fungicides	Herbicides
Organochlorines	Hexachlorobenzene	Chlorophenoxy compounds
Organophosphates	Organomercurials	Bipyridyl derivatives
Carbomate esters	Pentachlorophenol	
Pyrethroids	Phthalimides	
Botanical insecticides	Dithiocarbamates	
Fumigants	Rodenticides	
Phosphine	Zinc phosphide	
Ethylene dibromide	Fluroacetic acid and derivatives	
Dibromochloropropane	a-napthyl thiourea (ANTU) Anticoagulants	

fertilisers are significant challenges. Consumer demand for safe food with less chemical residues has led researchers to focus on greener approaches to detoxification or a reduction of pesticide use to a level safe for human consumption.

Ensuring food production with minimal residual pesticide levels is paramount to consumer protection. A comprehensive review of food processing with respect to pesticide residue dissipation was reported by

Kaushik et al. (2009). Washing of fruits and vegetables with or without chlorine has been shown to lower pesticides residue (Kaushik et al. 2009; Ong et al. 1996). Home preparation by washing with different concentrations of acetic acid, sodium chloride and tap water, as well as refrigeration and stir-frying for different times can remove a portion of residual pesticides in vegetables (Zhang et al. 2007). This chapter outlines the application of ozone for the degradation of pesticide residues found in food. The proposed mechanisms for degradation of pesticides, including organophosphates and organochlorinated compounds, are also discussed.

13.2 Types of pesticides

Pesticides can be classified by target organism, chemical structure and physical state, and as inorganic, synthetic or biological (Council on Scientific Affairs 1997). Classification of pesticides is generally based on their structure. Organophosphates, organochlorine compounds, carbamates and pyrethroids (Table 13.2) are the most widely used (USEPA 2009).

13.3 Fates of pesticides

Pesticides are released into the environment through runoff, leaching, absorption, chemical degradation, microbial breakdown, photodegradation, volatilisation, spray drift and agriculture crops (Nollet and Rathore 2009). Chemical degradation mechanisms of pesticides are well known, for example hydrolysis, photolysis and oxidation (Castillo et al. 1997). A research study showed that pesticides decompose more rapidly inside plastic-covered greenhouses than inside glass greenhouses, since glass screens out much of the ultraviolet (UV) light (Kleier 1994; Katagi 2004). Similarly, pesticides on foliage are more easily broken down by sunlight than when they are contained in the soil. However, most pesticides are resistant to chemical and photochemical degradation under typical environmental conditions (Nollet and Rathore 2009).

13.3.1 Degradation processes of pesticides

Degradation of chemical pesticides requires biotic and abiotic processes. Biotic processes include biodegradation and metabolism. Sorption and degradation are the second most critical elements that determine the rate of decay of pesticides in soils (Kah et al. 2007). However, most chemical pesticides are nonbiodegradable, persist in the environment and bioaccumulate in adipose tissues of living organisms. They are persistent, lipophilic and hydrophobic, which makes it easy for them to accumulate in sediment and biological tissues.

Table 13.2 Some examples of pesticides and their structures.

Type of pesticide	Chemical structure	Example
Organochlorine pesticides	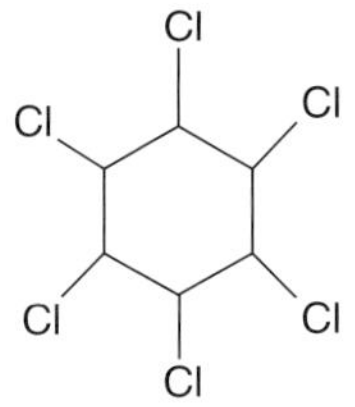	Hexachlorocyclohexane Lindane
	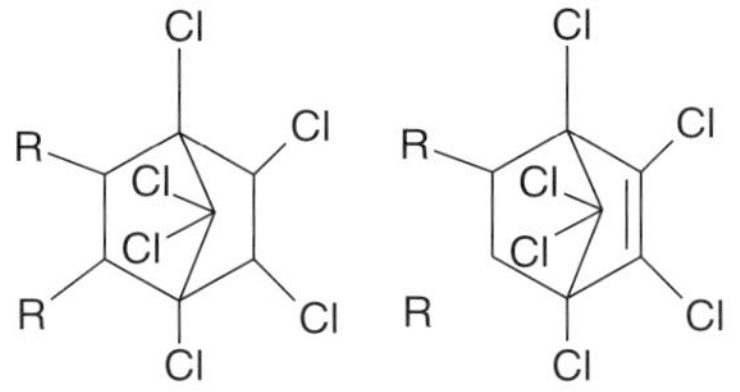	Toxaphene Chlordane
	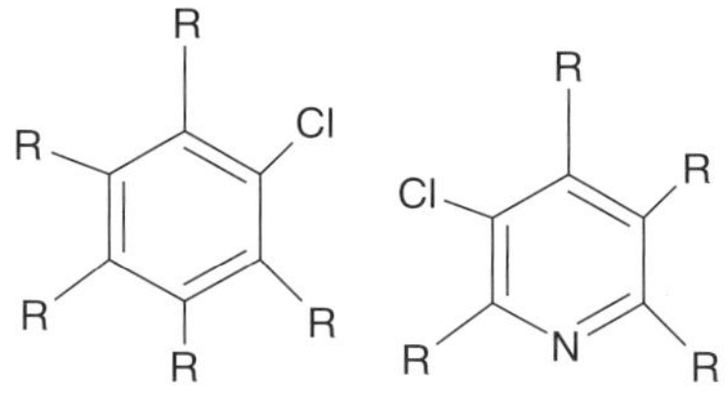	Isochlorthion Chlorpyrifos
Organophosphorus pesticides	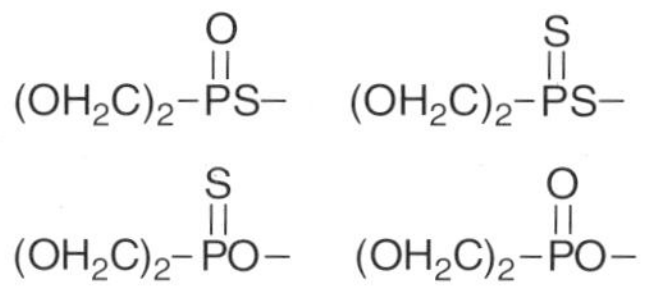	Temephos Mevinphos Omethoate Phenthoate
Pyrethroid pesticides	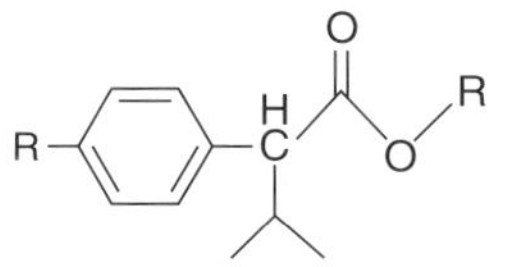	Fenvalerate

R denotes either H or another functional group.

Commercially, nanofiltration (Chen et al. 2004), activated carbon filtration (Foo and Hameed 2010), reverse osmosis (Bonné et al. 2000), distillation, adsorption (Gupta et al. 2006; Tepuš et al. 2009), photocatalytic degradation (Devipriya and Yesodharan 2005), photodegradation (Tanaka and Reddy 2009), ionising irradiation (Lepine 1991), gamma irradiation, biodenitrification reactors (Aslan and Turkman 2006), electrolysis adsorption (Vlyssides et al. 2005), microwaving (Salvador et al. 2002), electrochemical oxidation (Arapoglou et al. 2003) and ozonation (Ikeura et al. 2011a; Wu et al. 2007c) are all reported methods used to remove or reduce levels of pesticide residues.

13.3.2 Ozonation of pesticides

Organic contaminant removal, including of pesticides, pharmaceuticals and endocrine disruptors, is one of the major objectives in water treatment. The effectiveness of the treatments commonly used in Spanish drinking water plants to degrade 44 pesticides including alachlor, aldrin, ametryn, atrazine, chlorfenvinfos, chlorpyrifos, p,p′-DDD, o,p′-DDT, p,p′-DDT, desethylatrazine, 3,4-dichloroaniline, 4,4′-dichlorobenzophenone, dicofol, dieldrin, dimethoate, diuron, α-endosulphan, endosulphansulphate, endrin, α-HCH, β-HCH, γ-HCH, δ-HCH, heptachlor, heptachlor epoxide A, heptachlor epoxide B, hexachlorobenzene, isodrin, 4-isopropylaniline, isoproturon, metholachlor, methoxychlor, molinate, parathion methyl, parathion ethyl, prometon, prometryn, propazone, simazine, terbuthylazine, terbutryn, tetradifon and trifluralin systematically detected in the Ebro River Basin was investigated by Ormad et al. (2008). The results indicated that oxidation removed 70% of the pesticides. Ozonation in combination with a subsequent coagulation-flocculation-decantation did not improve the efficiency of the processes; however, a combination of ozonation with an activated-carbon absorption process gave rise to 90% removal of the studied pesticides. The authors concluded that oxidation was the most efficient method among the techniques studied for degradation of the majority of the studied pesticides. A similar study financed by the European Commission and reported by Maldonado et al. (2006) also indicated that complete pesticide mineralisation is hard to accomplish, and that large amounts of the oxidant are required to lower the organic content of the solutions. The possibility of ozonation cannot be ruled out if partial degradation is the expected goal.

Different organic contaminants require different treatment approaches with respect to ozonation. In municipal water treatment systems, progestogen endocrine disruptors reacted far slower with ozone than phenolic estrogens (Broséus et al. 2009). The same study also reported that ozone is effective for removing trace organic contaminants from water, with ozone doses typically applied in drinking water treatment; ozonation removed over 80% of caffeine, pharmaceuticals and endocrine disruptors with a "Ct" (concentration × exposure time) value of about 2 mg/min/L, with pesticides the most recalcitrant compounds to oxidise. Ozone was also applied for pesticide degradation in wastewater (Ballesteros Martín et al. 2011).

13.4 Degradation mechanisms

13.4.1 Kinetics

Ozonation of organic compounds including pesticide residues occurs by complex mechanisms involving mass transfer and a variety of possible chemical reactions. These reactions may be direct reactions of ozone with the target compound or its intermediates and (indirect) radical reactions between hydroxyl radicals (produced through ozone decomposition catalysed mainly by the hydroxide ion (HO^-) and organics (Mishchuk et al. 2008; Cullen et al. 2009). The rate constants of ozone with four groups of pesticides (4 phenolic, 8 organonitrogen, 8 phenoxyalkylacetic and 4 heterocyclic N-pesticides) were determined at pH7.5, ionic strength of ~10^{-3}M, and 100µM $NaHCO_3$ (Hu et al. 2000). The rate constants were found to vary widely according to the type and structure of the pesticides (Yao and Haag 1991). Hu et al. (2000) observed that the highest rate constant was 27600M/s and the lowest was 61.8M/s.

The logarithm of the rate constants was found to increase linearly with increase of the occupied molecular orbital (ε_{HOMO}) values, except for the phenoxyalkylacetic group. The term ε_{HOMO} is used to estimate the kinetic parameters of oxidation of organic compounds. When all examined pesticides were included in the correlation analysis, the rate constants gave a statistically unsatisfactory correlation with a coefficient of 0.92, but showed a similar trend of increasing reactivity with increasing ε_{HOMO} values of pesticides. The results of the correlation analysis suggested that the reactivity of pesticides with ozone follows the frontier orbital theory and can be estimated by ε_{HOMO}. Rate constants at 20°C and pH 7 were determined to be 11.9, 0.004 and 191.6M/s for organophosphorus pesticides chlorpyrifos, chlorfenvinfos and diazinon, respectively (Acero et al. 2008). The degradation kinetics of pesticides is influenced largely by pH of the medium and temperature. Temperature and pH have significant effects on the reactivity of ozone.

Different systems have been employed to study the degradation mechanisms of pesticides through ozonation alone and/or coupled with other oxidation methods such as hydrogen peroxide (H_2O_2), photo-fenton/ozone (PhFO) and TiO_2-photocatalysis/ozone (PhCO). PhFO- and PhCO-coupled systems were used as advanced oxidation processes (AOPs) for the degradation of biorecalcitrant pesticides including alachlor, atrazine, chlorfenvinfos, diuron, isoproturon and pentachlorophenol (Farré et al. 2005). These are considered Priority Hazardous Substances in the European Commission Water Framework Directive. The degradation processes of the different pesticides, which occurred through oxidation of the organic molecules by generated HO° radicals, followed first- and zero-order kinetics when PhFO and PhCO were applied, respectively. Javier-Benitez et al. (2002) also reported that the degradation of carbofuran (2,3-dihydro-2,2-dimethylbenzofuran-7-yl methylcarbamate, a priority pollutant) follows pseudo-first-order kinetics with degradation under ozone, UV radiation and Fenton's reagent treatment.

Most laboratory investigations of pesticide degradation kinetics employ ultrapure water and high-purity pesticide standards (e.g. Acero et al. 2008); however, under field conditions the full story may be very different. The presence of impurities or organic matter will compete with target pesticides and may greatly increase the running cost and duration for pesticide degradation. Under certain cases, impurities in the system are beneficial for degradation. Okawa et al. (2004) reported that degradation of 2,4-dichlorophenol by ozone in acetic acid was increased by 2.6 times with the inclusion of humic substances (natural organic matter). A comprehensive review paper on pesticide chemical oxidation has been summarised by Chiron et al. (2000). The presence of a low concentration of surfactant could improve the removal of atrazine by increasing the dissolution of ozone and the indirect generation of hydroxyl free radicals (Chu et al. 2006). Remediation time for trichloroethylene removal was reduced by more than 29% by applying acetic acid solutions saturated with ozone. However, increasing the ozone concentration was the best method for reducing remediation time, followed by increasing acetic acid concentration and increasing flow rate (Alcantara-Garduño et al. 2008).

13.4.2 Intermediates and oxidation products

Pesticides are complex molecules, but their degradation intermediates can be more toxic than their parental forms. In most cases, target pesticides cannot be mineralised solely by ozone, but are transformed into intermediates, and some of the products formed do not react further with ozone (Gottschalk et al. 2000). One of the health concerns of using oxidants to degrade pesticides is the formation of toxic intermediates. The structures of pesticides play an important role in byproduct toxicity; some degradation intermediates can be more toxic than their parental forms, while ozone byproducts of other pesticides may be less toxic. Oxidation products of organophosphorus pesticides (diazinon, methyl-parathion, parathion, chlorpyris) include oxons (such as diazinon oxon, methyl paraoxon, paraoxon, chlorpyris-oxon), and picric acid, phosphoric acid and nitrophenol were identified (Gunther et al. 1970; Laplanche et al. 1984; Pond et al. 1995; Wu et al. 2009). The oxon forms of organophosphorus pesticides were more potent acetylcholinesterase inhibitors than the parental organophosphorus pesticides (USEPA 2001). On the organism level, the degradation products of isoproturon are also more toxic to *Daphnia magna* than the parent compound (Mansour et al. 1999).

However, many studies suggested that most toxic intermediates can also be detoxified by ozonation. Some studies indicated that byproducts formed from the oxidation of organophosphorus pesticides and organochlorine pesticides were less toxic than the parental pesticides, evaluated with gap junction intercellular communication (GJIC) assay *in vitro* (Masten et al. 2001; Wu et al. 2007b). The oxidation intermediates of diazinon were found to be relatively safe, with no bioaccumulation and toxicity in fish, and the contamination of fish and other aquatic organisms by the oxidation products of diazinon and other organophosphorus insecticides in the environment was low (Tsuda et al. 1997). The catalytic ozonation of dichlorvos

(DDVP) using microporous silicate can successfully decrease its effect on total cytotoxicity (Fujita et al. 2002). Studies to date show that the pesticides can be detoxified by ozonation, but it is important to apply an excessive ozone dose and proper ozonation conditions so as to remove the potential accumulation of oxons (Wu et al. 2009).

Ozone is capable of causing ring cleavage in aqueous solutions (Turhan and Uzman 2008) for phenol removal. Removal of aniline was studied by Pierpoint et al. (2003) by radioisotope-labelled pesticides and soil columns exposed to 0.6% ozone gas (wt). The initial products included nitrosobenzene and nitrobenzene, which were then oxidised to CO_2. Similarly, trifluralin was oxidised to $^{14}CO_2$. Complete degradation is possible after extended ozonation, such as removal of the pesticide deltamethrin in an ozonation period of 210 minutes, as reported by Lafi and Al-Qodah (2006). Also using radioactive labelling, $^{14}CO_2$ production contributed up to 20% in the degradation process of chlorophenylurea pesticides (Amir Tahmasseb et al. 2002). There may be different pathways for mineralisation and for production of different intermediates. The degradation rate for chlorphenylurea pesticides was slower by a factor of 2.5, giving rise to additional pathways of hydroxylation of the phenyl ring and carbinolamine intermediate (Amir Tahmasseb et al. 2002).

Gas chromatography/electron capture detection is commonly used to determine degradation products, such as for organochlorine compounds after ozone and ozone/H_2O_2 treatment (Ormad et al. 1997). Degradation products only accounted for 70–80% of the initial concentration of methyl pirimiphos after ozonation (Chiron et al. 1998). Inhibition of bioluminescence from *Vibrio fischeri* appears to be one of the most versatile indicators of toxic compounds generated in oxidative degradation processes involving pesticides. However, detection of neurotoxic compounds, which can be generated in AOPs of some organophosphorus pesticides, requires acetylcholinesterase (AChE) bioassays that rely on inhibiting AChE activity. Bavcon Kralj et al. (2007) reported successful cases of the AChE flow-injection analysis (FIA)-thermal lens spectrometric (TLS) bioassay for toxicity testing in degradation of organophosphates by AOPs.

Chelme-Ayala et al. (2010) proposed a mechanism for the degradation of bromoxynil pesticide in aqueous solution with ozone (Figure 13.2). They suggested that bromoxynil may degrade via hydroxylation, leading to the formation of 3,5-dibromo-2,4-dihydroxybenzonitrile, or that debromination may proceed via 3-bromo-4,5-dihydroxybenzonitrile, 3-bromo-4-hydroxybenzonitrile, generating 4-hydroxybenzonitrile. Reaction pathways and mechanisms of photodegradation of pesticides were extensively reviewed by Burrows et al. (2002). They concluded that photodegradation of most pesticides by direct sunlight is minimal, which results in persistence of these pesticides in the environment.

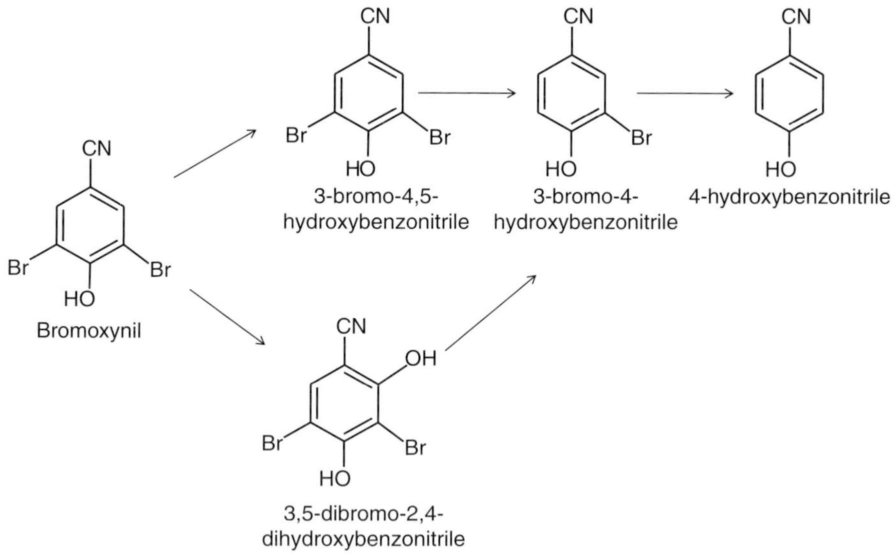

Figure 13.2 Proposed mechanism for the degradation of bromoxynil via conventional ozonation. (Reprinted from *Chemosphere*, Volume 78, Issue Number 5, Pamela Chelme-Ayala, Mohamed Gamal El-Din, Daniel W. Smith, Kinetics and mechanism of the degradation of two pesticides in aqueous solutions by ozonation, 557–562, 2010, with permission from Elsevier.)

Mechanisms of photodegradation could involve the following (Burrows et al. 2002):

(1) Direct photolysis leading to homolysis, heterolysis or photoionisation of the pesticide.
(2) Photosensitised photodegradation due to the absorption of light by a molecule.
(3) Photocatalytic degradation due to cyclic photoprocesses with the regeneration of the catalyst until all the substrate is destroyed.
(4) Direct reaction with the hydroxyl radical (HO°).

13.5 Ozone application for pesticide residues in fruits and vegetables

As oxidation is the major degradation process for common pesticides, it is logical that application of common oxidants, including ozone, is expected to be promising for the removal of pesticide residues. Ozonation may be a safe and promising process for the removal of pesticides on fruit and vegetable surfaces under domestic and industrial conditions. The effectiveness (50–100%) of ozonated and chlorinated water dips in the dissipation of azinphos-methyle, captan and formetanate hydrochloride on fresh and processed apples was reported by Ong et al. (1996). It is of concern that application of high levels of oxidants to foodstuffs may destroy some nutritional elements. The feasibilities of using low levels of dissolved ozone

Table 13.3 Pesticide removal after rinsing vegetables (*Brassica rapa*) with tap water and ozonated water at 24 °C (n = 3).

Pesticide	Contact time (min)	Tap water	Ozonated water	
		Removal (%)	(1.4 mg/L)	(2.0 mg/L)
Diazinon	15	24.9±2.3	33.8±1.8	44.5±5.0
	30	27.3±0.6	40.5±2.5	53.4±3.4
Methyl parathion	15	16.4±3.0	26.6±3.2	28.6±2.4
	30	30.0±3.1	39.6±2.6	47.9±5.7
Parathion	15	19.2±7.4	28.1±3.6	30.4±9.1
	30	30.5±5.1	52.1±4.7	55.3±4.0
Cypermethrin	15	25.5±0.6	33.4±2.3	53.5±8.7
	30	30.8±5.7	54.3±0.3	61.1±6.0

(1.4–2.0 mg/L) for the removal of the four pesticide residues (diazinon, methyl parathion, parathion, cypermethrin) on vegetable surfaces (*Brassica rapa*) was confirmed by Wu et al. (2007b). Their study indicated that more than 60% of cypermethrin and 27–55% of diazinon, methyl parathion and parathion was removed from vegetables (Table 13.3). Rinsing at 1.4 mg/L initial dissolved ozone for 15 minutes removed 27–34% of residual pesticide. Rinsing at a higher concentration of initial dissolved ozone (2.0 mg/L) increased the efficiency of pesticide removal to 30–54%. Extended rinsing at 2.0 mg/L initial dissolved ozone for 30 minutes increased the pesticide removal efficiency to 45–61%. In addition, the removal efficiency of pesticides depended on the dissolved ozone levels and temperature.

Wu et al. (2007c) observed that at 14 °C, only 36.2, 24.8, 19.7 and 44.3% of residual diazinon, methyl-parathion, parathion and cypermethrin, respectively, remained in vegetables. However, at 24 °C with similar dissolved ozone concentration (2 mg/L) the reduction efficiency of the same respective pesticides increased to 53.4, 47.9, 55.3 and 61.1%, respectively, in 30 minutes. Similarly, Ong et al. (1996) observed that the rate of degradation of the pesticides generally increased with increased pH and temperature. Ong et al. (1996) observed that about 53% of azinphos-methyl pesticide residue was removed with the water wash and 75% with ozone-containing water (0.25 mg/L) compared to unwashed apples. Hwang et al. (2001) also studied the effectiveness of various wash treatments (chlorine, chlorine dioxide, hydrogen peroxyacetic acid and ozone) on the removal of the mancozeb and ethylenethiourea (ETU) on and in fresh and processed apples. They observed 56–97% decreases in mancozeb residue and complete removal of ethylene-thiourea after ozone (3 ppm) washing.

In another study, Gabler et al. (2010) determined the effect of ozone fumigation on the residues of fenhexamid, cyprodinil, pyrimethanil and pyraclostrobin. They observed a significant reduction in the pesticide residue levels due to ozone fumigation of about 68.5, 75.4, 83.7 and 100.0%, respectively, after a single fumigation with 10 000 μl/L ozone for 1 hour.

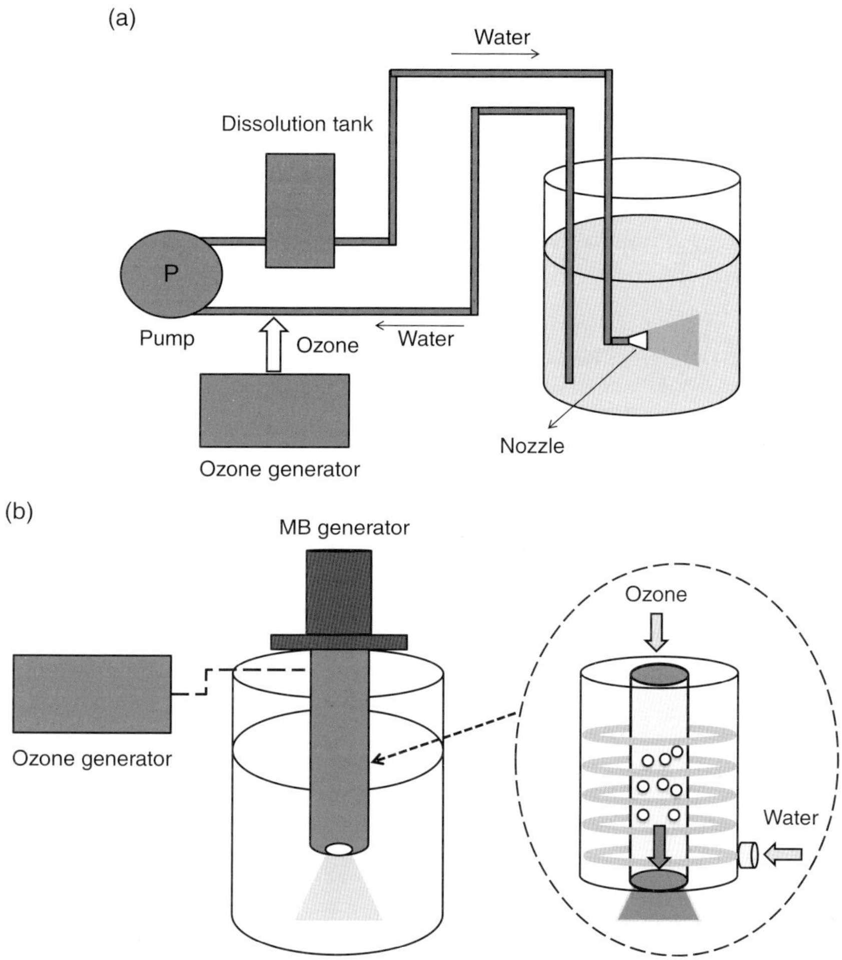

Figure 13.3 Schematic diagrams of equipment used for the generation of microbubbles (<50 μm). (a) The decompression-type microbubble generator (a sufficient amount of gas is dissolved in water under a 3–4 atmospheric pressure to cause a supersaturated condition). (b) The gas–water circulating-type microbubble generator (gas is introduced into the water vortex, and the formed gas bubbles are broken into microbubbles by breaking the vortex). (Reprinted from *Journal of Food Engineering*, Volume 103, Issue Number 3, H. Ikeura, F. Kobayashi, M. Tamaki, Removal of residual pesticide, fenitrothion, in vegetables by using ozone microbubbles generated by different methods, 345–49, 2011, with permission from Elsevier.)

However, they did not observe any significant effect on the residues of iprodione and boscalid. Degradation of pesticide residues owning to ozone is attributed to pesticide type. Both iprodione and boscalid, studied by Gabler et al. (2010), have systemic activity and might have protection from grape tissues. Iprodione is reported to degrade at a relatively slow rate (Hu et al. 2000; Gabler et al. 2010).

Ikeura et al. (2011a) investigated the effectiveness of an ozone microbubble (OMB) generator (Figure 13.3) for the removal of residual fenitrothion from lettuce, cherry tomatoes and strawberries. They observed that the decompression type was more effective than the gas–water circulation type in removing the residual pesticide in vegetables, which could be attributed to the larger number of small OMBs that could more easily infiltrate into vegetables than with the gas–water circulation type. Moreover, OMB treatment solutions generate hydroxyl radicals that are highly effective at decomposing organic molecules, including pesticide residues (Ikeura et al. 2011b; Sumikura et al. 2007; Takahashi et al. 2007).

13.6 Current status and opportunities

13.6.1 Ozone concentrations

As most residual pesticides found on fruit and vegetables are water-based, cleansing by ozone-containing water has two primary benefits: oxidative degradation, and conventional removal of soil and pesticides through simple dissolution and physical removal. Different ozone-based systems and gas/liquid phase combinations were compared by Tizaoui et al. (2008). Their study determined the removal of a reactive dyestuff, orange RO16, and 2-chlorophenol in relation to degradation performance and ozone consumption. The systems investigated included (1) liquid/gas-ozone (LGO): ozone was applied as it was produced in the gas phase; (2) liquid/solid-ozone (LSO): ozone was adsorbed on particulate silica-based material and then applied to water; (3) liquid/liquid-ozone (LLO): ozone was dissolved in a water-immiscible solvent and then applied to water; and (4) a photocatalytic system using a titanium dioxide catalyst (PHC). The reports indicated that the LSO system offered the possibility of using long contact times for slow ozone reactions and that the LLO system is most suitable for fast ozone reactions. However, both systems offer the prospect of more efficient use of ozone by extracting specific pollutants away from the water phase to the solid or solvent phases. Pflieger et al. (2009) also reported that ozonolysis of alachlor, trifluralin and terbutylazine adsorbed on silica particles under atmospheric conditions led to a long lifetime (more than 8 months) under 40 ppb of ozone. For the above reasons, using ozone gas for residual pesticide removal seems less promising.

In addition to pesticide removal, ozone has numerous applications in the food supply chain. Ozone gas is promising for microbial control of dried fruit, vegetables and crops: for example, microbial flora and degradation of alfatoxin in dried figs (Zorlugenç et al. 2008) and date fruits (Habibi Najafi and Haddad Khodaparast 2009), and insect pest control (*Tribolium* spp.) (Sousa et al. 2008). Ozone can be applied under vacuum for fresh-cut cantaloupe microbial control (Selma et al. 2008). However, certain levels of residual pesticide removal could not be removed through ozone fumigation.

Corona discharge for ozone generation is the most commonly employed and technologically feasible method for ozone gas generation; other methods are readily available but are generally expensive and are restricted for small-scale and laboratory production. Although ozone-containing water can be produced and directed by some alternative methods, mixing ozone with water through a Venturi tube or nanobubbling is the main method of application technology for production of ozone-containing water. Dissolved ozone levels are unlikely to exceed 1.0 mg/L by Venturi injection if ozone is produced by passing dried ambient air (oxygen content 20.9%) through a corona discharge device for ozone production. In principle, passing higher-purity oxygen (e.g. >90%, from an oxygen cylinder or molecular sieve concentrator) can produce an oxygen-enriched air stream with higher ozone concentration. If it is passed to a Venturi tube for ozone-containing water production, a higher level of dissolved ozone of about 5–10 mg/L can be achieved.

Since water with ozone micro- or nanobubbles has a dual function – water washing and direct ozone gas oxidation – micro- or nanobubble ozone generators are readily available in the market. The standardised indigo blue method is less feasible for dissolved ozone determination for water with fine bubbles, as the bubbles greatly affect the light path for colourimetric determination. Less-contaminated surface water has an oxidation reduction potential (ORP) level of around 200 mV. Higher ORP levels, such as 500 mV, can instantly kill many common microbes. It is also expected that it can remove a great portion of residual pesticides on fruits and vegetables.

13.6.2 Physical nature of plants affects degradation efficacy

As discussed above, degradation of pesticides by ozone must allow it to come into contact with the target pesticide. This physical phenomenon makes it less feasible for pesticide removal on fruit and vegetables with rough and hairy surfaces during ozone washing. However, inclusion of agitation may enhance penetration of ozone for effective disinfection.

13.6.3 Future trends

Advances in electronics will make ozone production simpler and more affordable (an ozone device powered by a 1.5 V battery is readily available on the market). These advances, combined with the US Food and Drug Administration (FDA) approval for use of ozone as an antimicrobial agent for direct contact with foods, will ensure that ozonation becomes a safe and promising process for the removal of pesticides from fruit and vegetable surfaces under commercial and domestic conditions. In an era of high energy demand, global concerns for energy conservation and enhanced environmentally friendly technology, ozone is set to become a more favourable option for pesticide removal; Tizaoui et al. (2008) estimated that a photocatalytic system using titanium dioxide catalyst showed the lowest rates and highest energy consumption by a factor of up to 400 times as compared to other ozone-based systems. Further research and development on ozonation technology should focus on:

- the development of improved devices for ozone gas production, to be employed for the production of ozone-containing water;
- investigating the feasibility of direct production of ozone-containing water without prior production of ozone gas;
- producing a higher purity of ozone gas without generation of other gaseous (e.g. nitrogen oxides) or soluble byproducts (e.g. nitrates).

References

Acero, J.L., Real, F.J., Javier-Benitez, F. and González, A. (2008b) Oxidation of chlorfenvinphos in ultrapure and natural waters by ozonation and photochemical processes, *Water Research*, **42**(12): 3198–206.

Alcantara-Garduño, M.E., Okuda, T., Tsai, T.Y., Nishijima, W. and Okada M. (2008) Experimental and mathematical evaluation of trichloroethylene removal from saturated soil using acetic acid with saturated ozone, *Separation and Purification Technology*, **60**(3): 299–307.

Amir Tahmasseb, L., Nélieu, S., Kerhoas, L. and Einhorn, J. (2002) Ozonation of chlorophenylurea pesticides in water: reaction monitoring and degradation pathways, *The Science of the Total Environment*, **291**(1–3): 33–44.

Arapoglou, D., Vlyssides, A., Israilides, C., Zorpas, A. and Karlis, P. (2003) Detoxification of methyl-parathion pesticide in aqueous solutions by electrochemical oxidation, *Journal of Hazardous Materials*, **98**(1–3): 191–9.

Aslan, S. and Turkman, A. (2006) Nitrate and pesticides removal from contaminated water using biodenitrification reactor, *Process Biochemistry*, **41**(4): 882–6.

Ballesteros Martín, M.M., Sánchez Pérez, J.A., García Sánchez, J.L., Casas López, J.L. and Malato Rodríguez, S. (2011) Effect of pesticide concentration on the degradation process by combined solar photo-Fenton and biological treatment, *Water Research*, **43**(15): 3838–48.

Bavcon Kralj, M., Trebše, P. and Franko, M. (2007) Applications of bioanalytical techniques in evaluating advanced oxidation processes in pesticide degradation, *TrAC Trends in Analytical Chemistry*, **26**(11): 1020–31.

Benbrook, C. (2009) *Impacts of Genetically Engineered Crops on Pesticide Use: The First Thirteen Years*, The Organic Center.

Bonné, P.A.C., Beerendonk, E.F., Van Der Hoek, J.P. and Hofman, J.A.M.H. (2000) Retention of herbicides and pesticides in relation to aging of RO membranes, *Desalination*, **132**(1–3): 189–93.

Broséus, R., Vincent, S., Aboulfadl, K., Daneshvar, A., Sauvé, S., Barbeau, B. and Prévost, M. (2009) Ozone oxidation of pharmaceuticals. endocrine disruptors and pesticides during drinking water treatment, *Water Research*, **43**(18): 4707–17.

Burrows, H.D., Canle, L.M., Santaballa, J.A. and Steenken, S. (2002) Reaction pathways and mechanisms of photodegradation of pesticides, *Journal of Photochemistry and Photobiology B: Biology*, **67**(2): 71–108.

Castillo, M., Domingues, R., Alpendurada, M.F. and Barceló, D. (1997) Persistence of selected pesticides and their phenolic transformation products in natural waters using off-line liquid solid extraction followed by liquid chromatographic techniques, *Analytica Chimica Acta*, **353**(1): 133–42.

Chelme-Ayala, P., El-Din, M.G. and Smith, D.W. (2010) Kinetics and mechanism of the degradation of two pesticides in aqueous solutions by ozonation, *Chemosphere*, **78**(5): 557–62.

Chen, S.S., Taylor, J.S., Mulford, L.A. and Norris, C.D. (2004) Influences of molecular weight, molecular size, flux, and recovery for aromatic pesticide removal by nanofiltration membranes, *Desalination*, **160**(2): 103–11.

Chiron, S., Rodriguez, A. and Fernandez-Alba, A. (1998) Application of gas and liquid chromatography–mass spectrometry to the evaluation of pirimiphos methyl degradation products in industrial water under ozone treatment, *Journal of Chromatography A*, **823**(1–2): 97–107.

Chiron, S., Fernandez-Alba, A., Rodriguez, A. and Garcia-Calvo, E. (2000) Pesticide chemical oxidation: state-of-the-art, *Water Research*, **34**(2): 366–77.

Chu, W., Chan, K.H. and Graham, N.J.D. (2006) Enhancement of ozone oxidation and its associated processes in the presence of surfactant: degradation of atrazine, *Chemosphere*, **64**(6): 931–6.

Colles, A., Koppen, G., Hanot, V., Nelen V., Dewolf, M-C., Noël, E., Malisch, R., Kotz, A., Kypke, K., Biot, P., Vinkx, C. and Schoeters, G. (2008) Fourth WHO-coordinated survey of human milk for persistent organic pollutants (POPs): Belgian results, *Chemosphere*, **73**(6): 907–14.

Council on Scientific Affairs (1997) Educational and informational strategies to reduce pesticide risks, *Preventive Medicine*, **26**(2): 191–200.

Cullen, P.J., Tiwari, B.K., O'Donnell, C.P. and Muthukumarappan, K. (2009) Modelling approaches to ozone processing of liquid foods, *Trends Food Sci Technol*, **20**(3–4): 125–36.

Devipriya, S. and Yesodharan, S. (2005) Photocatalytic degradation of pesticide contaminants in water, *Solar Energy Materials and Solar Cells*, **86**(3): 309–48.

Farré, M.J., Franch, M.I., Malato, S., Ayllón, J.A., Peral, J. and Doménech, X. (2005) Degradation of some biorecalcitrant pesticides by homogeneous and heterogeneous photocatalytic ozonation, *Chemosphere*, **58**(8): 1127–33.

Foo, K.Y. and Hameed, B.H. (2010) Detoxification of pesticide waste via activated carbon adsorption process, *Journal of Hazardous Materials*, **175**(1–3): 1–11.

Fujita, H., Sakai, Y., Sakoda, A. and Suzuki, M. (2002) Catalytic ozonation of an organophosphorus pesticide using microporous silicate and its effect on total toxicity reduction, *Water Science and Technology*, **46**: 35–41.

Furlani, P.R.Z., Marcilio, K.M., Leme, F.M. and Tfouni, S.A.V. (2011) Analysis of pesticide residues in sugarcane juice using QuEChERS sample preparation and gas chromatography with electron capture detection, *Food Chemistry*, **126**(3): 1283–7.

Gabler, F.M., Smilanick, J.L., Mansour, M.F. and Karaca, H. (2010) Influence of fumigation with high concentrations of ozone gas on postharvest gray mold and fungicide residues on table grapes, *Postharvest Biology and Technology*, **55**(2): 85–90.

Gottschalk, C., Libra, J.A. and Saupe, A. (2000) *Ozonation of Water and Wastewater: A Practical Guide to Understanding Ozone and its Application*, Weinheim, Germany: Wiley-VCH Verlag GmbH & Co. KGaA.

Grube, A., Donaldson, D., Kiely, T. and Wu, L. (2011) Pesticides industry sales and usage 2006 and 2007 market estimates. US Environmental Protection Agency. Retrieved from http://www.epa.gov/opp00001/pestsales/07pestsales/market_estimates2007.pdf. Last accessed 27 July 2011.

Gunther, F.A., Ott, D.E. and Ittig, M. (1970) Oxidation of parathion to paraoxon II by use of ozone, *B Environ Contam Tox*, **5**: 87–94.

Gupta, V.K., Suhas, Ali, I., Saini, V.K. (2006) Adsorption of 2,4-D and carbofuran pesticides using fertilizer and steel industry wastes, *Journal of Colloid and Interface Science*, **299**(2): 556–63.

Habibi Najafi, M.B., Haddad Khodaparast, M.H. (2009) Efficacy of ozone to reduce microbial populations in date fruits, *Food Control*, **20**(1): 27–30.

Hu, J.Y., Morita, T., Magara, Y. and Aizawa, T. (2000) Evaluation of reactivity of pesticides with ozone in water using the energies of frontier molecular orbitals, *Water Research*, **34**(8): 2215–22.

Hwang, E., Cash, J.N. and Zabik, M.J. (2001) Postharvest treatments for the reduction of mancozeb in fresh apples, *J Agric Food Chem*, **49**: 3127–32.

Ikeura, H., Kobayashi, F. and Tamaki, M. (2011a) Removal of residual pesticide, fenitrothion, in vegetables by using ozone microbubbles generated by different methods, *Journal of Food Engineering*, **103**(3): 345–9.

Ikeura, H., Kobayashi, F. and Tamaki, M. (2011b) Removal of residual pesticides in vegetables using ozone microbubbles, *Journal of Hazardous Materials*, **186**(1): 956–9.

Javier-Benitez, F., Acero, J.L. and Real, F.J. (2002) Degradation of carbofuran by using ozone: UV radiation and advanced oxidation processes, *Journal of Hazardous Materials*, **89**(1): 51–65.

Kah, M., Beulke, S. and Brown, C.D. (2007) Factors influencing degradation of pesticides in soil, *Journal of Agricultural and Food Chemistry*, **55**(11): 4487–92.

Katagi, T. (2004) Photodegradation of pesticides on plant and soil surfaces, *Reviews of Environmental Contamination & Toxicology*, **182**: 1–195.

Kaushik, G., Satya, S. and Naik, S.N. (2009) Food processing a tool to pesticide residue dissipation: a review, *Food Research International*, **42**(1): 26–40.

Kleier, D.A. (1994) Environmental effects on the photodegradation of pesticides, in *Comparing Glasshouse & Field Pesticide Performance, Vol II*, BCPC Monograph No. 59.

Lafi, W.K. and Al-Qodah, Z. (2006) Combined advanced oxidation and biological treatment processes for the removal of pesticides from aqueous solutions, *Journal of Hazardous Materials*, **137**(1): 489–97.

Laplanche, A., Martin, G. and Tronnard, E. (1984) Ozonation schemes of ogranophosphorus pesticides: application in drinking water treatment, *Ozone: Science & Engineering*, **6**: 207–19.

Lepine, F.L. (1991) Effects of ionizing radiation on pesticides in a food irradiation perspective: a bibliographic review, *Journal of Agricultural and Food Chemistry*, **39**(12): 2112–18.

Li, Z. (2006) Soil quality deteriorating in China, threatening public health and ecosystems. Retrieved from http://www.worldwatch.org/node/4419.

Maldonado, M.I., Malato, S. and Pérez-Estrada, L.A. (2006) Partial degradation of five pesticides and an industrial pollutant by ozonation in a pilot-plant scale reactor, *Journal of Hazardous Materials*, **138**(2): 363–9.

Mansour, M., Feicht, E.A., Behechti, A., Schramm, K.W. and Kettrup, A. (1999) Determination photostability of selected agrochemicals in water and soil, *Chemosphere*, **39**(4): 575–85.

Masten, S.J., Tian, M., Upham, B.L. and Trosko, J.E. (2001) Effect of selected pesticides and their ozonation by-products on gap junctional intercellular communication using rat liver epithelial cell lines, *Chemosphere*, **44**(3): 457–65.

Mishchuk, N.A., Goncharuk, V.V. and Vakulenko, V.F. (2008) Theoretical analysis of physicochemical processes occurring during water treatment by ozone and ultraviolet radiation, *Advances in Colloid and Interface Science*, **139**: 62–73.

Nollet, L.M.L. and Rathore, H.S. (2009) *Handbook of Pesticides: Methods of Pesticide Residues Analysis*, Boca Raton, FL: Taylor & Francis.

Okawa, K., Nakano, Y., Nishijima, W. and Okada, M. (2004) Effects of humic substances on the decomposition of 2,4-dichlorophenol by ozone after extraction from water into acetic acid through activated carbon, *Chemosphere*, **57**(9): 1231–5.

Ong, K.C., Cash, J.N., Zabik, M.J., Siddiq, M. and Jones, A.L. (1996) Chlorine and ozone washes for pesticide removal from apples and processed apple sauce, *Food Chemistry*, **55**(2): 153–60.

Ormad, P., Cortes, S., Puig, A. and Ovelleiro, J.L. (1997) Degradation of organochlorine compounds by O_3 and O_3/H_2O_2, *Water Research*, **31**(9): 2387–91.

Ormad, M.P., Miguel, N., Claver A., Matesanz, J.M. and Ovelleiro, J.L. (2008) Pesticides removal in the process of drinking water production, *Chemosphere*, **71**(1): 97–106.

Pflieger, M., Monod, A. and Wortham, H. (2009) Kinetic study of heterogeneous ozonolysis of alachlor, trifluralin and terbuthylazine adsorbed on silica particles under atmospheric conditions, *Atmospheric Environment*, **43**(35): 5597–603.

Pierpoint, A.C., Hapeman, C.J. and Torrents, A. (2003) Ozone treatment of soil contaminated with aniline and trifluralin, *Chemosphere*, **50**(8): 1025–34.

Polder, A., Skaare, J.U., Skjerve, E., Løken, K.B. and Eggesbø, M. (2009) Levels of chlorinated pesticides and polychlorinated biphenyls in Norwegian breast milk (2002–2006), and factors that may predict the level of contamination, *Science of the Total Environment*, **407**(16): 4584–90.

Pond, A.L., Chambers, H.W. and Chambers, J.E. (1995) Organophosphate detoxication potential of various rat tissues via A-esterase and aliesterase activities, *Toxicol Letter*, **78**(3): 245–52.

Salvador, R., Casal, B., Yates, M., Martín-Luengo, M.A. and Ruiz-Hitzky, E. (2002) Microwave decomposition of a chlorinated pesticide (Lindane) supported on modified sepiolites, *Applied Clay Science*, **22**(3): 103–13.

Schwarzenegger, A., Adams, L.S. and Warmerdam, M.A. (2008) *Summary of Pesticide Use Report Data 2007*, California Environmental Protection Agency, Department of Pesticide Regulation.

Selma, M.V., Allende, A., Lopez-Galvez, F., Conesa, M.A. and Gil, M.I. (2008) Disinfection potential of ozone, ultraviolet-C and their combination in wash water for the fresh-cut vegetable industry, *Food Microbiology*, **25**(6): 809–14.

Sousa, A.H., Faroni, L.R.D'A., Guedes, R.N.C., Tótola, M.R. and Urruchi, W.I. (2008) Ozone as a management alternative against phosphine-resistant insect pests of stored products, *Journal of Stored Products Research*, **44**(4): 379–85.

Sumikura, M., Hidaka, M., Murakami, H., Nobutomo, Y. and Murakami, T. (2007) Ozone microbubble disinfection method for wastewater reuse system, *Water Sci Technol*, **56**: 53–61.

Takahashi, M., Chiba, K. and Li, P. (2007) Formation of hydroxyl radicals by collapsing ozone microbubbles under strongly acidic conditions, *J Phys Chem B*, **111**: 11 443–6.

Tanaka, K. and Reddy, K.S.N. (2002) Photodegradation of phenoxyacetic acid and carbamate pesticides on TiO2, *Applied Catalysis B: Environmental*, **39**(4): 305–10.

Tepuš, B., Simonič, M. and Petrinić, I. (2009) Comparison between nitrate and pesticide removal from ground water using adsorbents and NF and RO membranes, *Journal of Hazardous Materials*, **170**(2–3): 1210–17.

Tizaoui, C., Bickley, R.I., Slater, M.J., Wang, W.-J., Ward, D.B. and Al-Jaberi, A. (2008) A comparison of novel ozone-based systems and photocatalysis for the removal of water pollutants, *Desalination*, **227**(1–3): 57–71.

Tsuda, T., Kojima, M., Harada, H., Nakajima, A. and Aoki, S. (1997) Acute toxicity, accumulation and excretion of organophosphorous insecticides and their oxidation products in killifish, *Chemosphere*, **35**(5): 939–49.

Turhan, K. and Uzman, S. (2008) Removal of phenol from water using ozone, *Desalination*, **229**(1–3): 257–63.

USEPA (2001) The incorporation of water treatment effects on pesticide removal and transformation in Food Quality Protection Act (FQPA) drinking water assessments, Office of Pesticide Programs, **50**.

USEPA (2009) Types of pesticides. Retrieved from http://www.epa.gov/opp00001/about/types.htm.

Vlyssides, A., Arapoglou, D., Mai, S. and Barampouti, E.M. (2005) Electrochemical detoxification of four phosphorothioate obsolete pesticides stocks, *Chemosphere*, **58**(4): 439–47.

Wu, J., Luan, T., Lan, C., Hung Lo, T.W. and Chan, G.Y.S. (2007a) Removal of residual pesticides on vegetables using ozonated water, *Food Control*, **18**(5): 466–72.

Wu, J.G., Luan, T.G., Lan, C.Y., Lo, W.H. and Chan, G.Y.S. (2007b) Efficacy evaluation of low-concentration of ozonated water in removal of residual diazinon, parathion, methyl-parathion and cypermethrin on vegetable, *Journal of Food Engineering*, **79**(3): 803–9.

Wu, J., Lin, L., Luan, T., Chan, G.Y.S. and Lan, C. (2007c) Effects of organophosphorus pesticides and their ozonation byproducts on gap junctional intercellular communication in rat liver cell line, *Food and Chemical Toxicology*, **45**(10): 2057–63.

Wu, J., Lan C. and Chan, G.Y.S. (2009) Organophosphorus pesticide ozonation and formation of oxon intermediates, *Chemosphere*, **76**(9): 1308–14.

Yao, C.C.D. and Haag, W.R. (1991) Rate constants for direct reactions of ozone with several drinking water contaminants, *Water Research*, **25**(7): 751–73.

Zhang, Z.Y., Liu, X.J. and Hong, X.Y. (2007) Effects of home preparation on pesticide residues in cabbage, *Food Control*, **18**(12): 1484–7.

Zorlugenç, B., Zorlugenç, F.K., Öztekin, S. and Bülend Evliya, I. (2008) The influence of gaseous ozone and ozonated water on microbial flora and degradation of aflatoxin B1 in dried figs, *Food and Chemical Toxicology*, **46**(12): 3593–7.

14 Modelling Approaches for Ozone Processing

Vasilis P. Valdramidis, P.J. Cullen and B.K. Tiwari

Nomenclature

- α_{AB}: stoichiometric yield ratio (mL O_3 consumed/mol organic matter);
- ε_G: gas hold up (dimensionless);
- δ: time for the first decimal reduction of the Weibull model (minutes);
- Δx: distance between two measuring points (m);
- ΔP: pressure difference (Pa);
- λ: coefficient of specific lethality (mg/L/min);
- ρ_L: density of the liquid (kg/m^3);
- ψ: ozone consumption or absorption rate (mg/L/s);
- A_b: surface area of the bubble (m^2);
- E: enhancement factor (–);
- C_L^e: dissolved ozone concentration in equilibrium with the ozone gas (mg/L);
- C_L: dissolved ozone concentration (mg/L);
- C_{ALb}: dissolved ozone concentration in liquid (mg/L);
- C_{BLb}: concentration of organic matter in liquid (mg/L);
- D: instantaneous ozone demand (mg/L);
- g: 9.81 m/s^2
- K_{AB}: ozonation rate constant of organic matter in liquid (per minute);
- K_d: self-decomposition rate constant of ozone (per minute);
- KL_a: overall mass transfer coefficient (per minute);
- k': (pseudo) first-order kinetic constant (per minute);
- k: overall reaction kinetic constant (per minute);
- k_{max}: microbial inactivation rate (per minute);
- k^*: ozone decomposition (per minute);
- N_t: number of microorganisms at time t (CFU/mL);
- N_o: initial number of microorganisms (CFU/mL);
- N_c: increased initial number of microorganisms to compensate initial lag phase (CFU/mL);
- OD: ozone dose (mg/L);
- OD: initial applied ozone dose (mg/L);

Ozone in Food Processing, First Edition. Edited by Colm O'Donnell, B.K. Tiwari, P.J. Cullen, and Rip G. Rice.

- O_R: residual ozone (mg/L);
- O: ozone applied (mg/L);
- p: parameter related to the shape of the Weibul inactivation curve (–);
- SD: Sauter mean diameter (m);
- S_l: shoulder length (minutes);
- V_b: surface volume of the bubble (m^3).

14.1 Introduction

Mathematical models in food processing aim to describe the different aspects of the processes and their dynamics quantitatively using theoretical analysis and experimental results. In addition to using these models to understand and describe a process, they can also be applied for control and optimisation purposes. In general, mathematical models offer new possibilities to manage the increasing complexity of the studied technologies and explore solutions in a short time, while reducing costs of operation. The chosen or constructed models should address universal concepts, indentify level of detail, necessary simplifications, and uncertainty, for example due to lack of data or inherited variability, especially when dealing with biological systems like food products. Depending on the mechanistic knowledge upon which models are built, they can be subdivided into deductive or inductive (Hills 2001). Inductive kinetic models have as a starting point the available data, while deductive models start with the general laws – that is, (bio)chemical/physical – and use them to build realistic mathematical expressions.

The powerful tool of quantitative analysis also has applications in ozone processing of foods. Ozonation of food products is a physical process that involves chemical, physical and microbiological changes. In cases of microbial kinetics, most of the current modelling approaches are built on inductive models, although some parameters of the selected model structures can have a biological interpretation. However, physical and chemical kinetics can be described by the use of deductive models. The physical processes include mass transfer and hydrodynamics, while the chemical processes include all direct and indirect reactions of ozone.

This chapter reviews modelling approaches to be considered for ozone processing of food products and builds upon well established developments in the field of ozone processing of water and wastewater. Microbial modelling, reaction kinetic modelling during ozonation and modelling of ozone operation set-ups are outlined.

14.2 Modelling approaches for microbial inactivation

Use of kinetic models that describe the microbial responses during ozone treatment is a quantitative approach to designing and optimising an ozone treatment focusing on the production of safe food products. These models can be further exploited to quantitatively describe the influence of processing conditions on food safety. Consequently, the effects of intrinsic, extrinsic and/or processing factors on the resulting microbial proliferation in food products or food model systems can be evaluated. Ideally these models should be parsimonious, flexible and built upon parameters based on the physiological mechanism of inactivation (Van Impe et al. 2004).

The bactericidal effects of ozone have been studied on a variety of organisms, including Gram-positive and Gram-negative bacteria, as well as spores and vegetative cells (Restaino et al. 1995). The antimicrobial efficacy of ozone against food-related microorganisms has been studied for Gram-positive bacteria: *Listeria monocytogenes*, *Staphylococcus aureus*, *Bacillus cereus*, *Enterococcus faecalis*; Gram-negative bacteria: *Pseudomonas aeruginosa* and *Yersinia enterocolitica*; yeasts: *Candida albicans* and *Zygosaccharomyces bacilli*; and spores of *Aspergillus niger* (Restaino et al. 1995). For the mechanisms of antimicrobial action, refer to Chapter 4 of this book.

In microbiological studies involving ozone treatment, modelling of microbial inactivation has been reported by numerous researchers (Bialka et al. 2008; Selma et al. 2007). Use of ozone as an antimicrobial agent causes difficulty in predicting ozone reactions in the presence of complex food materials. Similarly, it is more difficult to predict the effect of ozone treatment on microbes in the presence of organic matter (Cho et al. 2003) than in model fluids or microbiological media. However, kinetic models are a fast and economical way to assess microbial inactivation and may be used to predict the influence of ozonation control parameters such as treatment time, gas flow rate and ozone concentration when the process is performed at an ozone bubble column (Tiwari et al. 2008).

In recent years, the numerous nonlinear regressions that have been developed and proposed for various food preservation techniques have been reviewed by Geeraerd et al. (2005) and reported by among others Anderson et al. (1996), Augustin et al. (1998), Baranyi and Pin (2001), Peleg and Cole (1998), Peleg (2003), Geeraerd et al. (2005) and Valdramidis et al. (2006, 2007). These nonlinearities could include shoulder and tailing effects of the microbial kinetics (Raso et al. 2000; Smelt et al. 2002). The occurrence of a shoulder in the survival curve is often related to sublethal injury, multitarget inactivation, cell clumping or activation phenomena for spores (Mañas and Pagan 2005; Geeraerd et al. 2000). In the case of ozonation, the occurrence of tail could be attributed to inherent resistance or resistance acquired during ozonation, while the formation of a shoulder or a tail during ozone treatment could be due to rapid reactivity between ozone and the produced radicals and microorganisms, or to environmental factors, such as the presence of organic material (Hunt and Mariñas 1997; Restaino et al. 1995). Gujer and von Gunten (2003) reported that during ozone treatments some microorganisms are characterised by a minimum ozone exposure in order to achieve the desired inactivation.

A summary of proposed models for microbial inactivation using ozone is given in Table 14.1. Some of these modelling approaches are further

Table 14.1 Modelling approaches for microbial inactivation kinetics during ozone processing. (Adjusted by Cullen et al. (2009) with permission from Elsevier.)

Ozone process	Food product	Microorganism	Model and parameters	Reference
Gaseous ozone	Buffer solution	*E. coli* 0157:H7	First-order $\ln\left(\frac{N_t}{N_0}\right) = -k \cdot t$	Hunt and Mariñas (1997)
Ozone gas	Media	*Cryptosporidium parvum*	Chick–Watson model $\ln\left(\frac{N_t}{N_0}\right) = -k \cdot C_{avg} \cdot t$ $C_{avg} = \frac{C_o - C_t}{2}$	Li et al. (2001)
Gaseous ozone	Apple cider	*E. coli* 0157:H7	First-order $\ln\left(\frac{N_t}{N_0}\right) = -k \cdot t$	Steenstrup and Floros (2004)
Gaseous ozone	Different media	*Listeria innocua*	Gompertz model $N_t = \alpha + \gamma \exp[-\exp(-\beta(t-\mu))]$ where N_t = log CFU/mL, t = elapsed time, μ = the inflection point, β = slope parameter, γ=range and α = final CFU/mL	Fan et al. (2007)
Ozone-containing water	Shredded lettuce and water	*Shigella sonnei*	Chick–Watson: $\ln\left(\frac{N_t}{N_0}\right) = -k \cdot OD$ Modified Chick: $\ln\left(\frac{N_t}{N_0}\right) = -k\frac{(OD_0 - D)}{k^*}[1-\exp(-k^* t)]$	Selma et al. (2007)

			Modified Chick–Watson: $\ln\left(\frac{N_t}{N_0}\right) = -k\frac{(OD_0 - D)^q}{q \cdot k^*}[1-\exp(-q \cdot k^* \cdot t)]$ Modified multiple target: $\ln\left(\frac{N_t}{N_0}\right) = \log_e\left[1-(1-\exp)\left(\frac{k \cdot (OD_0 - D) \cdot (\exp(-k^* \cdot t) - 1)}{k^*}\right)^n\right]$ where OD = ozone dose, OD_0 = instantaneous ozone demand, k^* = ozone decomposition rate at time t and k, p, q are model constants	
Aqueous ozone Gaseous ozone	Raspberries and strawberries	*E. coli* O157:H7 and *Salmonella enterica*	Weibull model $\ln\left(\frac{N_t}{N_0}\right) = -\frac{1}{2.303} \cdot \left(\frac{t}{\alpha}\right)^\beta$ First-order	Bialka et al. (2008)
Ozone gas	Orange/apple juice	*E. coli* ATCC 25922 and NCTC 12900 *L. monocytogenes* ATCC 7644, *L. monocytogenes* NCTC 11994 and *L. innocua* NCTC 11288	Weibull model $\log_{10}(N) = \log_{10}(N_0) - \left(\frac{t}{\delta}\right)^p$	Patil et al. (2009, 2010a)
Ozone gas	Apple juice	*E. coli* ATCC 25922 and NCTC 12900	Geeraerd et al. (2000) model $\log_{10}(N) = \log_{10}(N_0) - \frac{k_{\max}(t)}{\ln(10)} + \frac{\log_{10} e^{(k_{\max} \cdot S_l)}}{\left(1+e^{(k_{\max} \cdot S_l)} - 1\right)} \cdot e^{(-k_{\max} \cdot t)}$	Patil et al. (2010a)

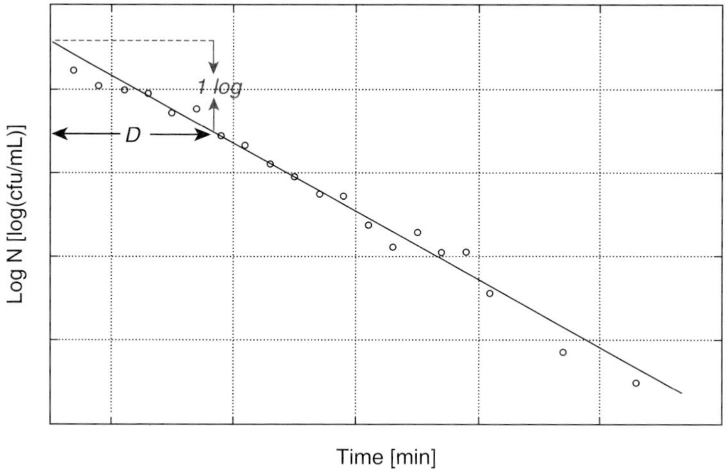

Figure 14.1 Graphical representation of the decimal reduction time, *D*, in a fit of the loglinear model on microbial inactivation data.

discussed in this section. The inactivation of pathogenic microorganisms in water during ozonation has been described by the use of first-order kinetics with respect to ozone concentration (Hunt and Mariñas 1997). In the general case of first-order kinetics, a linear relationship between the logarithm of the microbial population and time is considered. This reads as follows:

$$\frac{dN_t}{dt} = -k \cdot N_t \text{ with } k = \frac{\ln 10}{D} \tag{14.1}$$

or in static conditions:

$$\log \frac{N_t}{N_o} = -k' \cdot t \tag{14.2}$$

where k (where $k' = k \cdot \log(\exp(1))$) is the first-order inactivation constant and D is the decimal reduction time in minutes (i.e. the time needed to achieve 1 log reduction). A representative example is given in Figure 14.1.

Rennecker et al. (1999) proposed a delayed Chick–Watson model to characterise the minimum ozonation time and concentration for inactivation of *Cryptosporidium parvum* oocysts. According to the Chick–Watson model, microbial inactivation follows the following relationship (which is a loglinear model – see Equation 14.2 – under constant disinfectant concentration):

$$\ln \frac{N_t}{N_o} = \lambda \cdot C_L \cdot t \tag{14.3}$$

Ozone exposure (OE) in a batch reactor can be described by a delayed Chick–Watson model for inactivation in a batch process:

$$\frac{N_t}{N_o} = \frac{N_c}{N_o} \cdot \exp(-k \cdot (OE - OE_{lag}) \tag{14.4}$$

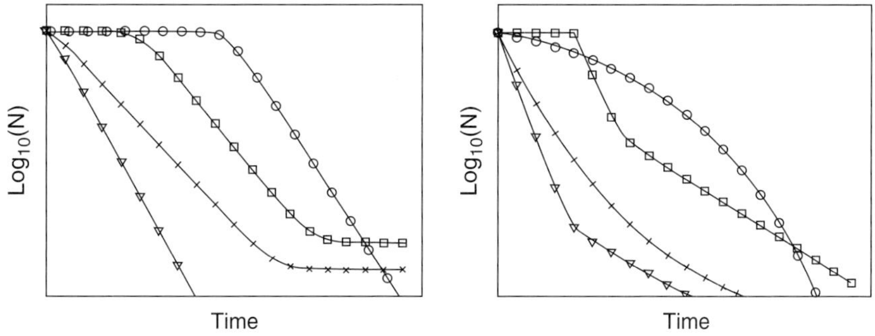

Figure 14.2 Inactivation curves described by the GInaFiT freeware tool. Left plot: linear (Δ), linear with tailing (×), sigmoidal-like (□), linear with a preceding shoulder (○). Right plot: biphasic (Δ), concave (×), biphasic with a shoulder (□), convex (○). (Reprinted from *International Journal of Microbiology*, Volume 102, Issue Number 1, A.H. Geeraerd, V.P. Valdramidis, J.F. Van Impe, GInaFiT, a freeware tool to assess non-log-linear microbial survivor curves, 2010, with permission from Elsevier.)

Selma et al. (2007) employed the Chick–Watson (Haas et al. 1995), modified Chick (Kaymak 2003), modified Chick–Watson (Cho et al. 2003) and modified multiple target (Kaymak 2003) kinetic models to determine the influence of ozone concentration, reaction time and ozone demand on inactivation of *Shigella sonnei* inoculated in water. The ozone demand (*OD*) factor allows comparison of the efficacy of different ozone treatments where concentration and treatment time are different (Selma et al. 2007). *OD* can be determined by the following expression:

$$OD = \frac{OD_0 - D}{k^*} \cdot [1 - \exp(-k^* \cdot t)] \tag{14.5}$$

The rate of ozone decomposition (k^*) can be estimated by measuring the concentration of ozone applied and the residual ozone in the fluid after time t and using the following first-order equation (Kaymak 2003):

$$O_R = (O_i - D) \cdot \exp(-k^* \cdot t) \tag{14.6}$$

The instantaneous ozone demand, D, can be defined as the minimum dose required, which may be determined experimentally and is dependent on the microorganism under investigation.

A useful tool for evaluating several nonlinearities during microbial inactivation is the GInaFiT tool (Geeraerd et al. 2005). Estimation of the inactivation parameters by the use of this can permit the calculation of the time required to obtain a specific log reduction, as defined by regulatory authorities (e.g. the US Food and Drug Administration (FDA) requires a 5 log reduction) or the producer. This tool can be applied for the description of eight different inactivation curves (see Figure 14.2).

Some specific examples of dealing with nonlinearities of the microbial kinetics during ozone treatments are further described in this section. Studies by Patil et al. (2010a,b) have shown that two types of model – the Weibull model and the shoulder–loglinear model – seem to describe more accurately the microbial kinetics of *Escherichia coli* ATCC 25922 and NCTC 12900, *L. monocytogenes* and *L. innocua* during ozonation when inoculated in orange or apple juice (see also Table 14.1). The parameterisation of the Weibull equation employed for these studies is as follows (Mafart et al. 2002):

$$\log_{10}(N_t) = \log_{10}(N_0) - \left(\frac{t}{\delta}\right)^p \tag{14.7}$$

The Weibull distribution corresponds to a concave upward survival curve if $p<1$ and concave downward if $p>1$ (van Boekel 2002). Bialka et al. (2008) also proposed the Weibull model to describe the inactivation kinetics of *E. coli* O157:H7 and *Salmonella enterica* in strawberries and raspberries during ozonation.

The estimated parameters δ and p were used to calculate a desired log reduction. The time required to obtain an x log reduction (t_{xd}) was calculated as follows:

$$t_{xd} = \delta \cdot (x)^{\frac{1}{p}} \tag{14.8}$$

The shoulder–loglinear model (Geeraerd et al. 2000) reads:

$$\log_{10}(N_t) = \log_{10}(N_0) - \frac{k_{\max}\cdot(t)}{\ln(10)} + \frac{\log_{10} e^{(k_{\max}\cdot S_l)}}{\left(1 + e^{(k_{\max}\cdot S_l)} - 1\right)} \cdot e^{(-k_{\max}\cdot t)} \tag{14.9}$$

The numerical values of S_l, $\log_{10}(N_0)$ and k_{max} were used to calculate a desired log reduction. The time required to obtain an x log reduction (t_{xd}) was calculated using Equation 14.10 (Valdramidis et al. 2005). For a case study designed to meet the FDA requirement of 5-log reduction, x is equal to 5:

$$t_{xd} = S_l + (x)\cdot\frac{\ln(10)}{k_{\max}} \tag{14.10}$$

An illustrative representation for the calculation of this value is given in Figure 14.3.

Estimated parameters of these modelling approaches can be further exploited for quantifying the effect of intrinsic (pH, solid contents) and extrinsic parameters (for example, flow rate, concentration, temperature). For example, when the efficacy of continuous gaseous ozone treatment in a bubble column reactor was studied for apple juice, a correlation between the t_{5d} values for *E. coli* strains and the pH of the apple juice was established (Figure 14.4). This correlation was described by an exponential equation:

$$t_{5d} = a \cdot \mathrm{e}^{k\cdot pH} \tag{14.11}$$

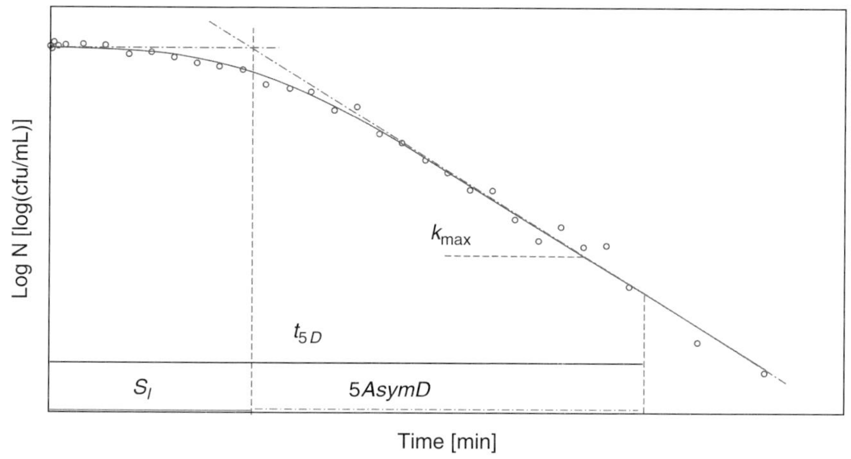

Figure 14.3 Visualisation of the mathematical factors of Equation 14.10 when x = 5.

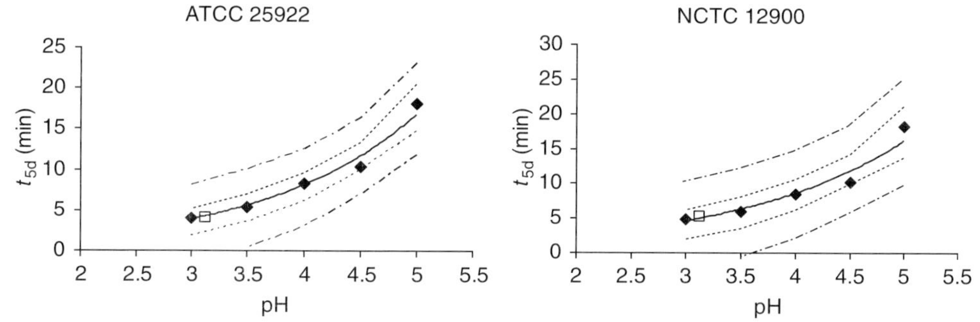

Figure 14.4 Achieved t_{5d} of *E. coli* ATCC 25922, *E. coli* NCTC 12900 in apple juice of different pH as described by $t_{5d} = 0.42 \cdot \exp(0.74 \cdot pH)$ and $t_{5d} = 0.69 \cdot \exp(0.63 \cdot pH)$, respectively. (–) fit of Equation 14.11; (·–·) 95% prediction bounds; (–) 95% confidence bounds. (□) validation point for apple juice of pH 3.13; (♦) observed data values at controlled experimental conditions. (Reprinted from Patil et al. 2010a.)

where a and k are the parameters and t_{5d} is the time required for a 5-log reduction (Figure 14.4). Similarly, polynomial relationships of the microbial Weibull parameter δ (for the same microorganism) were correlated with ozone flow rate and temperature in order to identify the optimal conditions of a similar experimental set-up for a liquid model system of TSB (tryptic soya broth) (Figure 14.5).

The microbial quantitative analysis during any kind of treatment is dependent on the chosen target microorganisms, which can induce adaptation responses, and the environmental stresses applied by expressing specific sets of genes on exposure to acid, salt, heat, cold, reactive oxygen species, starvation and so on. Therefore, it is of great importance to evaluate

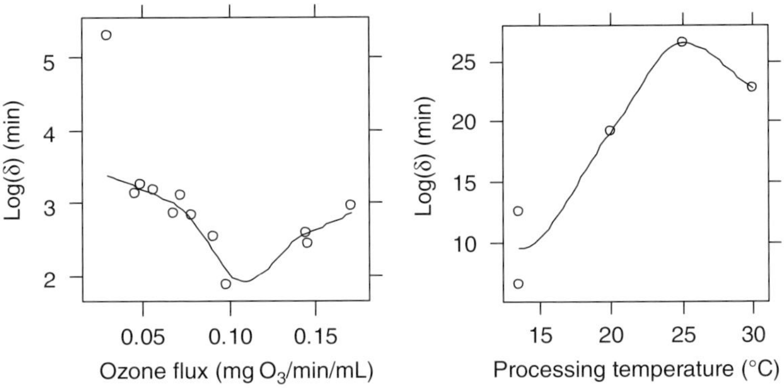

Figure 14.5 Relationship of the log of the characteristic time δ (a) with the ozone processing conditions at ambient temperature and (b) with the processing temperature at 0.09 mg O_3/min/mL. The continuous line represents a regression of a polynomial equation on the experimental data. (Reprinted from *Journal of Applied Microbiology*, Patil, S., Cullen, P.J., Kelly, B. et al., Extrinsic control parameters for ozone inactivation of *Escherichia coli* using a bubble column, 830–837, 2010, with permission from John Wiley & Sons.)

the efficiency of ozone preservation treatments using appropriate resistant strains based on the microbial modelling approaches that are built.

14.3 Chemical reaction kinetics

The mechanism of a reaction is an hypothesis about the sequence in which molecular events in a certain reaction take place. At the level of molecules, ions, atoms and radicals, the events are discrete (van Boekel 2009). According to van Boekel (2009), these individual occurrences should be tackled as probabilities which appear to follow a deterministic law, because the number of particles involved is incredibly high. In general, correspondence between stoichiometry and kinetics holds only for elementary reactions.

Ozone is unstable in aqueous solution, and its effectiveness as a disinfectant depends upon the rate at which it decomposes. The decomposition mechanism is very complex and is still not totally unveiled. Nevertheless, to be able to design an efficient ozonation system, it is imperative to determine a working equation for the kinetics of ozone decomposition. This equation should include the main variables affecting this rate. It should be mentioned that limited information is available on the reaction kinetics of systems in which the organic matter is excessive, as in food products. Ozonation of organic compounds present in foods is a complex mechanism involving a variety of possible chemical reactions. These reactions may be direct reactions of ozone with the target compound or its intermediates or radical reactions between hydroxyl radicals produced through ozone decomposition catalysed mainly by the hydroxide ion (OH^-) and organics (Mishchuk et al. 2008). To understand the complexity of the reactions, approaches involving the dissolution of ozone in distilled water are useful. Mishchuk et al. (2008) developed a theoretical model to describe the process involved. Dissolution of ozone sparged through a diffuser involves equilibration of gaseous ozone concentration (O_{cG}) and the layer of water adjacent to the ozone bubble (O_{cL}). As this occurs instantaneously, the distance (d) between the bubble and the liquid is almost zero. This process depends upon the nature of the dissolving substance, denoted by the Henry constant (H).

$$O_{cL}(d \to 0) = HO_{cG} \tag{14.12}$$

Ozone diffusion in water in the vicinity of the bubble or at some distance from it which after a certain time satisfies the above relationship can be used to determine the Henry constant by the following relationship:

$$H = \frac{O_{cL}^{Max}}{O_{cG}} \tag{14.13}$$

Ozone degradation of organic compounds present in food might be due to either direct reaction with ozone or an indirect reaction of secondary oxidators (see also Chapter 4).

Degradation of organic compounds also is reported to be due to various intermediate radical formations, leading to electrophilic and nucleophilic reactions occurring with aromatic compounds that are substituted with an electron donor (e.g. HO^-) having a high electron density on the carbon compounds in the ortho and para positions. Chemical reactions involve the breakage of bonds and the formation of new ones (Luo 2005). Lovato et al. (2009) presented a predominant kinetic mechanism for ozone breakdown in pure water of neutral to acidic conditions. The mechanism presented in Box 14.1 for the formation and transformation of radicals during ozonation can be considered for the development of a set of 12 differential equations (Box 14.2) describing the reaction kinetics. This illustrated example indicates that reactions involving ozone and food components will be even more complex, with simultaneous degradation and formation of new compounds. The development of a complete set of reaction kinetics incorporating the organic matter is a very challenging procedure.

Various organic compounds present in fruit juices, such as carotenoids, anthocyanins, ascorbic acid and amino acids, may be oxidised according to any of the mechanisms described in previous chapters. According to Xue et al. (2008), ozone plays an important role not only in the degradation process of organic dye but also in the formation of other high-reactive species, such as $\bullet OH$, $HO^{2\bullet}$, $\bullet O_2^-$ and $\bullet O_3^-$, which facilitates degradation. A similar oxidative degradation of ascorbic acid in the presence of oxygen has been reported by Kennedy et al. (1992) and Robertson and Saminego-Esguerra (1986).

The absorption of ozone and parallel reactions involving various oxidants present in the fluid have been reported to follow first- or second-order

Box 14.1 Mechanism of reaction kinetics of a pure water system, as presented by Lovato et al. (2009). (Reprinted from Lovato et al. (2009), with permission from Elsevier).

(1) $O_3 + HO^- \rightarrow HO_2^- + O_2$

(2) $O_3 + HO_2^- \rightarrow HO_2^\bullet + O_3^{\bullet-}$

(3) $HO_2^\bullet \rightarrow O_2^{\bullet-} + H^+$

(4) $O_2^{\bullet-} + H^+ \rightarrow HO_2^\bullet$

(5) $O_3 + O_2^{\bullet-} \rightarrow O_3^{\bullet-} + O_2$

(6) $O_3^{\bullet-} + H^+ \rightarrow HO_3^\bullet$

(7) $HO_3^\bullet \rightarrow O_3^{\bullet-} + O_2$

(8) $HO_3^\bullet \rightarrow HO^{\bullet-} + O_2$

(9) $O_3 + HO^\bullet \rightarrow HO_4^\bullet$

(10) $HO_4^\bullet \rightarrow HO_2^\bullet + O_2$

(11) $HO_2^- \rightarrow H^+ + H_2O_2$

(12) $H_2O_2 \rightarrow HO_2^- + H^+$

(13) $HO_4^\bullet + HO_4^\bullet \rightarrow H_2O_2 + 2O_3$

(14) $HO_4^\bullet + HO_3^\bullet \rightarrow H_2O_2 + O_2 + O_3$

(15) $HO^\bullet + H_2O_2 \rightarrow HO_2^\bullet + H_2O$

(16) $HO^\bullet + HO_2^- \rightarrow HO_2^\bullet + HO^-$

(17) $HO^\bullet + HO^\bullet \rightarrow H_2O_2$

(18) $HO_2^\bullet + HO_2^\bullet \rightarrow H_2O_2 + O_2$

kinetics for both ozone and individual constituents (Cheng et al. 2003). This can be generalised as follows:

$$a_n A - B_n \rightarrow P_n$$

where a_n is the stoichiometric ratio of the reaction between the ozone (A) and the compound (B), leading to the formation of the reaction product (P).

At equilibrium, the rate of degradation of a food component due to ozone has been shown to follow *pseudo*-first-order (Erol and Özbelge, 2007) and first-order reactions, with respect to both organic component and ozone concentration (Masten and Hoigné 1992; Saunders et al. 1983; Trapido et al. 1997). Zimeri and Tong (1999) also reported a degradation mechanism of epigallocatechin gallate in the presence of dissolved oxygen within a model liquid solution as pseudo first-order.

Box 14.2 System of 12 ordinary differential equations for the reactant species present in a pure water system of neutral to acidic conditions, as presented by Lovato et al. (2009). (Reprinted from Lovato et al. (2009), with permission from Elsevier).

$$\frac{dC_{O_3}}{dt} = -k_1C_{O_3}C_{HO^-} - k_2C_{O_3}C_{HO_2^-} - k_5C_{O_3}C_{O_2^{\bullet-}} - k_9C_{O_3}C_{HO^\bullet} + 2k_{13}C^2_{HO_4^\bullet} + k_{14}C_{HO_4^\bullet}C_{HO_3^\bullet}$$

$$\frac{dC_{HO_2^-}}{dt} = k_1C_{O_3}C_{HO^-} - k_2C_{O_3}C_{HO_2^-} - k_{11}C_{HO_2^-}C_{H^+} + k_{12}C_{H_2O_2} - k_{16}C_{HO^\bullet}C_{HO_2^-}$$

$$\frac{dC_{HO_2^\bullet}}{dt} = k_2C_{O_3}C_{HO_2^-} - k_3C_{HO_2^\bullet} - k_4C_{O_2^{\bullet-}}C_{H^+} + k_{10}C_{HO_4^\bullet} + k_{15}C_{OH^\bullet}C_{H_2O_2} + k_{16}C_{OH^\bullet}C_{HO_2^-} - k_{18}C^2_{HO_2^\bullet}$$

$$\frac{dC_{O_3^{\bullet-}}}{dt} = k_2C_{O_3}C_{HO_2^-} + k_5C_{O_3}C_{O_2^{\bullet-}} - k_6C_{O_3^{\bullet-}}C_{H^+} - k_7C_{HO_3^\bullet}$$

$$\frac{dC_{O_2^{\bullet-}}}{dt} = k_3C_{HO_2^\bullet} - k_4C_{O_2^{\bullet-}}C_{H^+} - k_5C_{O_3}C_{O_2^{\bullet-}}$$

$$\frac{dC_{HO_3^\bullet}}{dt} = k_6C_{O_3^{\bullet-}}C_{H^+} - k_7C_{HO_3^\bullet} - k_8C_{HO_3^\bullet} - k_{14}C_{HO_4^\bullet}C_{HO_3^\bullet}$$

$$\frac{dC_{OH^\bullet}}{dt} = k_8C_{HO_3^\bullet} - k_9C_{O^3}C_{HO^\bullet} - k_{15}C_{HO^\bullet}C_{H_2O_2} - k_{16}C_{HO^\bullet}C_{HO_2^-} - k_{17}C^2_{HO^\bullet}$$

$$\frac{dC_{H^{\bullet+}}}{dt} = k_3C_{HO_2^\bullet} - k_4C_{O_2^{\bullet-}}C_{H^+} - k_6C_{O_3^{\bullet-}}C_{H^+} + k_7C_{HO_3^\bullet} - k_{11}C_{HO_2^-}C_{H^+} + k_{12}C_{H_2O_2}$$

$$\frac{dC_{OH^-}}{dt} = -k_1C_{O_3}C_{HO^-} + k_{16}C_{HO^\bullet}C_{HO_2^-}$$

$$\frac{dC_{HO_4^\bullet}}{dt} = k_9C_{O_3}C_{HO^\bullet} - k_{10}C_{HO_4^\bullet} - k_{13}C^2_{HO_4^\bullet} - k_{14}C_{HO_4^\bullet}C_{HO_3^\bullet}$$

$$\frac{dC_{O_2}}{dt} = k_1C_{O_3}C_{OH^-} + k_5C_{O_3}C_{O_2^{\bullet-}} + k_8C_{HO_3^\bullet} + k_{10}C_{HO_4^\bullet} + k_{14}C_{HO_4^\bullet}C_{HO_3^\bullet} + k_{18}C^2_{HO_2^\bullet}$$

$$\frac{dC_{H_2O_2}}{dt} = k_{11}C_{HO_2^-}C_{H^+} - k_{12}C_{H_2O_2} + k_{13}C^2_{HO_4^\bullet} + k_{14}C_{HO_4^\bullet}C_{HO_3^\bullet} - k_{15}C_{HO^\bullet}C_{H_2O_2} + k_{17}C^2_{HO^\bullet} + k_{18}C^2_{HO_2^\bullet}$$

The degradation rate of a compound at any time can be represented by the following relationship:

$$\frac{dA_t}{dt} = k[C_{O3}]^m[A_t]^n \tag{14.14}$$

where A_t is the concentration of organic compound at time t, $[C_{O3}]$ is the dissolved ozone concentration during treatment, k is the reaction rate constant, n is the reaction order and m is the reaction order with respect to ozone. In Equation 14.14, when the dissolved ozone concentration is in excess compared to the concentration of the organic component, m = 0 (Zimeri and Tong 1999). In such a reaction, if the ozone concentration remains constant during the course of the kinetic study, Equation 14.14 reduces to:

$$C_{O3(t)} = C_{O3(s)} \text{ (in equilibrium)}$$

$$-\frac{dA_t}{dt} = kA_t \tag{14.15}$$

Rearranging and integrating Equation 14.15 yields:

$$-\int \frac{d(A_t)}{A_0} = \int k dt \tag{14.16}$$

$$\ln\left(\frac{A_t}{A_0}\right) = -kt \tag{14.17}$$

The equation assumes complete degradation of any organic content after prolonged treatment. A special case of first-order model is the fractional conversion model (van den Broeck et al. 2000), which can be applied when a fraction of compound under study remains at residual levels. The fractional conversion model was employed by Zimeri and Tong (1999) to model the degradation kinetics of epigallocatechin gallate as a function of pH and dissolved oxygen in a model liquid solution. Tiwari et al. (2010) employed this approach to model the kinetics of degradation of ascorbic acid and the anthocyanin content of ozonated strawberry juice. Figure 14.6 shows the relative change in ascorbic acid and anthocyanins for the fitted fractional conversion model. The fractional conversion model predicts well when a fraction of compound with non-zero equilibrium concentration will remain at the end of a kinetic study. Hence the following equations can be employed:

$$\ln\left(\frac{A_0 - A_t}{A_0 - A_\infty}\right) = -K_F t \tag{14.18}$$

$$f = \frac{A_0 - A_t}{A_0 - A_\infty} \tag{14.19}$$

A plot of the logarithm of $(1 - f)$ against time is linear and the rate constant (K_F) is the negative of the slope (Levenspiel 1972):

$$\ln(1 - f) = \ln\left(\frac{A_0 - A_t}{A_0 - A_\infty}\right) = -K_F t \tag{14.20}$$

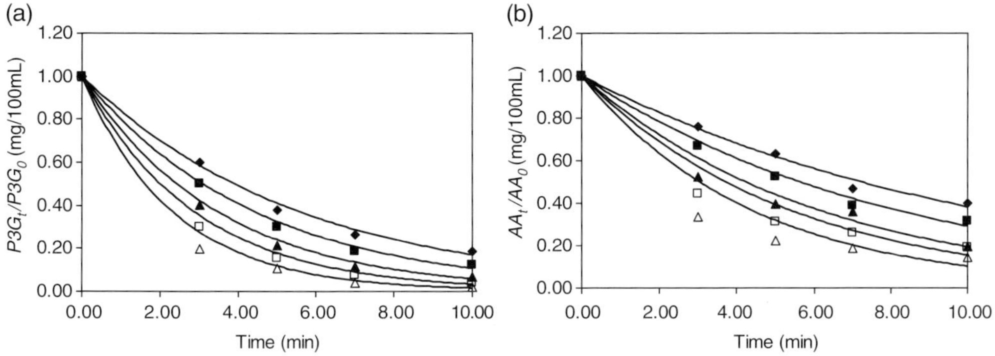

Figure 14.6 Changes in (a) anthocyanin content and (b) ascorbic acid content (mg/100 mL of strawberry juice) with treatment time (minutes) when described with a fraction conversion model at a constant flow rate of 0.0625 L/min and an ozone concentration of (♦) 1.6% w/w, (□) 3.2% w/w, (Δ) 4.8% w/w, (■) 6.4% w/w and (▲) 7.8% w/w. (Reprinted from *Food Chemistry*, Volume 113, Issue Number 4, B.K. Tiwari, C.P. O'Donnell, Effect of ozone processing on anthocyanins and ascorbic acid degradation of strawberry juice, 2010, with permission from Elsevier.)

Rearranging Equations 14.18 and 14.20 results in following expression:

$$A_t = A_\infty + (A_0 - A_\infty)e^{-K_F t} \tag{14.21}$$

From the above equation it may be observed that when $A_\infty \to 0$, the equation tends to conform to a first-order kinetic model (Equation 14.15). This approach may be used to model the degradation kinetics at a constant temperature. The rate of degradation dependence can be further modelled using the Arrhenius relationship with respect to temperature.

14.4 Modelling ozonation processes

Ozone processing of food products is carried out either by the use of gaseous treatments or by washing with ozone-containing water. The mass transfer modelling in a bubble column for ozonation of liquid products or in other set-ups for production of ozone-containing water (Dhillon et al. 2009) is based upon the same principles. Applications exist for both solid and liquid foods. In this section the modelling approaches for different ozonation processes are reviewed.

14.4.1 Modelling ozone bubble columns

Bubble columns are utilised as multiphase contactors and reactors in various food, chemical, petrochemical, biochemical and metallurgical industries (Degaleesan et al. 2001). A typical bubble column reactor is a cylindrical vessel with a gas diffuser to sparge ozone in a gaseous state into either a liquid phase or a liquid–solid dispersion (Figure 14.7). Many

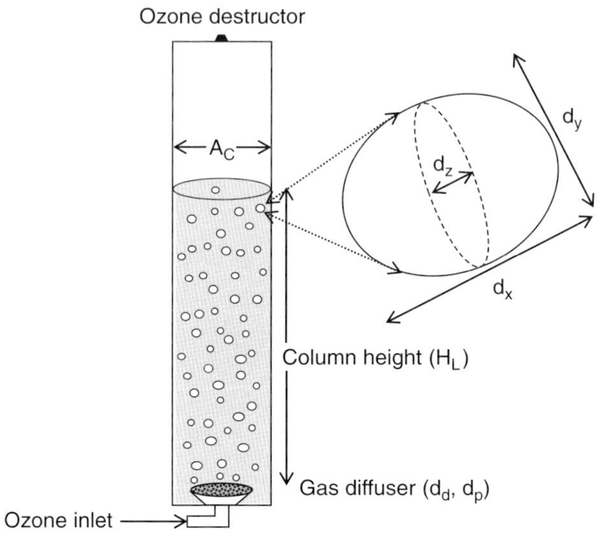

Figure 14.7 Bubble column reactor. (Reprinted from *Trends in Food Science & Technology*, Volume 20, Issue Number 3–4, P.J. Cullen, B.K. Tiwari, C.P. O'Donnell, K. Muthukumarappan, Modelling approaches to ozone processing of liquid foods, 2010, with permission from Elsevier.)

empirical correlations have been proposed to estimate design parameters for a bubble column based on the physical and chemical properties of the material under investigation, including experimental conditions. In liquid treatment applications, ozonation is limited by the mass transfer process governing the overall performance of ozone contactors (Zhou et al. 1994; Gamal El-Din and Smith 2002). The design parameters for a bubble column are: gas–liquid-specific interfacial area, individual mass transfer coefficient, flow behaviour, bubble size and distribution, and coalescence of bubbles (Zhao et al. 2004).

As ozone gas solubility and diffusivity within liquids are much lower than in the gas phase, the gas diffusion through the liquid film becomes the rate-limiting step of the mass transfer process (Valentin 1967). When gaseous ozone is sparged into a liquid phase, agitation occurs, inducing turbulent shear stresses. This causes the liquid film to become thinner. Consequently, higher rates of diffusion through the liquid film occur, resulting in an increased local mass transfer coefficient (k_L).

In a bubble column, ozone gas interacts with liquid food, where ozone is consumed, followed by a chemical reaction involving oxidation. The overall reaction rate is governed by two steps: (1) the mass transfer from the gas phase to the liquid phase and (2) the chemical reaction in the liquid phase (Benbelkacem and Debellefontaine 2003). Literature (Gamal El-Din and Smith 2003) shows that the design and modelling of ozone bubble columns are based on determination of overall mass transfer coefficient

(K_{La}), gas hold up (ε_G) and Sauter mean diameter (SD), defined as the diameter of a bubble that has the same volume/surface area, which can be determined as follows:

$$d_S = \sqrt{A_b / \pi} \tag{14.22}$$

$$d_v = \sqrt[3]{\left(6V_b / \pi\right)} \tag{14.23}$$

The functions d_s and d_v are usually measured directly using image analysis. Individual bubble diameter may be determined by Equation 14.24, assuming that the bubble shape is ellipsoidal (Figure 14.7). This three-dimensional technique may be simplified by assuming that the shortest length, d_x, and width, d_z, of the bubble are of equal length, thus reducing this measurement to a two-dimensional approach (Baawain et al. 2007):

$$d_b = \sqrt[3]{d_x d_y d_z} = \sqrt[3]{d_x d_y^2} \tag{14.24}$$

The Sauter mean diameter for a bubble is:

$$SD = d_v^3 / d_s^2 \tag{14.25}$$

14.4.2 Overall mass transfer coefficient

The overall mass transfer coefficient (KL_a) is used to describe the consumption of ozone gas into the liquid or solid–liquid phase, depending upon the oxidising material – for example, pulp, suspended soluble solids in fruit juice – defined as follows:

$$\psi = K_{La}\left(C_L^e - C_L\right) \tag{14.26}$$

Dissolved ozone concentration in water systems was determined using an ozone analyser, an indigo colourimetric method (Wu et al. 2007) or the iodometric method (Gottschalk et al. 2000). KL_a usually depends upon operating conditions and liquid characteristics such as surface tension, viscosity and density. The mass transfer of ozone gas into liquid phase is also influenced by temperature and pH. The half-life of ozone in distilled water at 20 °C is about 20–30 minutes (Khadre et al. 2001).

Although the determination of KL_a is useful for design purposes, it is important in many cases to determine the local mass transfer coefficient (K_L) in order to evaluate ozone absorption with chemical or gas–liquid reactions (Akita and Yoshida 1974). The gas bubbles' specific interfacial area (a) is required, which is equal to the ratio between the bubbles' surface area (A) and the volume of the dispersed phases (V). However, due to the difficulties associated with determining A, the value of a can be calculated using the following relationship:

$$a = 6\varepsilon_G / SD \tag{14.27}$$

SD can be determined as described above, whereas gas hold up (ε_G) can be determined by measuring pressure change within a bubble column using a pressure transducer and the following relationship:

$$\varepsilon_G = 1 - \frac{\Delta P}{\rho_L g \Delta x} \tag{14.28}$$

Ozone gas supply passing through the diffuser will be a mixture of ozone and air. Ozone is always produced in diluted form, from either air or high-purity oxygen, as it is not possible to produce 100% pure ozone (Tizaoui et al. 2008). In addition, ozone is sparingly soluble in water (i.e. 10 times less than chlorine and 130 times less than chlorine dioxide). Thus, low mass transfer and reaction rates are expected. Oxygen mass transfer is often neglected in ozone dissolution models. However, this information is necessary to study ozone dissolution in a bubble column. The mass transfer of ozone (A) and oxygen (O) from the gas to liquid phase can be described by the two-film model (Danckwerts 1970). According to the theory of surface renewal, oxygen-based KL_a is converted to ozone-based KL_a. As ozone is dissolved in water, it may be consumed via self-decomposition ($2O_3 \rightarrow 3O_2$) and oxidation with organic matter (B) present in the liquid. Chen et al. (2003) proposed the following pseudo-first-order and second-order reaction rate expressions (Chang et al. 2001):

$$\frac{dC_{ALb}}{dt} = K_d C_{ALb} - \alpha_{AB} K_{AB} C_{ALb} C_{BLb} \tag{14.29}$$

$$\frac{dC_{BLb}}{dt} = -K_{AB} C_{ALb} C_{BLb} \tag{14.30}$$

$$\frac{dC_{OLb}}{dt} = -\frac{3K_d C_{ALb}}{2} \tag{14.31}$$

With ozone consumption and oxygen formation, the mass transfer rates of ozone and oxygen may be enhanced or retarded (Chen et al. 2002). The ratio of the mass transfer rates of ozone and oxygen (with ozone consumption and oxygen formation) to those without may be designated by the enhancement factor of ozone consumption (E_A) and the retarding factor of oxygen formation (R_{FO}), respectively. Danckwerts (1970) defined E_A as:

$$E_A = \frac{\text{rate of gas absorption with chemical reactions}}{\text{rate of maximum pure physical gas absorbtion}} \tag{14.32}$$

The enhancement factor is a function of the reactivity of organic matter present in the fluid, the ozone diffusivity in the liquid phase, the local mass transfer coefficient, KL_a and feed gas flow rate (Gamal El-Din and Smith 2002). This enhancement factor decreases during ozonation and reaches a plateau after certain times, as reported by Zhou and Smith (1997, 2000).

Danckwerts' (1970) surface theory can be used to correct KL_a from oxygen to ozone base (Sherwood et al. 1975; Gamal El-Din and Smith 2002):

$$\frac{KL_aO_3}{KL_aO_2} = \sqrt{\frac{D_{O_3}}{D_{O_2}}} \tag{14.33}$$

where D_{O3} and D_{O2} are the molecular diffusivities of ozone and oxygen gas in water, respectively.

In summary, during ozonation of liquid food or water, the dissolved concentrations of ozone and oxygen increase with processing time, while the concentration of organic matter present in the liquid and the enhancement factor of ozone decrease with time (Chen et al. 2003).

14.5 Conclusions

The timeline course in processing of food and food products contains information about the underlying mechanism that the process has during its application. It is therefore essential to describe the kinetics that take place during these processes. Solution of the resulting equations shows whether the hypothesised mechanisms are consistent and contribute to the further optimisation, design and control of the applied process.

Mathematical models incorporating various independent factors governing ozone processing were reviewed in this chapter to describe the microbial kinetics, the biochemical reactions and the physical phenomena. These are the main mathematical tools for facilitating and enhancing the control of both quality and safety parameters of ozonated foods.

As food products vary extensively in organic matter, one of the biggest challenges will be to fully characterise the chemical reactions, the mass transfer and the hydrodynamics that take place during the ozonation of specific food products. Finally, development of different processing set-ups may also require the application of alternative modelling approaches to describe the processes.

References

Akita, K. and Yoshida, F. (1974) Bubble size, interfacial area, and liquid-phase mass-transfer coefficient in bubble columns, *Industrial and Engineering Chemistry Process Design and Development*, **13**(1): 84–91.

Anderson, W.A., McClure, P.J., Baird-Parker, A.C. and Cole, M.B. (1996) The application of a log-logistic model to describe the thermal inactivation of *Clostridium botulinum* 213B at temperatures below 121.1 °C, *Journal of Applied Bacteriology*, **80**: 283–90.

Augustin, J.C., Carlier, V. and Rozier, J. (1998) Mathematical modeling of the heat resistance of *L. monocytogenes*, *Journal of Applied Microbiology*, **84**: 185–91.

Baawain, M.S. Gamal El-Din, M. and Smith, D.W. (2007) Artificial neural networks modeling of ozone bubble columns: mass transfer coefficient, gas hold-up, and bubble size, *Ozone: Science and Engineering*, **29**: 343–52.

Baranyi, J. and Pin, C. (2001) A parallel study on bacterial growth and inactivation, *Journal of Theoretical Biology*, **210**: 327–36.

Benbelkacem, H. and Debellefontaine, H. (2003) Modeling of a gas–liquid reactor in batch conditions: study of the intermediate regime when part of the reaction occurs within the film and part within the bulk, *Chemical Engineering and Processing*, **42**(10): 723–32.

Bialka, K.L., Demirci, A. and Puri, V.M. (2008) Modeling the inactivation of *Escherichia coli* O157:H7 and *Salmonella enterica* on raspberries and strawberries resulting from exposure to ozone or pulsed UV-light, *Journal of Food Engineering*, **85**(3): 444–9.

Chang, C.Y, Chen, Y.H., Li, H., Chiu, C.Y., Yu, Y.H., Chiang, P.C., Ku, Y. and Chen, J.N. (2001) Kinetics of decomposition of polyethylene glycol in electroplating solution by ozonation with UV radiation, *Journal of Environmental Engineering*, **127**(10): 908–15.

Chen, Y.H., Chang, C.Y., Chiu, C.Y., Huang, W.H., Yu, Y.H., Chiang, P.C., Ku, Y. and Chen, J.N. (2002) Dynamic model of ozone contacting process with oxygen mass transfer in bubble columns, *Journal of Environmental Engineering-ASCE*, **128**(11): 1036–45.

Chen, Y.H., Chang, C.Y., Chiu, C.Y., Yu, Y.H., Chiang, P.C., KU, Y. and Chen, J.N. (2003) Dynamic behavior of ozonation with pollutant in a countercurrent bubble column with oxygen mass transfer, *Water Research*, **37**: 2583–94.

Cheng, J., Yang, Z.R., Chen, H.Q., Kuo, C.H. and Zappi, M.E. (2003) Simultaneous prediction of chemical mass transfer coefficients and rates for removal of organic pollutants in ozone absorption in an agitated semi-batch reactor, *Separation and Purification Technology*, **31**: 97–104.

Cho, M., Chung, H. and Yoon, J. (2003) Disinfection of water containing natural organic matter by using ozone-initiated radical reactions, *Applied Environmental Microbiology*, **69**: 2284–91.

Danckwerts, P.V. (1970) *Gas–Liquid Reactions*, NewYork, NY: McGraw-Hill, Inc.

Degaleesan, S., Dudukovic, M. and Pan, Y. (2001) Experimental study of gas-induced liquid-flow structures in bubble columns, *AIChE J*, **47**: 1913–31.

Dhillon, B., Wiesenborn, D., Wolf-Hall, C. and Manthey, F. (2009) Development and evaluation of ozonated water system for antimicrobial treatment of durum wheat, *Journal of Food Science*, **74**: E396–E403.

Erol, F. and Özbelge, T.A. (2007) Catalytic ozonation with non-polar bonded alumina phases for treatment of aqueous dye solutions in a semi-batch reactor, *Chemical Engineering Journal*, **139**: 272–83.

Fan, L., Song, J., McRae, K.B., Walker, B.A. and Sharpe, D. (2007) Gaseous ozone treatment inactivates *Listeria innocua in vitro*, *Journal of Applied Microbiology*, **103**: 2657–63.

Gamal El-Din, M. and Smith, D.W. (2002) Ozonation of kraft pulp mill effluents: process dynamics, *Journal of Environmental Engineering Sciences*, **1**: 45–57.

Gamal El-Din, M. and Smith, D.M. (2003) Mass transfer analysis in ozone bubble columns, *Journal of Environmental Engineering and Science*, **2**(1): 63–76.

Geeraerd, A.H., Herremans, C.H. and Van Impe, J.F. (2000) Structural model requirements to describe microbial inactivation during a mild heat treatment, *International Journal of Food Microbiology*, **59**: 185–209.

Geeraerd, A.H., Valdramidis, V.P. and Van Impe, J.F. (2005). GInaFiT, a freeware tool to assess non-log-linear microbial survivor curves, *International Journal of Food Microbiology*, **102**: 95–105.

Gottschalk, C, Libra, J.A. and Saupe, A. (2010) *Ozonation of Water and Waste Water*, 2 edn, Weinheim: John Wiley & Sons-VCH Verlag GmbH & Co. KGaA.

Gujer, W. and von Gunten, U. (2003) A stochastic model of an ozonation reactor, *Water Research*, **37**: 1667–77.

Haas, C.N., Joffe, J., Anmangandla, U., Hornberger, J.C., Heath, M., Jacangelo, S.J. and Glicker, J. (1995) *Development and Validation of Radiational Design Methods of Disinfection*. Denver, CO: AWWA Research Foundation.

Hills, B.P. (2001) The power and pitfalls of deductive modeling, in Tijskens, L.M.M., Hertog, M.L.A.T.M. and Nicolai, B.M. (eds) *Food Process Modeling*, Cambridge, UK: Woodhead Publishing Limited, pp. 3–18.

Hunt, N.K. and Mariñas, B.J. (1997) Kinetics of *Escherichia coli* inactivation with ozone, *Water Research*, **31**: 1355–62.

Kaymak, B. (2003) Effects of initial microbial density on disinfection efficiency and explanatory mechanisms, Philadelphia, PA: Electronic Theses. Retrieved from http://dspace.library.drexel.edu/handle/1860/195.

Kennedy, J.F., Rivera, Z.S., Lloyd, L.L., Warner, F.P. and Jumel, K. (1992) L-ascorbic acid stability in aseptically processed orange juice in TetraBrik cartons and the effect of oxygen, *Food Chemistry*, **45**: 327–31.

Khadre, M.A., Yousef, A.E. and Kim, J. (2001) Microbiological aspects of ozone applications in food: a review, *Journal of Food Science*, **6**: 1242–52.

Levenspiel, O. (1972) *Chemical Reaction Engineering*, 2 edn, New York, NY: John Wiley & Sons, Inc.

Li, H.B., Finch, G.R., Smith, D.W. and Belosevic, M. (2001) Sequential inactivation of *Cryptosporidium parvum* using ozone and chlorine, *Water Research*, **35**: 4339–48.

Lovato, M.E., Martin, C.A. and Cassano, A.E. (2009) A reaction kinetic model for ozone decomposition in aqueous media valid for neutral and acidic pH, *Chemical Engineering*, **146**: 486–97.

Luo, Y.R. (2005) *Handbook of Bond Dissociation Energies in Organic Compounds*, Beijing, China: Science Press.

Mafart, P., Couvert, O., Gaillard, S. and Leguerinel I. (2002) On calculating sterility in thermal preservation methods: application of the Weibull frequency distribution model, *International Journal of Food Microbiology*, **72**: 107–13.

Mañas, P. and Pagán, R. (2005) Microbial inactivation by new technologies of food preservation, *Journal of Applied Microbiology*, **98**: 1387–99.

Masten, S.J. and Hoigné, J. (1992) Comparison of ozone and hydroxyl radical-induced oxidation of hydrocarbons in water, *Ozone: Science & Engineering*, **14**: 197–214.

Mishchuk, N.A., Goncharuk, V.V. and Vakulenko, V.F. (2008) Theoretical analysis of physicochemical processes occurring during water treatment by ozone and ultraviolet radiation, *Advances in Colloid and Interface Science*, **139**: 62–73.

Patil, S., Bourke, P., Frias, J., Tiwari, B.K. and Cullen, P.J. (2009) Inactivation of *Escherichia coli* in orange juice using ozone, *Innovative Food Science & Emerging Technologies*, **10**: 551–7.

Patil, S., Valdramidis, V.P., Cullen, P.J., Frias, J. and Bourke, P. (2010a) Ozone treatment of apple juice at different pH levels; effects on the microbial responses of *Escherichia coli.*, *Food Microbiol*, **27**: 835–40.

Patil, S., Valdramidis, V.P., Cullen, P.J., Frias, J. and Bourke, P. (2010b) Ozone inactivation of *Listeria monocytogenes* and *Listeria innocua* in orange juice using a bubble column, *Food Control*, **21**(13): 1723–30.

Peleg, M. (2003) Microbial survival curves: interpretation, mathematical modeling and utilization, *Comments on Theoretical Biology*, **8**: 357–87.

Peleg, M. and Cole, M.B. (1998) Reinterpretation of microbial survival curves, *Critical Reviews in Food Science and Nutrition*, **38**: 353–80.

Raso, J., Alvarez, I., Condon, S. and Sala, F.J. (2000) Predicting inactivation of *Salmonella senftenberg* by pulsed electric fields, *Innovative Food Science and Emerging Tecnology*, **1**: 21–9.

Rennecker, J.L., Mariñas, B.J., Owens, J.H. and Rice, E.W. (1999) Inactivation of *Cryptosporidium parvum* oocysts with ozone, *Water Research*, **33**: 2481–8.

Restaino, L., Frampton, E., Hemphill, J. and Palnikar, P. (1995) Efficacy of ozonated water against various food related microorganisms, *Applied Environmental Microbiology*, **61**: 3471–5.

Robertson, G.L. and Saminego-Esguerra, C.M. (1986) Effect of initial dissolved oxygen levels on the degradation of ascorbic acid and browning of lemon juice during storage, *Journal of Food Science*, **51**: 184–7, 192.

Saunders, F., Gould, J.P. and Southerland, C.R. (1983) The effect of solute competition on ozonolysis of industrial dyes, *Water Research*, **17**: 1407–19.

Selma, M.V., Beltrán, D., Allende, A., Chacón-Vera, E. and Gil, M.I. (2007) Elimination by ozone of *Shigella sonnei* in shredded lettuce and water, *Food Microbiology*, **24**(5): 492–9.

Sherwood, T.K., Robert, L.T. and Charles, R.W. (1975) *Mass Transfer*, New York, NY: McGraw-Hill.

Smelt, J.P.P.M., Hellemons, J.C., Wouters, P.C. and van Gerwen, S.J.C. (2002) Physiological and mathematical aspects in setting criteria for decontamination of foods by physical means, *International Journal of Food Microbiology*, **78**: 57–77.

Steenstrup, L.D. and Floros, J.D. (2004) Inactivation of *E. coli* 0157:H7 in apple cider by ozone at various temperatures and concentrations, *Journal of Food Processing Preservation*, **28**: 103–16.

Tiwari, B.K., Muthukumarappan, K., O'Donnell, C.P. and Cullen, P.J. (2008) Modeling colour degradation of orange juice by ozone treatment using response surface methodology, *Journal of Food Engineering*, **88**: 553–60.

Tiwari, B.K., O'Donnell, C.P., Patras, A., Brunton, N. and Cullen, P.J. (2010) Effect of ozone processing on anthocyanins and ascorbic acid degradation of strawberry juice, *Food Chemistry*, **113**: 1119–26.

Tizaoui, C., Bickley, R.I., Slater, M.J., Wang, W.-J., Ward, D.B. and Al-Jaberi, D.B. (2008) A comparison of novel ozone-based systems and photocatalysis for the removal of water pollutants, *Desalination*, **227**: 57–71.

Trapido, M., Veressinina, Y., Hentunen, J.K. and Hirvonen, A. (1997) Ozonation of chlorophenols: kinetics, by-products and toxicity, *Environmental Technology*, **18**: 325–32.

Valdramidis, V.P., Geeraerd, A.H., Bernaerts, K. and Van Impe, J.F. (2005) An alternative approach for non log-linear thermal microbial inactivation: modelling the number of log cycles reduction with respect to temperature, *Food Technology and Biotechnology*, **43**(4): 321–7.

Valdramidis, V.P., Geeraerd, A.H., Bernaerts, K. and Van Impe, J.F. (2006) Microbial dynamics versus mathematical model dynamics: the case of microbial heat resistance induction, *Innovative Food Science and Emerging Technologies*, **7**: 80–7.

Valdramidis, V.P., Geeraerd, A.H. and Van Impe, J.F. (2007) Stress-adaptive responses by heat under the microscope of predictive microbiology, *Journal of Applied Microbiology*, **103**: 1922–30.

Valentin, F.H.H. (1967) *Absorption in Gas–Liquid Dispersions: Some Aspects of Bubble Technology*, London, UK: E. and F.N. Spon Ltd.

van Boekel, M.A.J.S. (2002) On the use of the Weibull model to describe thermal inactivation of microbial vegetative cells, *International Journal of Food Microbiology*, **74**(1–2): 139–59.

van Boekel, M.A.J.S. (2009) *Kinetic Modeling of Reactions in Foods*, Boca Raton, FL: CRC Press.

van den Broeck, I., Ludikhuyze, L.R., Loey, A.M.V. and Hendrickx, M.E. (2000) Inactivation of orange pectinesterase by combined high-pressure and -temperature treatments: a kinetic study, *Journal of Agriculture and Food Chemistry*, **48**: 1960–70.

Van Impe, Poschet, F., Geeraerd, A.H. and Vereecken, K.M. (2004) Towards a novel class of predictive microbial growth models, *International Journal of Food Microbiology*, **100**: 97–105.

Wu, J.G., Luan, T.G., Lan, C.Y., Lo, W.H. and Chan, G.Y.S. (2007) Efficacy evaluation of low-concentration of ozonated water in removal of residual diazinon, parathion, methyl-parathion and cypermethrin on vegetable, *Journal of Food Engineering*, **79**: 803–9.

Xue, J., Chen, L. and Wang, H. (2008) Degradation mechanism of Alizarin Red in hybrid gas–liquid phase dielectric barrier discharge plasmas: experimental and theoretical examination, *Chemical Engineering Journal*, **138**: 120–7.

Zhao, W., Shi, H. and Wang, D. (2004) Modeling of mass transfer characteristics of bubble column reactor with surfactant present, *Journal of Zhejiang University Science*, **5**(6): 714–20.

Zhou, H. and Smith D.W. (1997) Process parameter development for ozonation of kraft pulp mill effluents, *Water Science and Technology*, **35**(2–3): 251–9.

Zhou, H. and Smith D.W. (2000) Ozone mass transfer in water and wastewater treatment: experimental observations using a 2D laser particle dynamics analyzer, *Water Research*, **34**(3): 909–21.

Zhou, H., Smith D.W. and Stanley S.J. (1994) Modeling of dissolved ozone concentration profiles in bubble columns, *Journal of Environmental Engineering-Asce*, **120**(4): 821–40.

Zimeri, J. and Tong, C.H. (1999) Degradation kinetics of (–) epigallocatechin gallate as a function of pH and dissolved oxygen in a liquid model system, *Journal of Food Science*, **64**: 753–8.

15 Health and Safety Aspects of Ozone Processing

Rip G. Rice

15.1 Introduction

When considering health and safety aspects of ozone for food processing, a larger number of factors come into play than first may meet the eye. The word 'health' calls to mind the well-being of human beings, including the workers involved in processing foods with ozone, which is simultaneously a very strong disinfectant and a very strong oxidising agent. Because of its oxidising ability, ozone has the capability to change the chemical structures of easily oxidised food molecules with which it comes in contact, thereby creating some 'byproducts' of ozone oxidation that were not part of the original foodstuff. The byproducts of such oxidations at any point in the food processing treatment may be detrimental, beneficial or have no effect on consumers, from the health and safety viewpoint. Further, the levels of some nutrients in foods (vitamins, proteins, lipids) can be changed during ozone processing (usually lowered, but not always), which may or may not be thought to affect the health and safety of consumers at first glance.

Likewise, the term 'safety' applies to food processing plant workers, and also to the equipment and instrumentation that is installed within the food processing plant to process the various food products. Such processing equipment used to generate, apply and process foods with ozone must be compatible with ozone, which must not be unknowingly converted into other materials that might then (1) damage processing equipment and/or (2) contaminate the foods being processed.

Concerns about the safe use of ozone derive from its properties, described in detail in Chapter 3 of this book. Ozone is a gas that is partially water-soluble and unstable (meaning that it has a short half-life). It is a very strong disinfecting and oxidising agent that is relatively new to the food processing industry, having been approved by the US Food and Drug Administration (FDA) as an antimicrobial food additive in 2001 (FDA 2001). The instability of ozone prevents it from being a material of commerce

Ozone in Food Processing, First Edition. Edited by Colm O'Donnell, B.K. Tiwari, P.J. Cullen, and Rip G. Rice.

that can be purchased in containers, shipped and stored onsite until used. Instead, its instability requires that ozone be produced onsite and used immediately for its intended purposes. Any excess ozone is then destroyed by converting it to oxygen, from which it was formed initially.

When ozone is added to water for disinfection and other purposes, its partial solubility means that it can degas from solution downstream of its point of initial contact with the water, thus becoming available for plant workers to breathe. Food process engineers experienced with ozone are quite familiar with this property, and routinely provide system designs that avoid worker exposures to ozone during processing.

15.2 Points of application of ozone during food processing

Specific health and safety aspects of ozone in food processing are direct functions of the presence of ozone at specific points in the processing plant. Because of ozone's great versatility as an oxidant/disinfectant, there are a great number of places within any food processing plant where it can be and is being utilised. These applications can be considered in the two primary categories of aqueous ozone and gaseous ozone phases. Wherever ozone is applied in a food processing plant there is a resultant safety responsibility.

15.2.1 Aqueous phase ozone applications

Ozone in aqueous solution can be used to process plant influent water and product water (such as juice products), to provide ozone-containing water for spray washing of incoming food products prior to processing, for treatment of process water (sometimes for reuse, sometimes prior to discharge), for spray washing food products, for sanitising plant equipment (clean-in-place, CIP) and for spray sanitation of floors and drains, as well as of food contact and non-food contact surfaces (surface sanitation).

Food transportation trucks can also benefit from spray washing of empty food containers and the truck interiors, not only to reduce levels of microorganisms present, but also to destroy odours, colours and flavours and prevent odour transfers between foods during shipments.

Ozone-containing water can be fed to an ice-making machine, where the small amount of ozone that off-gasses then gathers at the bottom of the ice storage chamber (the density of ozone gas is slightly higher than that of air) and its presence maintains the ice and chamber slime-free.

When ozone is applied to treat a food processing plant's influent or effluent waters or to treat food processing waters for reuse, the water/wastewater equipment is usually designed and operated as a mini-water/wastewater treatment plant. Such subunits normally will be an adjunct to, but not an integral part of, the food storage and food processing lines. Consequently, system equipment will be designed with all of the necessary controls to ensure that no ozone will escape to come in contact with humans in those subunit areas.

Potential health and safety issues with aqueous ozone applications

When ozone is used in water for spray washing of produce, plant equipment or floors and drains, the safety item of most concern is the exposure of plant personnel to quantities of ozone gas that might escape from the processing equipment, or from aqueous solution. When ozone is dissolved in aqueous solution, there is a tendency for it to degas from solution, due to agitation applied to the solution, particularly in a nozzle, and thus escape to the plant atmosphere. Methods to avoid contact of plant personnel with ozone are discussed in Section 15.4.

15.2.2 Gas phase ozone applications

In the gas phase, ozone is used commercially during the storage of raw foods arriving at the plant prior to processing, as well as in storage of processed foods prior to shipment from the plant. Ozone gas is also used in food transportation trucks: to minimise the growth of microorganisms, but also to control food odours and to destroy any ethylene (generated by the ripening of some foods), which otherwise will speed up the ripening process.

In some European food processing plants, gaseous ozone is also used to disinfect plastic packaging wrap as it is unwound just before surrounding a food item or a food container. Additionally, gaseous ozone is sometimes included in a modified air packaging (MAP) mixture so as to extend the microbiological reduction benefits of ozone during transportation or storage of packaged food items (Steffen et al. 2010).

Another unique application of ozone is to package processed food items using MAP mixtures containing higher oxygen concentrations than the ~21% found in ambient air. Wrapped packages so assembled are then subjected to UV-185 nm radiation, which upon contact with some of the oxygen inside the package generates some ozone within the package, again to prolong the microbiological reduction benefit of ozone during storage and transportation (Steffen et al. 2010).

15.3 Health and safety issues with ozone for food plant workers

15.3.1 Ozone exposure regulations

As with all strong oxidising agents, ozone is potentially harmful if humans are exposed to sufficient concentrations for sufficient time durations. In the USA, ozone (and other toxic gases) in commercial/industrial work places is regulated by the Occupational Safety and Health Administration (OSHA) of the US Department of Labor.

Table 15.1 Personal exposure effects and limits for gaseous ozone – USA (Rakness 2005, p. 9; from Compressed Gas Association 2001).

Observed effects	Ozone concentration, ppm_v
Threshold odour detection, normal person	0.01–0.04
Maximum 8-hour average personal exposure limit	0.1
Minor eye, nose and throat irritation; headache, shortness of breath	>0.1
Breathing disorders, reduction in oxygen consumption, lung irritation, severe fatigue, chest pain, dry cough	0.5–1.0
Headache, respiratory irritation, and possible coma, possibility of severe pneumonia at higher levels of exposure	1–10
Lethal to small animals within 2 hours	15–20
Lethal in a few minutes	>1700

Table 15.2 Personal exposure effects and limits for gaseous ozone – Canada (Rakness 2005, p. 9; from Workers Compensation Board of British Columbia 1991).

Observed effects	Ozone concentration, ppm_v
Detectable odour	0.01–0.04
TLV–TWA (threshold limit value – time-weighted average): 8-hour limit	0.1
Headache, shortness of breath	>0.1
TLV–STEL (threshold limit value – short-term exposure limit): 15-minute limit	0.3
Chest pain, dry cough, lung irritation, severe fatigue	0.6–1.0 (1–2 hours)
Immediately dangerous to life and health	10.0
Expected to be fatal	50 (30 minutes)

OSHA ozone regulations

The current permissible exposure limit (PEL) to ozone allowed by OSHA regulations is 0.1 ppm, time-weighted average over an 8-hour work day, 5 days per week. This level is rather high considering the fact that the olfactory senses of an average human being can detect ozone gas at levels as low as 0.01 ppm (0.02 mg/m^3).

The OSHA also has a short-term exposure limit (STEL) to ozone of 0.3 ppm (0.6 mg/m^3), defined as a 15-minute exposure to ozone, not to be exceeded more than four times per day (Pryor and Rice 2000).

Suggested personal exposure effects and limits for exposure to gaseous ozone compiled by the Compressed Gas Association (2001) are listed in Table 15.1 (Rakness 2005, p. 9). Similar data compiled by the Workers Compensation Board of British Columbia (1991) for exposure to gaseous ozone in Canada are listed in Table 15.2 (Rakness 2005, p. 9).

Uniform Fire Code regulation of ozone

Because of ozone's properties, ozone gas is not regulated or subjected to the same strict safety standards governing the use of other site-stored, compressed oxidising gases. However, despite the inherently less hazardous properties of ozone, minimum safety practices should be implemented in plants of any type using ozone, in order to prevent any excessive human exposures. Depending on the jurisdiction, many of the recommended safety requirements are mandated by the Uniform Fire Code (UFC), first adopted in 1994 (International Fire Code Institute 1994), which are summarised in this section (Pryor and Rice 2000).

Regulations pertaining to the transportation, storage, handling and use of other compressed gases (such as chlorine gas) are embodied in Article 80 of the UFC. Use of ozone is exempted from these provisions because there is never 'greater than 0.5 lb of ozone stored on-site at any one time'. In recognition of the potential for excessive exposure, *albeit* reduced, which always exists when ozone is being generated and used in an industrial or public environment, the Executive Committee of the National UFC approved a new standard (in 1994) for the safe use of commercial and industrial ozonation systems (International Fire Code Institute 1994). The most recent version was published in 2006 (Uniform Fire Code 2006).

Note: Before continuing, it is very important to understand that all versions of the UFC are written with the assumption that ozone is a compressed gas and will always be handled under pressure. With ozone under pressure in a pipeline, there is always the potential for a leak to occur, resulting in ozone escaping into the plant atmosphere, where workers can be exposed to it. For large-scale applications of ozone (requiring up to tons/day of ozone), this is a valid assumption. Unfortunately, applications which involve applying ozone under partial vacuum have not yet been considered or addressed by even the most recent version of the UFC regulations. In most food processing plants, ozone is applied under partial vacuum.

A distinct safety advantage of applying ozone under partial vacuum (as when using an eductor or injector contactor) is that there is no opportunity for ozone to leak from a faulty pipe or failed pump into the surrounding plant atmosphere. Under such failure circumstances, the system is shut down, not because of an ozone leak into the atmosphere, but because of a loss of vacuum, caused by a broken gas transfer line or pump, or a hydraulic system failure. At such times, the flow of electricity to ozone generators is immediately ceased, thereby immediately ceasing the production of ozone and preventing any worker exposure to potential 'ozone or oxygen leakage'.

On the other hand, because the UFC exists and is widely in effect, it is appropriate to understand these regulations, so as to be aware of their intent but also of their applicability or not to the specific method(s) of applying and using ozone in a food processing plant.

The UFC regulations are far-reaching in their scope and represent a significant consideration in the development of new-construction ozonation projects and the maintenance or upgrading of older, existing ozone systems.

Many aspects of the code changes require certain standard safety precautions that currently are in widespread use with other (*compressed*) toxic gases. These practices include installation of ambient air ozone monitors to continually measure ozone levels when ozone is used in enclosed spaces. Such monitors can be preset to flash alarm lights, sound an alarm and even start up exhaust fans if the monitor detects ozone at levels above the set point, which in many plants is at or just below the OSHA PEL of 0.1 ppm. All systems designed to produce and apply ozone should also include appropriate ozone-destruction units to prevent the possibility of unconsumed ozone from being released into plant atmospheres. Many other required safety precautions have been employed routinely in the past by most vendors of large ozonation systems, and include external interlocks on the ozone equipment, treatment of ozone ventilated or exhausted from ozone generator cabinets or treatment processes, and secondary containment of pressurised ozone gas-carrying lines.

For all nonresidential ozone generators which have maximum daily output capacities of greater than 0.5 lb of ozone, the UFC code specifies that ozone generators must be located in either: (1) unoccupied, ventilated rooms labelled with appropriate warnings and monitored continuously for ozone with an alarm and safety system; or (2) approved cabinets with ventilation. The cabinets must also meet standards appropriate for their use and NEMA (the National Electrical Manufacturers Association) standard 250 is noted as an approved design guide.

One exception to this requirement is if the 'generator is in an approved pressure vessel'. Any such pressure vessel approval within any jurisdiction will, in the absence of more definitive standards, require ASME (American Society of Mechanical Engineers) compliance. Ozone generators must also be appropriately labelled with specified warning stickers and must be seismically anchored per Uniform Building Code requirements.

Minimum ventilation standards require six changes of the room or cabinet air volume per hour. Ventilation exhaust from either a room or a cabinet must be directed to a treatment system designed to reduce the discharge concentration of the exhausted gas to 5 ppm. This is one-half of the level determined to be 'immediately dangerous to life and health' (IDLH = 10 ppm). If cabinet ventilation is employed, air intake into the ozone generator must be at a velocity greater than 200 feet per minute to ensure that no fugitive emissions escape from the generator. A high-ambient-ozone-concentration monitor interlock on the ozone generator may be used in lieu of ventilation exhaust treatment when ozone generators are in rooms.

Secondary containment of all valves, fittings, gauges and piping carrying ozone gas must be installed, except when welded stainless steel pipe is employed. This secondary containment must also be vented to an appropriate exhaust treatment system.

All materials that come into contact with ozone gas must be completely ozone-compatible in order to prevent ozone leaks. Acceptable materials include 304 stainless steel, PVC, glass, Hypalon, Teflon, Kel-F and so on.

An external interlock also must be provided for the ozone generator in order to ensure shutdown of the generator in the event of an external system failure or shutdown. The external interlocks must include as a minimum a failure of the ventilation treatment system, a failure of the ambient ozone concentration monitor (if being used in a room) and a failure of the process being treated (generally indicated by a pump on/off condition and/or a high ORP (oxidation-reduction potential) set point or dissolved ozone controller interlock).

Emergency shut-off switches must be on the ozone generator and within 10 feet of the primary exit if the generator is in a room.

These requirements were developed primarily for large-scale ozone generating equipment (tons/day) in which the ozone is moved throughout the plant under pressure, with the potential of leaking into the plant atmosphere. The concerns about worker exposure to ozone are equally applicable to smaller-sized ozone systems, such as those used in many food processing plants. However, the 'pressurised ozone' safety concerns become moot with smaller systems designed to apply and carry ozone under partial vacuum, such as those that are used in food processing plants.

Since adoption of the UFC in 1994, its requirements (Section 3705, Ozone Gas Generators) have been routinely applied in the USA to new plant construction projects, but not to the retrofitting of ozonation systems into existing food processing plants, and not to systems that apply ozone under partial vacuum.

15.3.2 Potential fire hazards from high-purity oxygen use

For reasons of economy and efficacy, it is usually wiser to feed oxygen-enriched air (sometimes containing up to 95% oxygen) to an ozone generator than to feed it dried air. On the other hand, concentrations of oxygen in air significantly higher than the ~21% normally present in ambient air can increase the ability of otherwise nonflammable materials to burn if accidentally ignited by a stray spark or flame.

Most high school chemistry courses include a laboratory experiment in which a sample of red-hot steel wool (which does not burn in ambient air) is inserted into a test tube containing nearly 100% oxygen. Immediately, the hot steel wool bursts into flame and burns, leaving only powdery iron oxide (rust) as a residue. The metallic steel wool has 'disappeared', by being oxidised to ferric oxide (rust).

Although breathing high concentrations of oxygen produced by passing ambient air through an oxygen concentrator does not represent a hazard to human health (in fact many infirm or elderly people who have difficulty breathing are fitted with devices leading to an oxygen cylinder to increase their intake of oxygen), there does exist the potential for fires if the concentrated oxygen is allowed to come in contact with materials that are considered to be nonflammable in ambient atmospheric air, as well as to increase the rate of combustion of flammable materials (paper, cloth, wood, rubber, etc.) in the presence of a spark or flame.

Note: When ozone is generated and distributed under pressure (as described by the UFC), there is always the potential for oxygen to escape along with ozone in the event of a leak. However, with ozone and oxygen mixtures applied under partial vacuum, there is no potential for ozone or oxygen leaks because of automatic system shutdown in the event of a loss of vacuum caused by a broken gas transfer line or pump, or a hydraulic failure.

A particular note of caution is warranted when using oxygen as the feed gas. It is strongly advised not be use carbon filters to convert any out-gassed ozone into oxygen before release into the environment. This is because the use of oxygen in conjunction with activated carbon in the off-gas catalytic converter is a serious hazard to human health and safety due to the risk of rapid build up of heat of decomposition of the ozone, with resulting explosion of the carbon canister if it is ignited while having oxygen passed through it. A thermal/catalytic ozone destruct system is recommended.

Further, as a result of the potential for fire in any application involving high-purity oxygen gas (including both compressed oxygen gas and oxygen produced onsite by oxygen concentrators), specific safety criteria have been embodied in the UFC (International Fire Code Institute 1994) pertaining to the proper and safe use of oxygen. The regulations include, among numerous additional requirements, specifications for the type and gauge of gas delivery lines, acceptable solders and brazing materials, pipe and equipment cleaning procedures, material compatibility and so on.

Although not specifically required in any regulations, the use of automatic, temperature-actuated halon fire extinguishers mounted internal to the ozone generator cabinet is also recommended in those food processing ozone applications employing oxygen as a feed gas. Further, an automatic oxygen shutoff valve should be employed in all oxygen-fed ozonation applications, whereby the shutoff valve to the oxygen source will close immediately when the ozone generator is turned off for any reason. This will prevent the continued feed of oxygen from fueling any potential fires in the system.

15.3.3 Safety history of ozone in commercial/industrial applications

Ozone has been in commercial use for the treatment of drinking water since 1906, when the city of Nice, France installed ozone to disinfect mountain spring water. This Mediterranean resort town has now used ozone continually for the treatment of its drinking water for over 100 years without incident, and today thousands of potable water plants throughout the world are also using this technology. Many of the most recent new-construction industrial plants, and even many older upgraded plants, use high-purity oxygen to generate ozone, without experiencing hazards, either from the ozone or from the high-purity oxygen. In these many drinking water plants, ozone is generated routinely in quantities ranging from grams per hour (small plants) up to tons per day (large municipal plants). Many other commercial/industrial applications for ozone also exist throughout the world, including pulp bleaching, kaolin bleaching, wastewater treatment and reuse, bottled water treatment, swimming pools, cooling towers, synthesis of nylon intermediates, air treatment, marine aquaria, aquaculture, food storage and processing plants, wineries and so on.

In the century that has passed since ozone was first installed in Nice, there has never been a reported death due to ozone exposure. Why? Because engineers were quick to recognise the potential danger to humans of ozone exposure. Consequently, processes involving ozone are routinely designed with appropriate precautions to avoid exposure of workers to ozone.

The situation is analogous to that of chlorine, also a very strong disinfectant and oxidising agent, and a chlorinating agent as well. This chemical was used as a poison gas during World War I, and many troops were killed on both sides of the trenches when exposed to it. But today, chlorine is an essential industrial chemical used safely in tons/day quantities for a variety of commercial/industrial processes, all as a result of attention to the safety of humans handling this strong disinfecting, oxidising and chlorinating material.

15.4 Avoiding worker exposure to ozone in food processing plants

15.4.1 General considerations

The very first item to define is the design of the ozone system that generates and delivers ozone to the desired point(s) in the food processing plant, followed closely by how and to where in the plant the produced ozone is applied and to where it might travel after its application. It is important to recognise before installing an ozone system how and where in a process a leak in ozone equipment or delivery piping might occur.

When purchasing an ozonation subsystem from a reputable supplier, one can have confidence that the supplier is motivated to provide a well-designed, reliable system that includes appropriate instrumentation and controls to guard against the accidental release of ozone during processing. However, food processing plant engineering staff should become involved early in the process of purchasing ozone equipment, so that they can learn about ozone and then cross-check the design of the ozonation system to be purchased with the point in the plant at which it is to be applied, to be sure that health and safety aspects of installing ozone have been taken into account and appropriate changes have been made, if necessary.

These objectives are attained by the ozone equipment supplier first selecting appropriate materials of construction that are compatible with ozone and are not slowly destroyed by it. An additional safety item is the installation of ambient air monitors at key locations in the plant where ozone leaks might occur, such as near the ozone generator, near the ozone contacting unit (aqueous applications), in pipes carrying ozone gas and so on. These monitors detect the presence of ozone in the plant air, and if/when the set point (say the OSHA PEL limit of 0.1 ppm) is exceeded, due to the accidental release of ozone as a result of some processing or equipment problem, the ambient air monitors/controllers can signal the ozone generator, causing the generator to cease production. At the same time, an alarm can be programmed to sound and red lights to flash, and plant exhaust fans can be triggered to increase the normal rate of room/plant air turnover. Plant maintenance personnel then can address the source of the specific problem, restarting the processing equipment when it has been corrected.

Although human olfactory senses can detect ozone at levels as low as 0.01–0.04 ppm, they can become desensitised during prolonged and continuous exposure to ozone. Consequently, once ozone is detected, appropriate safety actions should be taken. Olfactory senses will regain their initial sensitivity in the absence of ozone. In some processing plants located in urban areas that experience high outdoor ambient ozone levels due to air pollution, the first-alert sound and visual alarm point are often higher than the OSHA PEL of 0.1 ppm; for example, 0.2 ppm. In these cases, a higher monitor setting is used, because if the alarm sounds when an air pollution event occurs, staff may ignore the warning if and when an ozone leak actually occurs inside the plant (Rakness 2005, p. 9).

The 15-minute STEL limit of 0.3 ppm is normally used as the alarm condition for an ozone generator power shutdown, which ceases the production of ozone (Rakness 2005, p.9). Feed gas flow continues, however, in order to flush ozone from the pipes, with due consideration of potential fire dangers when oxygen is the primary feed gas.

The safety measures just recounted are primarily for those systems applying ozone under pressure. If systems are installed wherein ozone is applied under partial vacuum, then in the event of a leakage, plant air 'leaks' into the piping system rather than ozone and oxygen leaking into the plant air, thus eliminating many of the potential safety problems with larger ozone systems.

15.4.2 Specific plant safety measures

For the protection of food processing plant workers from exposure to low levels of leaked or degassed ozone, small low-cost badges and similar analytical instruments are available and are recommended by ozone equipment suppliers. These devices can be belt-mounted or clipped/pinned on each worker, and a glance at the meter from time to time shows the level of ozone in the ambient air, wherever the worker may be in the plant. These personal ozone monitoring devices are an extra level of protection for food plant workers, in addition to the larger wall-mounted ozone-ambient air monitors that are installed at various points in the plant where significantly higher ozone levels might be present (in case of some accident or equipment failure).

Plant workers normally do not smell ozone while working at ozone facilities. However, they will often be able to identify minor leaks via the odour of low-concentration ozone, which would not cause a shutdown of the ozone facilities. If and when these low-ozone-concentration leaks occur, they should be located and repaired as soon as practical (Rakness 2005, pp. 9–10).

Very large ozone systems, producing lbs/minute to tons/day of ozone, and some smaller systems as well, usually include an ozone off-gas destruction subsystem, through which excess ozone from the ozone/water contacting system is passed. Such ozone destruct units convert excess ozone to oxygen prior to discharging the off-gases to the ambient environment outside of the processing plant. In such cases, an ambient ozone monitor is installed at the exhaust outlet to ensure that the ozone destruct unit is performing efficiently.

In most of the food processing plants currently using ozone, the production rates are below ~1000 g/hr (a little over 2 lbs/hr) from oxygen-enriched air, and ozone is applied and distributed in aqueous solution under partial vacuum, so as to avoid the safety hazards of ozone and oxygen under pressure (B. Hamil, DEL Ozone, 2010, personal communication).

15.4.3 Controlling off-gas ozone at Fresher Than Fresh fish processing/packaging plant

The Fresher Than Fresh (FTF) company (Gastonia, NC) uses ozone in water for many washing applications (Figure 15.1) (Rice and Wrenn 2007). These applications include spray washing bulk fish containers, doorways, utensils, hoses, latex gloves, process machinery, workers' shoes, floor drains, washing fish prior to and during processing, and conducting plant wash-downs. Each of these washing applications involves agitation of the ozone-containing water solution, and thus poses the potential for significant ozone off-gassing, with the ozone escaping to plant ambient atmosphere at the washing points.

In order to reduce the amount of ozone off-gassing, FTF controls the ozone residual in water at the outlet of the ozone/water contactor at 2.5 mg/L (ppm). If this level is exceeded, ozone generation is programmed to shut down automatically. For the distances the ozone-water solutions are piped throughout the plant, this initial concentration of dissolved ozone

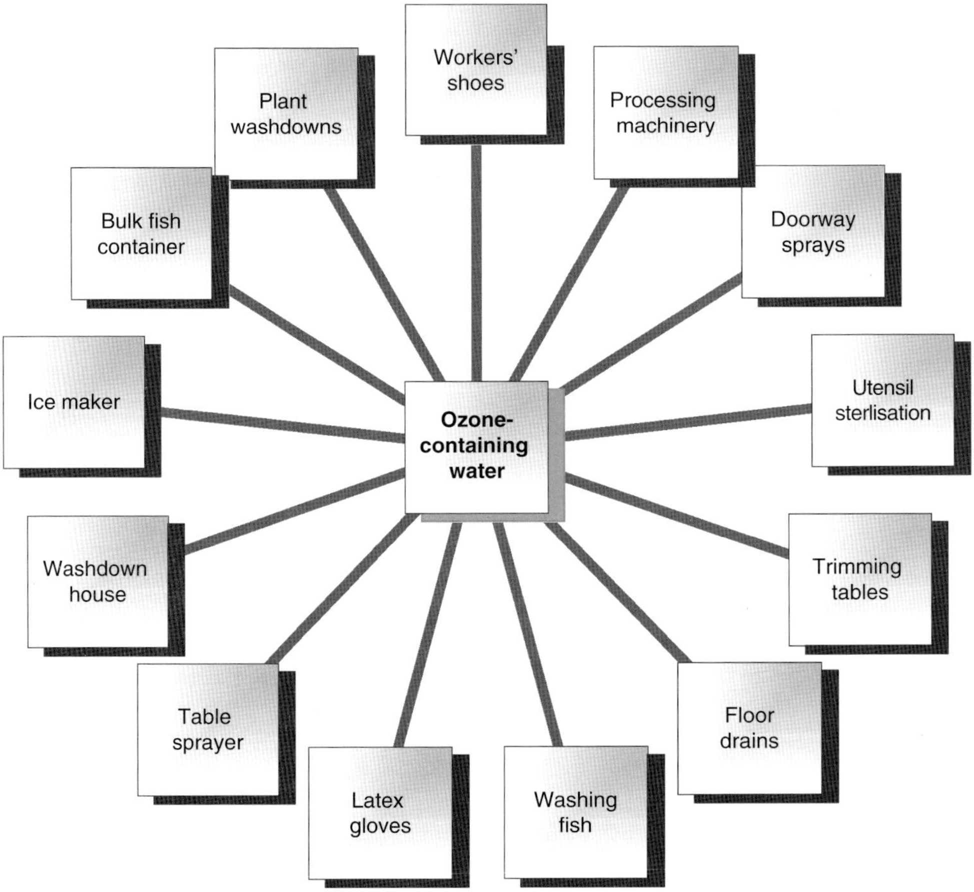

Figure 15.1 Ozone-containing water uses at Fresher Than Fresh.

ensures that only very small amounts of ozone off-gas at the various wash points into the plant ambient air. By the time ozone-containing water reaches its various points of use, the 2.5 ppm ozone concentration has decreased (due to autodecomposition of ozone in solution). As long as the dissolved ozone concentration at its point(s) of use in this plant does not exceed 1.5 ppm, off-gas ozone concentrations in the plant atmosphere never exceeded the OSHA PEL of 0.1 ppm (time-weighted average over an 8-hour working day) since the ozone system was installed in 2003. Nor have the plant workers been made uncomfortable by low-level fugitive ozone emissions (Rice and Wrenn 2007).

Because this plant processes fresh fish and other seafood products, there was usually a noticeable 'fishy odour' throughout the premises prior to the adoption of ozone treatment of plant process water. After installation of the ozone system, a sharp decrease in 'fishy odour' in the plant ambient air was noted, probably because ozone in the gas phase is well-known to destroy fishy odours.

The off-gassing potential of ozone from aqueous solutions during spraying can also be overcome by using equipment designed to employ a combination of techniques (Hamil 2007):

(1) Degassing all undissolved ozone in aqueous solution.
(2) Destroying the degassed excess ozone in an ozone destruction subunit.
(3) Lowering the water line pressure to the spray nozzle to ~19 psi (lbs/in^2).
(4) Using a low-pressure, high-flow, low-sheer spray nozzle.

Combining all these factors reduces the tendency of the remaining dissolved ozone to degas during spray washing when washing foods, cleaning in place or cleaning out of place, and other applications. This specially designed ozone washing equipment has been evaluated for both efficacy and worker safety by third-party testing, as described in Section 15.4.4.

15.4.4 Third-party evaluation of aqueous ozone spray wash equipment

In 2002, the Toxicology Group, a wholly owned company of NSF International (Ann Arbor MI), conducted detailed third-party efficacy and hazard assessments and analyses for DEL Agricultural (a subsidiary of DEL Ozone, San Luis Obispo, CA) and Air Liquide America. Two devices (DEL AGW-0500 Mobile Ozone Surface Sanitation System, AL SSS 0500 Mobile Ozone Surface Sanitation System, and the DEL AGW-1500G Mobile Recirculating Ozone Sanitation System, AL SSS 1500 Mobile Recirculating Ozone Sanitation System) are manufactured by DEL Ozone and marketed by these two firms for spray washing applications in food processing plants.

Both models are mobile. One provides a 10 gal/min water spray with a 3.0–3.5 ppm applied ozone dose, and is designed to sanitise equipment, walls, floors, drains, tables, conveyors, containers, tanks and barrels. The other, designed for CIP and COP (clean-out-of-place) processes, recirculates ozone-containing water at 35 gal/min with a 3.0 ppm applied ozone dose through tanks ranging in size from 50 to 2500 gallons. In any of these systems, the residual ozone dose that is applied as a spray is in the range of 1.5–2.0 ppm; and in the case of the recirculation system, the residual ozone dose is monitored and controlled at 2.0–2.5 ppm.

Third-party efficacy testing

The methods used for the efficacy tests were AOAC Official Methods 960.09, Germicidal and Detergent Sanitizing Action of Disinfectants, and 961.02, Germicidal Spray Products as Disinfectants (Boisrobert 2002). Ozone spray washing was conducted on samples of individual microorganisms listed in Table 15.3, which also shows the number of log reductions obtained for each microorganism tested. Each microorganism received an ozone dosage of 1.85–2.25 ppm from the spray nozzle, except for *Escherichia coli*, which received an ozone dosage of 2.1 ppm.

Table 15.3 Efficacy testing of mobile ozone surface washing system (Boisrobert 2002; cited in Pascual et al. 2007).

Microorganism	Log reduction
Trichophyton mentagrophytes (ATCC 9533)	6
Salmonella choleraesuis (ATCC 9533)	6
Staphylococcus aureus (ATCC 6358)	6
Pseudomonas aeruginosa (ATCC 15442)	6
Campylobacter jejuni (ATCC 33250)	4
Listeria monocytogenes (ATCC 7644)	4
Aspergillus flavus (ATCC 9296)	4
Brettanomyces bruxellensis (ATCC 10560)	4
Escherichia coli (ATCC 11229)	5

The results obtained (log reductions of 4–6 for the nine microorganisms tested) substantiate the efficacy of these two systems in sanitising previously cleaned nonporous surfaces, including processing equipment, which has come into contact with food (Pascual et al. 2007).

Third-party hazard analysis testing

The hazard assessment was conducted in accordance with the Hazard Communication Standard (HCS) as promulgated through the Occupational Safety and Health Act of 1970 and documented in the Code of (USA) Federal Regulations, Title 29, Chapter XVII, Part 1910, Section 1910.1200, addressing the physical and chemical hazards associated with the operation of the devices, and DEL Agricultural has included this information in its Owner's Manual in a manner consistent with the HCS.

Assessment of the two devices tested by the Toxicology Group, reported by Boisrobert (2002) and cited by Pascual et al. (2007), showed that the system has inherently low hazards due to its design and construction, which includes a number of safety features. The safety interlock, vacuum switch, flow pressure switch and ozone sensors provide an automated fail-safe mechanism to shut down the corona discharge ozone generation unit in the event the equipment fails to operate under preset parameters.

During the efficacy testing and validation, operator monitoring was conducted using a ChromAir passive ozone monitor (K&M Environmental, Santa Fe, NM). The efficacy testing was conducted within a laboratory bay of approximately 60 000 cubic feet. Monitoring results indicated that during the operation, ozone concentrations were at or below the OSHA PEL of 0.1 ppm on over 10 separate occasions.

The hazard analysis was based on an engineering review, DEL Agricultural internal test reports and in-house experience gained from

Table 15.4 Selected operational ozone monitoring results (Toxicology Group 2002). Results were collected using a ChromAir Passive Ozone Monitor (badge) from K&M Environmental.

Date	Ozone concentration (direct reading) ppm_v	Operating duration (minutes)	Time-weighted average (TWA) ppm_v
May 9, 2001	0.08	60	0.01
May 9, 2001	1.6[a]	60	0.2
May 10, 2001	0.05	75	0.08
June 20, 2001	>1.6[a]	60	Not calculated
June 20, 2001	0.7	65	0.1
July 13, 2001	0.08	60	0.01
July 13, 2001	0.8	Not known	Not calculated
August 8, 2001	0.08	120	0.02
August 8, 2001	0.08	120	0.02
August 13, 2001	0.2	60	0.02
August 15, 2001	0.2	60	0.02
August 17, 2001	0.2	90	0.04
August 17, 2001	0.6	90	0.1
August 21, 2001	0.5	90	0.09
August 21, 2001	0.3	90	0.06
August 31, 2001	0.3	60	0.04
August 31, 2001	0.3	60	0.04

[a]Based on data from the other operator during those same test runs, the data set as a whole, the unit's mode of operation and the lack of respiratory irritation noted by any operating personnel, these levels were considered falsely elevated, most likely due to a splash of ozone-enriched water directly on the badge.

operating the system during the determination of the efficacy of the AGW-0500/AL SSS 500 Mobile Ozone Surface Sanitation System (Boisrobert 2002; Pascual et al. 2007).

Over a period of four months, the Toxicology Group operated the AGW-0500 Mobile Ozone Surface Sanitation System to validate DEL Agricultural's antimicrobial claims. During use of the system, operators were monitored using passive ozone monitoring badges affixed near their breathing zone. As noted in Table 15.4, the levels detected by the badges were at or below acceptable levels in all but two cases (Operator 2 on both 9 May 2001 and 20 June 2001). Based on the data from the other operator during those same test runs, the data set as a whole, the unit's mode of operation and the lack of respiratory irritation noted by any operating personnel, these levels were considered falsely elevated, most likely due to a splash of ozone-enriched water directly on the badge. Although no monitoring was conducted to assess the peak exposures against the OSHA proposed STEL of 0.3 ppm (acceptable level for a 15-minute exposure), the data indicate that it is unlikely that there was any exceeding of the STEL.

It is the professional opinion of the Toxicology Group that the DEL Agricultural devices tested deliver a consistent applied ozone dose which meets the critical level required to ensure the antimicrobial efficacy claims while still maintaining exposures below the OSHA PEL. The product literature has provided sufficient information characterising the physical and chemical hazards associated with use of these devices, thereby allowing employers adequate guidance to put in place a hazard communication programme around the use of each device as required by the HCS.

Conclusions by the third party

The inherent low hazard due to the unit's design and construction, coupled with the safety features, monitoring data (noted in Table 15.4) and precautionary directions provided in the Owner's Manual are sufficient for the Toxicology Group to provide a professional opinion that the DEL Agricultural AGW-0500/AL SSS 500 Mobile Ozone Surface Sanitation System and AGW-1500G/AL SSS 1500 Mobile Recirculating Ozone Sanitation System pose no safety concerns, when operated under the prescribed conditions as set forth in the owner's manual.

15.5 Safety of foods processed with ozone

One primary benefit of employing ozone anywhere in a food processing plant is the diminishment of the levels of microorganisms present on the surfaces of foods. Achieving this benefit extends the shelf life of the specific foodstuff. Counterbalancing this benefit, however, is ozone's strong oxidation capability. This means that extended contact with ozone (particularly at higher concentrations) increases the chances that the foodstuff itself will react with ozone, thereby changing the initial chemical nature of the food. In such applications, therefore, it is important for the user to apply just enough ozone to attain the microorganism-lowering objective, while at the same time minimising any unwanted oxidative decomposition of the food itself caused by exposure to excessive amounts of ozone.

In 1997, a panel of food experts was convened by the Electric Power Research Institute (EPRI) to evaluate the available literature regarding processing of foods of all types with ozone. Among other aspects of ozone, the panel evaluated its nutrient impacts and potential toxicological effects on foods treated with it (EPRI 1997).

15.5.1 Nutrient impacts of ozone contact with foods

In his introduction to the 'Nutrient Impacts' chapter of EPRI (1997), Erdman (1997) discussed the following points of interest to food processors introduced to ozone for the first time:

Foods are processed for several purposes: to preserve the food and to extend its shelf-life, to increase its digestibility, to improve its palatability and texture, to prepare it for serving, to remove inedible parts, to destroy anti-nutritional factors, to create new types of foods, and to destroy toxins and eliminate microorganisms. Of foremost concern is the safety of the food, as consumed. Although most food preservation techniques decrease the overall nutrient content of the food to some degree, this is a necessary price to pay for safety. Thus choice of the most appropriate food processing techniques is always a question of balance between extent of preservation and safety of the food and the nutrient retention following processing (Erdman and Poneros-Schneier 1994).

There are three major causes of food spoilage: chemical changes, enzymatic changes, and, most relevant to an application of ozone, microbiological spoilage. The traditional food preservation methods utilised to reduce microbiological load include thermal processing, alteration of pH, use of chemical preservatives, use of microwaves and ionizing radiation, removal of water, or a combination of these techniques. Essentially all of these techniques will reduce, to a varied degree, the nutrient content of the preserved food.

Ozone is a strong oxidant and thus would be expected to cause alterations in nutrient levels in foods if high concentrations of this material are used for extended periods. However, ozone does not penetrate deeply into foods (Kuprianoff 1953), and any negative impact on nutrient content (caused by ozone) is limited to nutrients on the surface of the food.

Erdman (1997) then reviewed the available published literature on the effects of ozone on each major nutrient class (Sections 15.5.2, 15.5.3, 15.5.4 and 15.5.5). Although there were only a few studies available at the time, 'it appears that under properly controlled preservation conditions, ozone causes only minor losses of nutrient content, lower than (when processed by) some other processes' (Erdman 1997).

15.5.2 Impacts of ozone processing of foods on vitamin contents

The vitamins most labile in oxidizing conditions include vitamin C, vitamin B1 (thiamin), folate and the carotenoids. Often vitamin C and thiamin are utilised as indicator nutrients when monitoring the effects of food processing techniques upon stability of nutrients. If these compounds are retained to a high degree, then it is assumed that others will as well.

Two Japanese papers have evaluated the effects of ozone treatment on the thiamin and riboflavin (vitamin B2, another water soluble vitamin with more stability to heat than thiamin but less stability to light) retention in cereal and bakery products, peas, beans and whole spices.

Naito and Nanba (1987) treated 24 different kinds of foods, including cereal grains, cereal grain powders, peas, beans and whole spices with 0.5 to ca 50 ppm ozone (gas phase) at 10 °C for 1 hour. Riboflavin proved to be very stable during treatment, with over 90% retention of vitamin B2 in all food samples, even with treatments of 50 ppm ozone. Thiamin decomposition was detected in some food samples of flours and spices treated with 50 ppm ozone. Losses amounted to up to 40% in products with high surface areas, but minimal losses (~10%) were seen in whole grain and bean products.

Naito et al. (1989) also treated wheat flour with 0.5–50 ppm ozone for 6 hours to control airborne microorganisms prior to the production of Japanese noodles, and again found no change in riboflavin content. The shelf-life of the produce was substantially increased, although some thiamin content was lost.

Henry et al. (1996) studied the degradation kinetics of the carotenoids, β-carotene and lycopene *in vitro* after exposure to a continuous flow of oxygenated or ozone-containing water. Each carotenoid was adsorbed onto a solid surface and ultra-filtered water saturated with either oxygen or ozone was flowed over the surface at ambient temperature. As expected, ozone was a stronger oxidant than oxygen. Approximately 90% of the colour of lycopene and β-carotene were lost after 1 and 7 hours, respectively, with ozone, and in 2 and 7 hours for oxygen. Thus lycopene, the primary red pigment in tomatoes, is more susceptible than is β-carotene.

When potatoes were stored in an ozone-containing environment, there was a 1.2-fold ***higher*** content of vitamin C than in the control sample (Kolodyaznaya and Suponina 1975). This change probably reflects an effect of ozone on metabolism in the potatoes, as total sugars also decreased by 1.3–1.5-fold with a 3–6% increase in starch content. The mechanism of these changes most likely is due to ozone's known oxidizing effects on ethylene (a plant hormone that increases ripening) with an end result of slowing the ripening process (Erdman 1997).

15.5.3 Impacts of ozone processing of foods on protein contents

It is quite clear from studies of ozonolysis of water for drinking or for food processing that extensive use of this oxidant will chemically destroy a variety of amino acids (Rice and Gomez-Taylor 1986). Similarly, LaLecheur and Glaze (1996) showed that oxidation of serine in water produces a set of carbonyl and carboxylate containing byproducts that reflect both molecular ozone and hydroxyl free radical chemistries. Oxidative decarboxylation and nitrate formation is the preferred route of reaction of serine with molecular ozone, while ammonia formation indicated hydroxyl free-radical chemistry. The reaction sequences of ozone with glycine, alanine and phenylalanine, for example, have been delineated, but then progress of the reactions is the same as for chlorine, another strong oxidant used to purify water.

Kasei et al. (1994) used ozonolysis processing to prepare an ozonated casein. This harsh treatment completely destroyed all aromatic amino acids, except for a few of the phenylalanine residues, and reduced the true digestibility of the casein. Naitoh (1992) evaluated the effects of high levels of ozone (110–120 ppm) on the ozonation of amino acids in aqueous solution and found that the most labile amino acids were tryptophan, tyrosine, phenylalanine and methionine. The metabolic products produced during this process were not found to be mutagenic, whether or not the pure amino acids or mixtures of amino acids and glucose were ozonated.

In all of the above studies, extremely high levels of ozone were utilised. There is no evidence that exposure of foods to levels of ozone typically suggested for food preservation will cause any destruction of amino acid residues or reduction of protein quality (Erdman 1997).

15.5.4 Impacts of ozone processing of foods on lipid contents

Naitoh (1989) treated cereal powders, peas, beans, pulse products and cereal grains with 0.05 to ca 50 ppm ozone and found that up to 5 ppm ozone, oxidation of lipids rarely occurred, while at higher ozone levels (50 ppm and higher), considerable lipid oxidation was noted. Gorman et al. (1995) studied the effects of several types of disinfecting treatments, including 0.5% ozone-containing water on the TBA (thiobarbituric acid) content of beef following 29 days of storage at 4 °C. Sheldon and Brown (1986) compared the use of ozonated poultry carcasses with chilled water carcasses and reported higher TBA numbers in some tissues with ozone but not with others. Sensory panels could not detect differences between treatments of the broiler meat. Watanabe et al. (1994) used ozone (0, 0.03, and 0.1 ppm) during cultivation of nameko mushrooms and found that the ozone treatment increased palmitoleic acid contents and decreased the linoleic acid and PUFA/SAT FA ratios (PUFA = PolyUnsaturated Fatty Acids; SAT FA = Saturated Fatty Acids).

Ozone is known to react rapidly with unsaturated organic compounds (Rice et al. 1982); thus it would not be a surprise to find that at higher levels of exposure to ozone, some oxidation of PUFA and increase in peroxidation of fat had occurred. However, at usual levels of (ozone) treatment, no significant effects are expected to occur (Erdman 1997).

15.5.5 Toxicology aspects of ozone processing of foods

Two studies reported the feeding of ozone-treated casein to rats (Kasei et al. 1993, 1994). Casein was dissolved in 8 M urea, and then 0.3% ozone in an oxygen stream was bubbled into the solution at the rate of 100 L/h for 20 hours (a severe regimen). Ozonated casein was obtained after lyophilisation of the dialyzed solution. Chemical analyses showed significant losses of cystine, methionine, tyrosine, tryptophan, phenylalanine and histidine in the ozonated caseins. The concentration of each amino acid in the treated and untreated caseins was equalised by the addition of a suitable amount of the free amino acid. Groups of six rats were fed the experimental diets for a two-week period, with the caseins at 8%. A fourth group received 4% egg white and was included in the measurement of biological value and true digestibility of each experimental diet.

Growth and food intake was essentially the same for the ozonated casein groups and slightly less than the untreated casein group, but the difference was not significant. The biological value of the ozonated casein diet was not inferior to the native casein diet, but the true digestibility of the ozonated casein diet was significantly less than that of the native casein diet. Kidney, cecum and liver weights of rats fed the ozonated caseins were significantly greater than those fed the native casein diet. No significant differences in weight of other organs was observed. The enlarged cecum was considered to be caused by the lower digestibility of the ozonated casein – because the cecum of rats fed casein ozonated after predigestion with pepsin were smaller compared with those fed ozonated casein. The cause of kidney enlargement was not determined. A part of the liver enlargement of the ozonated casein-fed group was due to the accumulation of triglyceride.

Effects of amino acids on fat deposition in the liver of rats fed a lower protein diet are attributed to amino acid imbalances. Others have shown that an 8% casein diet with 0.3% methionine develops fatty livers as the plasma level of threonine and serine decreases. Liver fat accumulation is prevented by the addition of threonine.

The literature also suggests that the oxidation products formed when foodstuffs are treated with ozone are similar to those formed when water is treated with ozone. Although the data are somewhat limited, the available information does not suggest significant health problems. Results to support this position include:

(1) Repeated, long term inhalation studies with animals show that ozone is not a carcinogen (NTP 1995).
(2) No mutagenic products were detected after 18 different amino acids and 10 freeze-dried saccharides were treated for 1–5 hours with ozone (Naitoh 1992).
(3) By-products of the reaction of ozone with unsaturated fatty acids are primarily aldehydes, ketones, and hydrogen peroxide (Kozumbo et al. 1996).
(4) The biological value of ozonated casein was shown to be comparable to untreated casein, although the digestibility of an ozonated (casein) diet was less than that of a native casein diet. Adverse metabolic effects

in rats fed ozonated casein were shown to be due to a loss of certain amino acids, and not from an accumulation of toxic components.

(5) In Japan and Australia there are no quantitative restrictions on ozone as a food processing agent. The most recent publication of French Standards (France 1995) appears to represent an attempt to make the regulations consistent with the requirement of other member countries in the European Community (Newell 1997).

5.6 Conclusions

Ozone is a chemical that can provide many benefits to food processors, by itself and in combination with other food processing treatment procedures. However, its potentially hazardous nature must be understood and measures must be taken to ensure that plant workers are not exposed to breathing excessive levels of ozone. It is not a systemic poison, nor is it carcinogenic. Exposure to low concentrations of ozone for short periods of time does not cause damage to individuals, as long as the concentrations and exposure times are below currently regulated maximum exposure levels. In more than 100 years of ozone application in a variety of commercial-industrial manufacturing and processing plants, there has never been a death reported from exposure to ozone.

Equipment is commercially available to generate, apply, monitor and control ozone in both the aqueous and gaseous phases for a variety of applications within a food processing plant.

A panel of food experts convened by the EPRI in 1996/1997 reviewed the then existing scientific literature relating to ozone contact with foods and concluded that when used under Good Manufacturing Practices, ozone is safe for plant workers to use, and does not impart detrimental effects to foods processed with ozone.

The US FDA approved the use of ozone in 2001 as an antimicrobial food additive as long as it is applied according to Good Manufacturing Practices.

Third-party testing of commercially available equipment designed to produce ozone-containing water for spraying in food processing plants for a number of washing applications has been conducted under conditions that provide 4–6 log reductions of microorganisms typically found in food processing plants. Results of these tests show that plant workers can operate such equipment with minimal exposure to gaseous ozone, at levels that are below existing current OSHA maximum permissible exposure standards.

When oxygen is employed as the feed gas for generating ozone, users should be aware that flammability of many organic materials increases dramatically. Exposure of some organic materials of construction to high-purity oxygen may also increase the degradation of those materials. Therefore, when oxygen is used to prepare ozone – a common occurrence in many food processing plants – appropriate precautions should be taken to avoid potential fires that might be caused by stray sparks or flames in the event of oxygen leaks. Further, proper attention should be paid to oxygen- and ozone-resistant materials of construction when these two food processing aids are designed and installed in food processing plants.

The capability of applying ozone to an aqueous stream under partial vacuum is a true safety advantage, because such a procedure ensures that in the event of equipment leakage, plant air will leak into the moving water stream rather than ozone leaking out of the system into the plant air. When ozone is generated from oxygen-enriched air (from an oxygen concentrator), if hydraulic or pumping failure occurs, the ozone generating equipment is programmed to shut down immediately, at which time air feeding the oxygen concentrator is also ceased, thereby eliminating the potential for oxygen fires.

Safe handling and use of dangerous materials in commercial-industrial applications has been a reality for many decades. It is necessary that the potential hazards of such materials be understood and appropriate precautions be designed into the processes that use them.

Acknowledgments

The author is grateful for information and advice provided by Ms Beth Hamil, Vice President of Corporate Compliance, DEL Ozone, San Luis Obispo, CA, and by Mr Cameron Tapp, President/CEO of ClearWater Tech LLC, San Luis Obispo, CA during the construction of this chapter. Ms Hamil also granted the author permission to include data generated by third-party testing of disinfection efficacy and worker hazard analysis of certain commercial ozone aqueous spray washing equipment.

References

Boisrobert, C. (2002) US regulatory review of ozone use in the food industry, Presented at *Ozone III: Agricultural & Food Processing Applications of Ozone as an Antimicrobial Agent*, Fresno, CA.

Compressed Gas Association (2001) Safe handling of ozone-containing mixtures including the installation and operation of ozone-generating equipment, Product No. CGA P-34, Compressed Gas Association, Inc., Chantilly, VA.

EPRI (1997) Expert *Panel Report: Evaluation of the History and Safety of Ozone in Processing Food for Human Consumption, Vol. 1: Executive Summary*, Palo Alto, CA: Electric Power Research Institute, Final Report, TR-108026-VI-4827.

Erdman, J.W. Jr (1997) Nutrient impact (of ozone contact with foods), in EPRI (1997) Expert *Panel Report: Evaluation of the History and Safety of Ozone in Processing Food for Human Consumption, Vol. 1: Executive Summary*, Palo Alto, CA: Electric Power Research Institute, Final Report, TR-108026-VI-4827, Chapter 5).

Erdman, J.W. Jr and Poneros-Schneier, A. (1994) Factors affecting nutritive value in processed foods, in Shils, M.E., Olson, J.A. and Shike, M. (eds) *Modern Nutrition in Health and Disease*, Philadelphia, PA: Lee and Febiger, pp. 1569–78.

FDA (2001) Secondary direct food additives permitted in food for human consumption, *Federal Register*, **66**(123): 33 829–30.

France (1995) Decrees of 24 January and 23 January 1995 relative to various technological additives in human foods, *Journal Officiel de la République Française*, 4 Mar., p. 3457.

Gorman, B.M., Kochevar, S.L., Sofus, J.H., Morgan, J.G., Schmidt, G.R. and Smith, G.S. (1995) Shelf-life of beef tissue following spray-washing with chemical solutions and 35 °C or 74 °C water (Abstract), *IFT Annual Meeting 1995*, p.140.

Hamil, B. (2007) Ozone technology for winery applications, *Proc. Intl. Water Technology Conference & Ozone V*, Fresno, CA: California State Univ, 2–4 Apr.

Henry, L., Jaren, M., Catignani, G. and Schwartz, S. (1996) The effects of ozone and oxygen exposure on the degradation kinetics of β-carotene and lycopene, Presented at the *11th International Symposium on Carotenoids, Leiden, The Netherlands*, 18–23 Aug. (Abstract 42).

International Fire Code Institute (1994) *Uniform Fire Code*, Vol. 1, Appendix II-I.

Kasei, T., Iwashita, A. and Kiriyama, S. (1993) Growth is compromised in rats fed ozone-treated casein, *J Nutrition*, **123**: 893–9.

Kasei, T., Sakamaki, T. and Kiriyama, S. (1994) Modified process for preparation of ozonated casein and nutritional evaluation of the product with rats, *Biosci Biotech Biochem*, **58**(3): 451–4.

Kolodyaznaya, V.S. and Suponina, T.A. (1975) Storage of foods using ozone, *Kholodil'naya Tekhnika*, **6**: 39–41.

Kozumbo, W.J., Hanley, N.M., Agarwal, S., Thomas, M.J. and Madden, M.C. (1996) Products of ozonated arachidonic acid potentiate the formation of DNA single strand breaks in cultured human lung cells, *Environ Mol Mutagenesis*, **27**: 165–95.

Kuprianoff (1953) The use of ozone for the cold storage of fruit, *Z Kaltentechnik*, Vol. 10 (translation).

LaLecheur, R. and Glaze, W.H. (1996) Reactions of ozone and hydroxyl radicals with serine, *Environ Sci and Technology*, **30**: 1072–80.

Naito, S. and Nanba, Y. (1987) Studies on the utilization of ozone in food processing. Part IV. The effect of ozone treatment on thiamine and riboflavin contents of cereal grains, cereal grain powders, legumes and whole spices, *Nippon Skokuhin Kogyo Gakkaishi*, **34**(12): 794–800.

Naito, S., Sawada, Y. and Yamaguchi, N. (1989) Studies on utilization of ozone in food preservation: effects of ozone treatment on packaged Namamen Japanese raw noodle, *J Antibacterial and Antifungal Agents, Japan*, **17**(11): 517–26.

Naitoh, S. (1989) The influence of ozone treatment on lipids contained in cereal grains, cereal grain powders, peas, beans and pulse products (studies on utilization of ozone for food preservation, Part VI), *J Japanese Soc of Food Science and Technology*, **36**(11): 878–83.

Naitoh, S. (1992) Studies on the application of ozone in food processing: mutagenicity of ozone-treated amino acids and saccharides, *J Antibact Antifung Agents*, **20**(7): 365–72.

Newell, G.W. (1997) Safety issues (with ozone in food contact) – toxicology, in EPRI (1997) Expert *Panel Report: Evaluation of the History and Safety of Ozone in Processing Food for Human Consumption, Vol. 1: Executive Summary*, Palo Alto, CA: Electric Power Research Institute, Final Report, TR-108026-VI-4827, Chapter 4.

NTP (1995) *Toxicology and Carcinogenesis: Studies with Ozone and Ozone/NNK*, US Natl Toxicology Program, Natl Inst of Health, Pub. No. 95-3371.

Pascual, A., Llorca, I. and Canut, A. (2007) Use of ozone in food industries for reducing the environmental impact of cleaning and disinfection activities, *Trends in Food Science & Technology*, **18**: S29–S35.

Pryor, A. and Rice, R.G. (2000) Ozone toxicology and guidelines for safe use in food processing systems, *Ozone News*, **28**(4): 19–28.

Rakness, K.L. (2005) *Ozone In Drinking Water Treatment: Process Design, Operation, and Optimization*, 1 edn, Denver, CO: American Water Works Association.

Rice, R.G. and Gomez-Taylor, M. (1986) Occurrence of the byproducts of strong oxidants reacting with drinking water contaminants – scope of the problem, *Environ Health Perspectives*, **69**: 31–44.

Rice, R.G. and Wrenn, R.H. (2007) Improving fish quality by means of ozone at Fresher Than Fresh, Inc., Presented at *IOA-AINIA Conference on Ozone and the Food Industry*, Valencia, Spain.

Rice, R.G., Farquhar, J.W. and Bollyky, L.J. (1982) Review of the applications of ozone for increasing storage times of perishable foods, *Ozone: Science & Engineering*, **4**: 147–63.

Steffen, H., Duerst, M. and Rice, R.G. (2010) User experiences with ozone, electrolytic water (active water) and UV-C light (Ventafresh Technology) in production processes and for hygiene maintenance in a Swiss SUSHI factory, *Ozone: Science & Engineering*, **32**(1): 71–8.

Toxicology Group (2002) Hazard and safety assessment of DEL AGW-0500 Mobile Ozone Surface Sanitation System, AL SSS 0500 Mobile Ozone Surface Sanitation System, DEL AGW-1500G Mobile Recirculating Ozone Sanitation System, and AL SSS 1500 Mobile Recirculating Ozone Sanitation System.

Uniform Fire Code (2006) Annex G. G.4.3, 2006 edition; International Fire Code, Section 3705.3.2.

Watanabe, T., Takai, Y., Tanaka, K. and Suzuki, A. (1994) Effects of ozone exposure during cultivation of nameko mushroom (*Pholiota nameko*) on fatty acid composition of the fruit bodies, *Journal of the Japanese Society for Food Science and Technology Nippon Shokuhin Kogyo Gakkaishi*, **41**(10): 709–14.

Workers Compensation Board of British Columbia (1991) *Ozone: A Manual of Standard Practice*, Richmond, BC: Workers Compensation Board of British Columbia.

Index

Ozone in Food Processing, First Edition. Edited by Colm O'Donnell, B.K. Tiwari, P.J. Cullen, and Rip G. Rice.